TESTING FOR NORMALITY

STATISTICS: Textbooks and Monographs

D. B. Owen, Founding Editor, 1972–1991

1. The Generalized Jackknife Statistic, *H. L. Gray and W. R. Schucany*
2. Multivariate Analysis, *Anant M. Kshirsagar*
3. Statistics and Society, *Walter T. Federer*
4. Multivariate Analysis: A Selected and Abstracted Bibliography, 1957–1972, *Kocherlakota Subrahmaniam and Kathleen Subrahmaniam*
5. Design of Experiments: A Realistic Approach, *Virgil L. Anderson and Robert A. McLean*
6. Statistical and Mathematical Aspects of Pollution Problems, *John W. Pratt*
7. Introduction to Probability and Statistics (in two parts), Part I: Probability; Part II: Statistics, *Narayan C. Giri*
8. Statistical Theory of the Analysis of Experimental Designs, *J. Ogawa*
9. Statistical Techniques in Simulation (in two parts), *Jack P. C. Kleijnen*
10. Data Quality Control and Editing, *Joseph I. Naus*
11. Cost of Living Index Numbers: Practice, Precision, and Theory, *Kali S. Banerjee*
12. Weighing Designs: For Chemistry, Medicine, Economics, Operations Research, Statistics, *Kali S. Banerjee*
13. The Search for Oil: Some Statistical Methods and Techniques, *edited by D. B. Owen*
14. Sample Size Choice: Charts for Experiments with Linear Models, *Robert E. Odeh and Martin Fox*
15. Statistical Methods for Engineers and Scientists, *Robert M. Bethea, Benjamin S. Duran, and Thomas L. Boullion*
16. Statistical Quality Control Methods, *Irving W. Burr*
17. On the History of Statistics and Probability, *edited by D. B. Owen*
18. Econometrics, *Peter Schmidt*
19. Sufficient Statistics: Selected Contributions, *Vasant S. Huzurbazar (edited by Anant M. Kshirsagar)*
20. Handbook of Statistical Distributions, *Jagdish K. Patel, C. H. Kapadia, and D. B. Owen*
21. Case Studies in Sample Design, *A. C. Rosander*
22. Pocket Book of Statistical Tables, *compiled by R. E. Odeh, D. B. Owen, Z. W. Birnbaum, and L. Fisher*
23. The Information in Contingency Tables, *D. V. Gokhale and Solomon Kullback*
24. Statistical Analysis of Reliability and Life-Testing Models: Theory and Methods, *Lee J. Bain*
25. Elementary Statistical Quality Control, *Irving W. Burr*
26. An Introduction to Probability and Statistics Using BASIC, *Richard A. Groeneveld*
27. Basic Applied Statistics, *B. L. Raktoe and J. J. Hubert*
28. A Primer in Probability, *Kathleen Subrahmaniam*
29. Random Processes: A First Look, *R. Syski*
30. Regression Methods: A Tool for Data Analysis, *Rudolf J. Freund and Paul D. Minton*
31. Randomization Tests, *Eugene S. Edgington*
32. Tables for Normal Tolerance Limits, Sampling Plans and Screening, *Robert E. Odeh and D. B. Owen*
33. Statistical Computing, *William J. Kennedy, Jr., and James E. Gentle*
34. Regression Analysis and Its Application: A Data-Oriented Approach, *Richard F. Gunst and Robert L. Mason*
35. Scientific Strategies to Save Your Life, *I. D. J. Bross*
36. Statistics in the Pharmaceutical Industry, *edited by C. Ralph Buncher and Jia-Yeong Tsay*
37. Sampling from a Finite Population, *J. Hajek*

TESTING FOR NORMALITY

HENRY C. THODE, JR.
*State University of New York
at Stony Brook
Stony Brook, New York*

CRC Press
Taylor & Francis Group
Boca Raton London New York

CRC Press is an imprint of the
Taylor & Francis Group, an **informa** business

First published 2002 by Marcel Dekker, Inc.

Published 2019 by CRC Press
Taylor & Francis Group
6000 Broken Sound Parkway NW, Suite 300
Boca Raton, FL 33487-2742

© 2002 by Taylor & Francis Group, LLC
CRC Press is an imprint of Taylor & Francis Group, an Informa business

First issued in paperback 2019

No claim to original U.S. Government works

ISBN 13: 978-0-367-44710-6 (pbk)
ISBN 13: 978-0-8247-9613-6 (hbk)

**Visit the Taylor & Francis Web site at
http://www.taylorandfrancis.com**

**and the CRC Press Web site at
http://www.crcpress.com**

To my children,

Matthew, John, and Samantha

PREFACE

In the development of statistical sampling theory it has often happened that more than one test of a given hypothesis is available. Generally on theoretical grounds it is possible to specify which of these tests is the most efficient; but it may happen that owing to mathematical difficulties in putting the ideal test into working form or to practical difficulties arising from the extent of computation involved, the statistician will choose to employ a second best but simpler test.

E.S. Pearson, 1935

The Gaussian or normal distribution has long been the focal point of much of statistical study, for a number of reasons. Data often approximates a normal "bell-shaped" curve; the normal distribution is mathematically easy to work with; and many statistics, both estimates and test statistics, as well as some distributions become normal asymptotically.

For the purposes of this text, we are primarily concerned with the first reason stated above. Normality (or lack thereof) of an underlying data distribution can have an effect to a greater or lesser degree on the properties of estimation or inferential procedures used in the analysis of the data. In order to address these issues, formal as well as informal methods have been developed in order to ascertain the apparent normality of a data sample, so that the appropriateness of applying a statistical procedure to that sample can be determined. Fortunately (or unfortunately), a large number of methods for testing for normality have been developed, providing researchers with a wide range of choices; however, this can also result in a corresponding good deal of confusion.

My objective in writing this text was to present, as completely as possible, goodness of fit tests that were designed or could be used for determining whether a sample of observations could have come from a normal distribution. My intent was to focus on methodology and utility of the tests rather than in-depth theoretical issues, so that readers would be able to make a judgment as to which test(s) would be best for their particular circumstances, and could easily perform the test. I intended to make this text accessible to researchers with a minimal amount of theoretical statistical background, and the sections in which I have delved into theory can generally be ignored without impairing the practical application of the methods. To my knowledge, this text contains the broadest and most comprehensive set of material published to date on the single subject of testing for normality.

I also hoped that by presenting this material I would provide a better understanding of underlying distributional assumptions and the effects of violating those assumptions. However, although more goodness of fit methodology is focused on the normal than on any other distribution, not all underlying assumptions are necessarily those of normality. Equipped with as complete as possible a description of tests for normality, readers concerned with goodness of fit in regard to other null distributions may develop analogous tests based on what is written here: perhaps tests can be extended or expanded to improve tests for multivariate normality, the exponential distribution, or Weibull distribution, for example.

Although some historical background and theory are provided for many of the tests, the emphasis here is on the calculation and performance of the tests. I have omitted complete details on the formulation of some of the tests identified herein, mainly those less useful for practical purposes. These were mentioned purely for the purpose of completeness. Although I limited this work to the normal distribution, on occasion the applicability of certain tests to other null distributions has been mentioned. I believe the large bibliography will be invaluable to anyone who feels the need to obtain more details in these areas.

This text comprises four sections. The first section (Chapter 1) is introductory. The second (Chapters 2 through 8) addresses the issue of testing for univariate normality in complete samples (Chapters 2–7) and censored samples (Chapter 8). The third section (Chapters 9 and 10) covers the topic of testing for multivariate normality. The remainder of the text covers additional miscellaneous topics, including normal mixture distributions (univariate and multivariate, Chapter 11), robust estimation (Chapter 12) and computational issues (Chapter 13). Data sets used in the examples throughout the book are included in Appendix A and tables of critical values for most of the tests presented here are given in Appendix B.

Henry C. Thode, Jr.

Reference

Pearson, E.S. (1935). A comparison of β_2 and Mr. Geary's ω_n criteria. *Biometrika* 27, 333-352.

CONTENTS

TESTING FOR NORMALITY

CHAPTER 1

INTRODUCTION

"...it is not enough to know that a sample could have come from a normal population; we must be clear that it is at the same time improbable that it has come from a population differing so much from the normal as to invalidate the use of "normal theory" tests in further handling of the material."

E.S. Pearson, 1930

1.1 Why Test for Normality?

The topic of this text is the problem of testing whether a sample of observations comes from a normal distribution. Normality is one of the most common assumptions made in the development and use of statistical procedures. The problem has not suffered from lack of attention. In our review of the literature we found more tests than we ever imagined existed. This text, for instance, considers about forty formal testing procedures that have been proposed to test specifically for normality, as well as plotting methods, outlier tests, general goodness of fit tests and other tests that are useful in detecting non-normality in specialized situations. Further, the list is probably not exhaustive. For example, while the sample moment skewness and kurtosis statistics are commonly used as tests of normality, many such moment tests could be considered (Chapter 3). Geary (1947) considered the larger class of absolute moment tests and developed quite general results concerning their power. Thode (1985) used Geary's calculations to further refine the evaluation of these absolute moment tests and found that some of these tests had modestly better power under certain circumstances than the kurtosis test or Geary's test.

The objective of this text is to summarize the vast literature on tests of normality and to describe which of the many tests are effective and which are not. Some results are surprising. Such popular tests as the Kolmogorov-Smirnov or chi-squared goodness of fit tests have power so low that they should not be seriously considered for testing normality (D'Agostino, Belanger and D'Agostino, 1990). In general, the performance of moment tests and the Wilk-Shapiro test is so impressive that we recommend their use in everyday practice.

1.1.1 Historical Review of Research on Whether Assumption of Normality Is Valid or Important

Statistical procedures such as t-tests, tests for regression coefficients, analysis of variance, and the F-test of homogeneity of variance have as an underlying assumption that the sampled data come from a normal distribution. Of course, the assumption of normality in a statistical procedure requires an effective test of whether the assumption holds, or a careful argument showing that the violation of the assumption does not invalidate the procedure used. Much statistical research has been concerned with evaluating the magnitude of the effect of violations of this assumption on the true significance level of a test or the efficiency of parameter estimates.

Research evaluating the effects of violations of the assumption of normality upon standard statistical procedures date back before Bartlett's 1935 paper on the t-test. Fisher (1930) thought the problem of such importance that he developed results on the cumulants of the skewness and kurtosis statistics as tests of normality.

This problem has come to have the generic label of "robustness" in the current statistical literature. A review of the literature of robustness and tests of normality shows many contributions from the most outstanding theorists and practitioners of statistics. Pitman (1937a, 1937b) considered the problem of the sensitivity of the t-test and the one way analysis of variance from the point of view of using permutation theory to establish important results on the lack of sensitivity of the t-test and one way analysis of variance to violations of normality.

Geary (1947) used the Gram-Charlier system of distributions as the non-normal alternatives being sampled in order to determine the validity

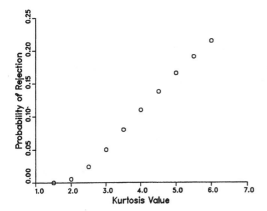

Figure 1.1 Actual significance level of z-test when underlying distribution has specified kurtosis value (from Geary, 1947).

of statistical procedures. The advantage of these distributions was that Geary could calculate the asymptotic moments of the absolute moment tests and from these the approximate distribution of the tests and hence could determine their power as tests for normality. He showed that for the two sample z-test and for testing the homogeneity of variance the effects of having a symmetric non-normal underlying distribution can seriously affect the true significance level of the test. For a value of 1.5 for the kurtosis of the alternative distribution, the actual significance level of the test is less than 0.0001, as compared to the level of 0.05 if the distribution sampled were normal. For a distribution with a kurtosis value of 6, the probability of rejection was 0.215. Figure 1.1 presents a graph of Geary's results for other values of the kurtosis under his chosen set of alternatives.

In contrast, for the t-test Geary determined that the distortion of the probability of rejection is slight if the underlying distribution is symmetric but non-normal, confirming Pitman's more general permutation distribution analysis. If the underlying distribution is skewed, marked changes to the probability of rejection can occur. For the two sample t-test Geary concluded that if the underlying distribution is the same for both populations, regardless of the type of non-normality, the changes in the probability that the null hypothesis is rejected are small. Large changes can occur if the distributions are different.

Box (1953) showed that Bartlett's test of homogeneity of variance was seriously compromised by non-normality and that it was foolish to use Bartlett's test as a pretest to determine whether the analysis of variance

might be invalid due to different variances. He summed up the effects of
ignoring non-normality when comparing two samples rather well:

> "So far as comparative test on means are concerned ... this prac-
> tice [of ignoring non-normality] is largely justifiable ... There is
> abundant evidence that these comparative tests on means are re-
> markably insensitive to general non-normality of the parent pop-
> ulation."

By "general non-normality" he meant "that the departure from normality, in particular skewness, is the same in the different groups ...". On
the other hand, robustness to non-normality "... is not necessarily shared
by other statistical tests, and in particular is not shared by the tests for
equality of variances ...".

Simulations by Pearson and Please (1975) confirmed the theoretical
results on the lack of robustness of the F-test for homogeneity of variance
for small samples. Using samples of size 10 and 25, they showed that non-
normality seriously affects the true significance level of the single sample
variance test and the two sample variance ratio, even for symmetric parent distributions. The one sample t-test is affected by skewness; the two
sample t-test is not affected greatly when the samples come from identical
populations.

Subrahmaniam, Subrahmaniam and Messeri (1975) looked at the effects on the true level of the test for one sample t-tests, analysis of variance
and analysis of variance for regression when the underlying distribution is a
location contaminated normal distribution. Their calculations were based
on samples of size 20 or less. There was little effect on any of the procedures when the contamination was small. For larger contaminations and
larger differences in component means, the effect on the probability of the
0.05-level t-test became quite large (probability of 0.16 for alternative with
contamination fraction 0.25, standardized difference in means of 1, and
sample size 20). This agrees with Geary's conclusion concerning the t-test.

Tukey (1960) started the more recent flood of consideration about the
problem of robustness in estimation against slight departures from normality. He showed the effects of non-normality on the estimation of location
and scale parameters of a distribution using an unbalanced mixture of two
normals with common mean and different variances. He called such a mixture a scale contaminated normal and concluded that if "... contamination
is a real possibility ... neither mean nor variance is likely to be a wisely
chosen basis for making estimates from a large sample". Tukey's work,
in fact, was the starting point for our work on the problem of testing for
normality: how large were the sample sizes needed to detect the contaminations that the researchers in robustness were examining (Thode, Smith

and Finch, 1983).

D'Agostino and Lee (1977) compared the efficiency of several estimates of location, including the sample mean, when the underlying distribution was either a Student's t or exponential power distribution. Both of these are symmetric families. The efficiency of the estimates was compared based on the kurtosis value of the underlying distribution. For the t distribution, the relative efficiency of the sample mean (which is 1 when the distribution is normal) only decreases to about 0.9 for a kurtosis value of 6 (corresponding to a t distribution with 6 degrees of freedom). For the exponential power distribution, however, the relative efficiency of the sample mean drops quickly and decreases to about 0.5 when the kurtosis value is 6 (the Laplace distribution).

1.1.2 Genetic Applications and Tests of Clustering

Another problem of interest to us in testing for normality arose from the problem of how one would test for a different type of mixture, the location-contaminated normal distribution (Thode, Finch and Mendell, 1988). Applications of this type are important in genetics, where the alternative often consists of a mixture of normal components with differences in the location parameters. The issue is that one can make an inference from the number of components in the mixture to aspects of the structure of the gene determining the variable being measured. For example, a mixture of two normal components suggests a simpler type of genetic model than a mixture of three normal components. This can be generalized to testing in cluster analysis, where each cluster is made up of observations from a normal distribution. In principle, then, a test of normality might have some promise as a tool in cluster analysis.

Thus a problem that can be focused on in this context is: when there is a null hypothesis that observations come from a single normal distribution, with an alternative that the observations come from several normal components, a test for normality can be used to support or reject the null hypothesis. The alternative may consist of a known or unknown number of components. The components may differ in either or both of the parameters of the normal distribution.

1.1.3 Comparative Evaluation of General Goodness of Fit Tests to Normality Tests

There are more tests designed specifically to assess normality than for any other particular distribution. The literature contains many tests that take advantage of special properties of the normal distribution. For example, the general absolute moment tests take advantage of specific relations among the moments of the normal distribution. The Wilk-Shapiro test compares an estimate of the standard deviation using a linear combination of the order statistics to the usual estimate.

Intuitively, such statistics should be more sensitive to certain alternatives than the completely general goodness of fit tests such as the χ^2 test or the Kolmogorov-Smirnov test. These procedures operate by using the cumulative distribution function to reduce the general problem to the specific one of testing the hypothesis of uniformity. In addition to the very common chi-squared test and the Kolmogorov-Smirnov test, there are many other general procedures. One of our objectives will be to consider such tests to determine which are effective for the specific problem of testing for normality. Such a comparison may well provide indications of the relative value of these general tests for testing other null hypotheses.

Relatively little work has been done in the field of testing for multivariate normality compared to that done for univariate normality. Small improvements in the ability to test for univariate normality may lead to larger improvements in the ability to handle the multivariate problem. We will survey the field to date and will present our own evaluations.

1.2 Hypothesis Testing for Distributional Assumptions

Suppose you have a random sample of n independent and identically distributed (iid) observations of a random variable X, labeled $x_1, x_2, \ldots, x_n$, from an unspecified density $f(x)$. The general goodness of fit problem consists of testing the null hypothesis

$$H_0 : f(x) = f_0(x)$$

against an alternative hypothesis. The probability density function (pdf) in the null hypothesis $f_0(x)$ has a specified distributional form. When the parameters are completely specified, the null hypothesis is called a simple hypothesis. If one or more of the parameters in H_0 are not specified, H_0 is called a composite hypothesis.

Depending upon the problem, the alternative may be completely specified (in the case of a simple null hypothesis) including the values of the parameters,

$$H_1 : \quad f(x) = f_1(x; \theta).$$

For composite null hypotheses it may consist of a class of distributions

$$H_1 : \quad f(x) = f_1(x),$$

(i.e., parameters not specified), or it may be completely general

$$H_1 : \quad f(x) \neq f_0(x).$$

Another general goodness of fit problem arises when the alternative is any distribution with a specified shape. Historically, non-normal alternatives are divided into three shape classes based on the comparison of their third and fourth standardized moments (denoted $\sqrt{\beta_1}$ and β_2, respectively) to those of the normal distribution. A distribution whose standardized third moment (skewness) is different from 0 is necessarily skewed. The value of the standardized fourth moment (kurtosis) for a normal distribution is 3, although a value of 3 does not necessarily indicate a normal distribution (e.g., Johnson, Tietjen and Beckman, 1980; Balanda and MacGillivray, 1988). Symmetric alternatives are often separated into those with population kurtosis less than or greater than 3.

In this text we consider tests of the composite hypothesis of normality

$$H_0 : \quad f(x) = \frac{1}{\sqrt{2\pi}\sigma} e^{\frac{-(x-\mu)^2}{2\sigma^2}}, \quad -\infty < x < \infty.$$

where both the mean (μ) and standard deviation (σ) are unknown. This is generally the case of interest in practice. In the past many general goodness of fit tests (specifically those based on the empirical distribution function [EDF tests]) required complete specification of the null parameters, to their disadvantage. Stephens (1974) and others improved the use of these tests by developing EDF tests for composite hypotheses. We examine tests derived for specific (except for parameters) alternatives, shape alternatives and the general alternative.

Some tests, such as likelihood ratio tests and most powerful location and scale invariant tests (Chapter 4), were derived for detecting a specific alternative to normality. These are based on the joint probabilities of the null and alternative distributions, given the values of the observations. The disadvantages of these tests are that many of them are not able to be calculated in closed form, critical values are rarely available, and they may not be efficient as tests of normality if in fact neither the null nor the specified alternative hypotheses are correct. One might consider rather a more general test which is useful in detecting an alternative of similar shape to the specified alternative. On the other hand, some likelihood ratio tests are useful in testing for a broader set of alternatives.

Shape tests are divided into two classes, tests for skewed alternatives and tests for non-normal symmetric alternatives. Most shape tests can further be broken down into directional and bidirectional tests. Directional tests for skewness are used when a left or right skewed distribution is known to be the alternative of concern. A bidirectional skewness test is used when a skewed alternative is of concern but the direction of skewness is not known.

For symmetric alternatives, directional tests are used when it is known that the alternative is assumed to be heavy- or light-tailed. A bidirectional test is used when it is assumed only that the alternative is symmetric and non-normal.

Omnibus tests are designed to cover all possible alternatives. They are not usually as powerful as specific or (directional or bidirectional) shape tests when the characteristics of the true alternative are correctly identified. These are usually single-tailed tests, e.g., the Wilk-Shapiro W test (Shapiro and Wilk, 1965) and the probability plot correlation test (Filliben, 1975). Combinations of directional tests have also been suggested as omnibus tests (e.g., D'Agostino and Pearson, 1973).

The tests which we will describe are also location and scale invariant, i.e., they have the property that a change in the location or scale of the observations do not affect the test statistic or the resulting test, i.e.,

$$T(x_1, x_2, \ldots, x_n) = T(kx_1 - u, kx_2 - u, \ldots, kx_n - u)$$

for constants k and u. This is a desirable property of a test since the parameters do not affect the shape of the normal distribution. For most statistical procedures, distribution assumptions which are made usually only concern shape.

Tests for normality that will be discussed come under one of three possibilities: for a specified H_0 and H_1, a test statistic t, a specified significance level α, and a constant k chosen appropriately, normality may be rejected if

(1) $t \leq k_{l,\alpha}$

This type of test is common for regression tests and some directional tests (e.g., tests for alternatives skewed to the left).

(2) $t \geq k_{u,\alpha}$

Likelihood ratio tests, tests for outliers, and some directional tests are usually of this form.

(3) $t \leq k_{l,\alpha/2}$ or $t \geq k_{u,\alpha/2}$

A two-tailed test is most often used for alternatives where the characteristics of the shape are specified but not the direction (e.g., symmetric alternative but it is not known if it has long or short tails).

In most cases the tests discussed in the text can be modified to deal with specified parameters by substituting the hypothesized values for the estimated values in the test statistic formula. However, adjustments to the critical values for the test must also be made. These types of tables are rarely available. One notable exception is Barnett and Lewis (1994) who provided tables of critical values for some outlier tests with 1, 2 or no parameters specified.

1.3 Symmetric Distributions and the Meaning of Kurtosis

A symmetric distribution is often called "heavy" or "long" tailed if the standardized fourth moment is greater than the normal value of 3 and called "light" or "short" tailed if the value of β_2 is less than three. Alternatively, these distributions are also sometimes called "peaked" or "flat", respectively, in relation to the normal distribution. The nomenclature for these two symmetric shape classes is misleading. There has been much discussion over whether β_2 describes peakedness, tail length/size, or the "shoulders" of a distribution. Dyson (1943) and Finucan (1964) showed that, for two symmetric densities $g(x)$ and $f(x)$ with mean 0 and equal variance, if

$$f(x) < g(x) \quad \text{for} \quad a <\mid x \mid< b$$

and

$$f(x) > g(x) \quad \text{for} \quad \mid x \mid< a \quad \text{or} \quad \mid x \mid> b$$

then $\mu_4(f) > \mu_4(g)$, i.e., if the value of the density $f(x)$ is lower than that of $g(x)$ in some interval between the mean and the tails (the "shoulders" of the distribution) and higher elsewhere, then $f(x)$ has the higher kurtosis. (For densities with 0 means and equal variance, μ_4 is essentially kurtosis, see Chapter 3.)

Darlington (1970) claimed that kurtosis is actually a measure of the probability mass around the two values $\mu \pm \sigma$ and therefore should be interpreted as a measure of bimodality around those two points. Following up on Darlington's work, Moors (1986) described kurtosis as a measure of dispersion around the two values $\mu \pm \sigma$, declaring the interpretation of bimodality to be false. Large values of β_2 occur with probability densities that have less probability mass around these two values (the "shoulders"). Since the mass must equal 1 the probability mass must be concentrated either near the mean μ or in the tails. By assimilating the results of Dyson, Finucan and Moors, one sees that a distribution with kurtosis higher than normal must have higher density values at μ (causing peakedness) and/or in the tails (causing heavy tails). Depending on how the mass is dispersed in the center and the tails of the distribution, different symmetric distributions may therefore have the same kurtosis value. Figure 1.2 shows the

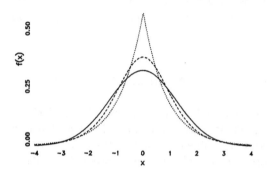

Figure 1.2 Comparison of the — normal ($\beta_2 = 3$), ···· Laplace ($\beta_2 = 6$) and - - - t_6 ($\beta_2 = 6$) densities.

density functions of the Laplace distribution and the t distribution with 6 degrees of freedom, both of which have kurtosis values of 6, compared to the normal distribution with the same mean and variance. As can be seen, the Laplace is considerably more peaked than the t distribution; it is difficult to see differences in the tails on this scale although both the t and Laplace distributions have heaver tails than the normal.

The use of kurtosis, and of skewness for that matter, as a measure of the degree of non-normality in a distribution is somewhat arbitrary, as others have proposed alternative measures of shape. Groeneveld and Meeden (1984) define a variety of other measures of distribution skewness and kurtosis; Crow and Siddiqui (1967), Uthoff (1968), Rogers and Tukey (1972), Filliben (1975), Ruppert (1987) and Moors (1988), among others, define other measures of skewness and departures from non-normal symmetry which are based on percentiles. Kendall and Stuart (1977) attribute the measure $(\bar{x} - x_m)/s$ as a measure of skewness in a sample to Karl Pearson, where $\bar{x}$, x_m and s are the sample mean, mode and standard deviation, respectively. Use of the median rather than the mode in this measure has also been suggested (Kendall and Stuart, 1977; Groeneveld and Meeden, 1984).

The definition of kurtosis and the presentation of alternatives has been reviewed in some detail by Balanda and MacGillivray (1988), who also provided a fairly comprehensive list of references on the meaning of kurtosis and some alternative measures to kurtosis. It is hoped that the reader will realize that kurtosis is not the final word in symmetric distributions;

for example, as mentioned before D'Agostino and Lee (1977) found quite different estimation efficiencies for the mean when sampled from two symmetric non-normal distributions even though they had equal population kurtosis values. Other results based purely on the kurtosis values of a single family of distributions such as those presented by Geary (1947) should also not be taken as universal.

However, at the risk of perpetuating a myth, in this text we will use β_2 to define the shape classes of symmetric distributions, and to impart a sense of the degree of the "non-normalness" of a symmetric distribution. We trust this will not cause much concern, since we will in general limit the use of β_2 as a division of symmetric distributions into two classes rather than as a quantitative value.

1.4 Objectives of This Text

Our main objective is to present, as completely as possible, a viable and valuable list of tests and procedures that can be used for assessing normality, including both univariate and multivariate normality. We compare tests in terms of power and ease of use.

Chapters 2 through 7 describe procedures for assessing univariate normality in complete samples. Chapter 2 describes probability plotting methods and regression and correlation tests. Chapter 3 contains tests based on sample moments, and moment-type tests. In Chapter 4 we present other tests specifically derived for testing normality. Chapter 5 describes general goodness of fit tests and their usefulness in testing specifically for normality. In Chapter 6 we present tests specifically designed to detect outliers. In Chapter 7 we summarize results of power studies and comparisons of the various univariate tests for normality presented in the preceding five chapters. Chapter 8 also focuses on univariate samples, but considers the case of censored data.

Chapters 9 and 10 are concerned with assessing normality in multivariate samples. Chapter 9 describes tests for multivariate normality, while Chapter 10 considers those tests designed to detect multivariate outliers. Chapter 11 focuses on a more specific problem, that of testing for mixtures of normal distributions, in both the univariate and multivariate cases.

Chapter 12 presents basic methods for robust estimation, which can be used in the event that data are determined not to be normal. Chapter 13 describes various computational issues in assessing normality. Appendices contain data sets used in the examples presented throughout the text, and tables of parameter and critical values for many of the procedures described.

References

Balanda, K.P., and MacGillivray, H.L. (1988). Kurtosis: a critical review. American Statistician 42, 111-119.

Barnett, V., and Lewis, T. (1994). **Outliers in Statistical Data,** 2nd ed. John Wiley and Sons, New York.

Bartlett, M.S. (1935). The effect of non-normality on the t-distribution. Proceedings of the Cambridge Philosophical Society 31, 223-231.

Box, G.E.P. (1953). Non-normality and tests on variances. Biometrika 40, 318-335.

Crow, E.L., and Siddiqui, M.M. (1967). Robust estimation of location. Journal of the American Statistical Association 62, 353-389.

D'Agostino, R.B., Belanger, A., and D'Agostino, Jr., R.B. (1990). A suggestion for using powerful and informative tests of normality. American Statistician 44, 316-321.

D'Agostino, R.B., and Lee, A.F.S. (1977). Robustness of location estimators under changes of population kurtosis. Journal of the American Statistical Association 72, 393-396.

D'Agostino, R., and Pearson, E.S. (1973). Tests for departure from normality. Empirical results for the distributions of β_2 and $\sqrt{\beta_1}$. Biometrika 62, 243-250.

Darlington, R.B. (1970). Is kurtosis really 'peakedness'? American Statistician 24, 19-22.

Dyson, F.J. (1943). A note on kurtosis. Journal of the Royal Statistical Society B 106, 360-361.

Filliben, J.J. (1975). The probability plot correlation coefficient test for normality. Technometrics 17, 111-117.

Finucan, H.M. (1964). A note on kurtosis. Journal of the Royal Statistical Society B 26, 111-112.

Fisher, R.A. (1930). The moments of the distribution for normal samples of measures of departure from normality. Proceedings of the Royal Society of London A 130, 16-28.

Geary, R.C. (1947). Testing for normality. Biometrika 34, 209-242.

Groeneveld, R.A., and Meeden, G. (1984). Measuring skewness and kurtosis. The Statistician 33, 391-399.

Johnson, M.E., Tietjen, G.L., and Beckman, R.J. (1980). A new family of probability distributions with application to Monte Carlo studies. Journal of the American Statistical Association 75, 276-279.

Kendall, M., and Stuart, A. (1977). **The Advanced Theory of Statistics, Vol. I.** MacMillan Publishing Co., New York.

Moors, J.J.A. (1986). The meaning of kurtosis: Darlington reexamined. American Statistician 40, 283-284.

Moors, J.J.A. (1988). A quantile alternative for kurtosis. The Statistician 37, 25-32.

Pearson, E.S. (1930). A further development of tests for normality. Biometrika 22, 239-249.

Pearson, E.S., and Please, N.W. (1975). Relation between the shape of population distribution and the robustness of four simple test statistics. Biometrika 62, 223-241.

Pitman, E.J.G. (1937a). Significance tests which may be applied to samples from any population. Supplement to the Journal of the Royal Statistical Society 4, 119-130.

Pitman, E.J.G. (1937b). Significance tests which may be applied to samples from any populations: III. The analysis of variance test. Biometrika 29, 322-335.

Rogers, W.H., and Tukey, J.W. (1972). Understanding some long-tailed symmetrical distributions. Statistica Neerlandica 26, 211-226.

Ruppert, D. (1987). What is kurtosis? American Statistician 41, 1-5.

Shapiro, S.S., and Wilk, M.B. (1965). An analysis of variance test for normality (complete samples). Biometrika 52, 591-611.

Stephens, M.A. (1974). EDF statistics for goodness of fit and some comparisons. Journal of the American Statistical Association 69, 730-737.

Subrahmaniam, K., Subrahmaniam, K., and Messeri, J.Y. (1975). On the robustness of some tests of significance in sampling from a compound normal distribution. Journal of the American Statistical Association 70, 435-438.

Thode, Jr., H.C. (1985). Power of absolute moment tests against symmetric non-normal alternatives. Ph.D. dissertation, University Microfilms, Ann Arbor MI.

Thode, Jr., H.C., Finch, S.J., and Mendell, N.R. (1988). Simulated percentage points for the null distribution of the likelihood ratio test for a mixture of two normals. Biometrics 44, 1195-1201.

Thode, Jr., H.C., Smith, L.A., and Finch, S.J. (1983). Power of tests of normality for detecting scale contaminated normal samples. Communications in Statistics - Simulation and Computation 12, 675-695.

Tukey, J.W. (1960). A survey of sampling from contaminated distributions. In I. Olkin, S.G. Ghurye, W. Hoeffding, W.G. Madow, and H.B. Mann, eds., **Contributions to Probability and Statistics**, Stanford Univ. Press, CA, 448-485.

Uthoff, V.A. (1968). Some scale and origin invariant tests for distributional assumptions. Ph.D. dissertation, University Microfilms, Ann Arbor MI.

CHAPTER 2

PLOTS, PROBABILITY PLOTS AND REGRESSION TESTS

"Graphical methods provide powerful diagnostic tools for confirming assumptions, or, when the assumptions are not met, for suggesting corrective actions. Without such tools, confirmation of assumptions can be replaced only by hope."

J.M. Chambers, W.S. Cleveland, B. Kleiner and P.A. Tukey, 1983

"There is no excuse for failing to plot and look."

J.W. Tukey, 1977

The importance of plotting can never be overemphasized in statistics. There are recommended plots for virtually every statistical method, for example, scatter plots, residual plots and diagnostic plots for regression. Plots are also important in the areas of goodness of fit and fitting distributions; they provide a sense of pattern and a level of detail not available in a single test statistic. Although formal testing procedures allow an objective judgment of normality (i.e., significance vs non-significance at some α level), they do not generally signal the reason for rejecting a null hypothesis, nor do they have the ability to compensate for masking effects within the data which may cause acceptance of a null hypothesis. Therefore, we begin our quest for normality by suggesting basic plotting procedures for the display of a single sample of data.

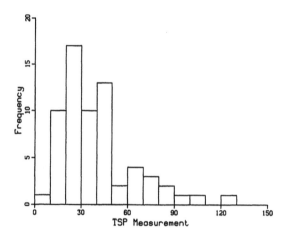

Figure 2.1 Histogram of 65 TSP measurements taken at a remote measuring site near the Navajo Generating Station.

Following a short description of these methods, we will describe plotting methods more specific to the assessment of normality, probability plots. Probability plots are graphical procedures in which sample order statistics are compared to some theoretical or expected values of the order statistics. This allows subjective judgments of distributional shape. In particular, normal quantile-quantile (Q-Q) and percent-percent (P-P) plots approximate straight lines if a sample comes from a normal distribution. Systematic nonlinearity in a probability plot provides an indication of the type of deviation from normality.

We will conclude this chapter with a description of regression, correlation and related tests for normality, included here because they are an obvious extension of probability plots. Regression or correlation tests are based on measures of linear correlation in probability plots. In contrast to probability plots, regression tests are formal procedures which can be used to objectively assess normality. However, the features of a set of data which cause the non-normality can not be determined solely on the basis of the test. It is therefore recommended that a test for normality, be it a regression test or other type of test, be done in conjunction with a raw data plot and a probability plot.

2.1 Raw Data Plots: Preliminary Plotting Procedures

It is strongly recommended that any statistical analysis always include a graphical inspection of the data (whether the assumption of normality is

```
          Cum.
Freq.    Freq.

   1        1       0  I  7
  10       11       1  I  1366778899
  17       28       2  I  00111234447788889
  10       38       3  I  0111235679
  13       51       4  I  0111222226677
   2       53       5  I  15
   4       57       6  I  4459
   3       60       7  I  244
   2       62       8  I  18
   1       63       9  I  1
   1       64      10  I  1
   0       64      11  I
   1       65      12  I  4
```

Figure 2.2 Stem-and-leaf plot of 65 TSP measurements.

of interest or not). For univariate data there are many types of raw data displays available, such as histograms, stem-and-leaf plots, and box plots. Since these are all fairly well known, we will limit our discussion of basic plotting procedures to a brief description of these three data displays. To some degree, each of these allows a subjective consideration of the symmetry and shape of a data density, as well as identification of other features, such as multimodality, outliers and data clusters. Moreover, sophisticated computer graphics and calculations are not required; for small and moderate sample sizes, they are easily done by hand.

Histograms or frequency distributions are displays of the distribution of a set of data. Distributional features which can be observed using this type of display include spread, symmetry, outliers and multimodality. A histogram of 65 air measurements (Data Set 1) of total suspended particulates (TSP) made at a remote measuring station in the vicinity of the Navajo Generating Station, Arizona, in 1974 is shown in Figure 2.1. From the histogram we can determine that these data are skewed, with one or more possible outliers at the high end of the data scale.

Stem-and-leaf plots (Tukey, 1977) provide the same information as histograms plus a further level of detail in the examination of data. Similar to histograms, they show the frequency of observations within bins, but also show the data values, allowing identification of the median, extremes and other sample percentiles directly from the plot, as well as groups of tied (or otherwise clustered) data values. Figure 2.2 is a stem-and-leaf plot of the TSP data; from this plot we can quickly identify the median (32) and fourths (22, 46) of the data set (Section 12.3.2), as well as values of the minimum and maximum.

Box plots (or box-and-whisker plots, Tukey, 1977) show less data de-tail than histograms or stem-and-leaf plots, but still give indications of the spread and symmetry of a data set, and are slightly more formal than his-tograms or stem-and-leaf plots in identifying outliers in a data set. Figure 2.3 is a box plot of the TSP data. The width of the box is the fourth spread (Section 12.3.2), which contains 75% of the observations. The bar inside the box indicates the location of the median. The whiskers extend to the observation in each tail of the data which is the furthest from the fourths but less than or equal to 1.5 times the fourth spread. Observations outside of the whiskers may be considered outliers (similar to "skipping", see Section 12.2.7), with different levels of "outlierness" depending on the distance from the fourths. Symmetry or skewness of the data can be as-certained from the location of the median within the box and the length of the whiskers on either side.

In Figure 2.3 we see that while the main body of the data is fairly symmetric, the few large values cause the sample to be skewed. The ob-servation with a value of 124 is identified as an outlier, while the four observations between 81 and 101 are considered "large" (Tukey, 1977).

Box plots are easily created from stem-and-leaf plots, where percentiles can be picked off quickly. Box plots can also be valuable for the simulta-neous display of several univariate sets of data.

2.2 Methods of Probability Plotting

The most common type of normal probability plot is the quantile-quantile $(Q\text{-}Q)$ plot. Percent-percent $(P\text{-}P)$ plots, plots of the empirical cumula-tive distribution function (e.g., Wilk and Gnanadesikan, 1968) and other variants of probability plots are also informative. In addition to providing insight to the evaluation of normality and/or type of deviation from nor-mality, some probability plots can provide estimates of the mean, standard deviation and percentiles of the distribution if the data are near normal.

2.2.1 Plotting Positions

Probability plots are plots of sample order statistics against some "ex-pected" values of the order statistics $(Q\text{-}Q$ plots), or against some uniform order statistic $(P\text{-}P$ plots). Regression tests are measures of the linearity in probability plots. The choice of a value for the abscissa of the plot is, or is dependent upon, an estimate of the empirical cumulative distribution

MINIMUM   MAXIMUM

```
            ----------------
--------I    x     I--------------------- * *    *        o
            ----------------
                                        81
 7     22   32      46                   88 91  101              124
```

Figure 2.3 Box plot of 65 TSP measurements.

function (ecdf) of the hypothesized null distribution. The ecdf estimates
used in plotting and regression tests are called the plotting positions. To
some extent, results of the plotting or testing procedure can be affected by
the choice of the plotting position.

Blom (1958) gave a general approximation for the means of order
statistics under any null distribution by using plotting positions, p_i, of
the form

$$p_i = \frac{i - \alpha_i}{n - 2\alpha_i + 1} \tag{2.1}$$

$0 \leq \alpha < 1$, noting that α_i is different for each n, i and null distribution.
Estimates of the mean order statistics can then be obtained by taking the
inverse cdf of the null distribution

$$E(X_{(i)}) = \Phi^{-1}(p_i)$$

where E is the expected value taken under the null (normal in this case)
distribution. Because of the small variations in the α_i in (2.1), for simplicity
he suggested using a constant value α, either $\alpha = 0.375$ or $\alpha = 0.5$, resulting
in the plotting positions

$$p_i = \frac{(i - 0.375)}{(n + 1/4)} \tag{2.2}$$

or

$$p_i = \frac{(i - 0.5)}{n}. \tag{2.3}$$

Blom's main concern was with the estimation of σ from the probability
plot rather than with goodness of fit. He suggested that (2.2) gives a
nearly unbiased estimate of σ whereas (2.3) gives a biased estimate of σ

Table 2.1 Plotting positions for probability plots.

Method	Plotting Position	Source
I	$\Phi(E(X_{(i)}))$	$E(X_{(i)})$ is tabulated (Harter, 1961) or calculated directly (Royston, 1982b)
II	$\Phi(med(X_{(i)}))$	
III	$\Phi(mode(X_{(i)}))$	
IV	$E(U_{(i)}) = i/(n+1)$	Kimball (1960); Filliben (1975)
V	$med(U_{(i)})$	
VI	$mode(U_{(i)}) =$ $(i-1)/(n-1)$	Filliben (1975); not useful for plotting at $i = 1$ or $i = n$
VII	$(i - 0.375)/(n + 0.25)$	Estimate of I (Blom, 1958)
VIII	$(i - 0.5)/n$	Estimate of I (Blom, 1958); attributed to Hazen (1930)
IX	$(i - 0.3175)/(n + 0.365)$	Estimate of V (Filliben, 1975)
X	$(i - 0.3)/(n + 0.4)$	Estimate of V (Benard and Bos-Levbach, 1953)
XI	$(i - 0.4)/(n + 0.2)$	Cunane (1978)
XII	$(3i - 1)/(3n + 1)$	BMDP (1983)
XIII	$(i - 0.44)/n + 0.12$	Gringorten (1963); Mage, (1982)
XIV	$(i - 0.567)/(n - 0.134)$	Larsen, Curran and Hunt (1980); Mage (1982)

with minimum mean square error. The plotting position (2.3) was also the choice of Hazen (1930).

One of the most common plotting positions is

$$p_i = \frac{i}{n+1} \tag{2.4}$$

which is the expected value of the distribution function of the standard normal variate, i.e., $p_i = E(\Phi^{-1}(Z_{(i)})) = E(U_{(i)})$ where $U_{(i)}$ are the order statistics for a uniform random variable. This plotting position is (2.1) with a constant value of $\alpha = 0$.

Kimball (1960) identified three objectives which might be considered when choosing a plotting position: (1) to determine if the sample comes from the null distribution; (2) as a short-cut method of obtaining an estimate of the standard deviation; and, (3) graphical extrapolation at one of the extremes. He recommended (2.2) as the plotting position for objectives (1) and (2); however, his choice was not based on the evaluation of formal regression tests.

Kimball (1960) and Mage (1982) indicated commonly used plotting positions (Table 2.1): most are based on (2.1) for some choice of constant α. Looney and Gulledge (1984, 1985a, 1985b) used various plotting positions in evaluations of probability plot correlation tests (Section 2.3.2).

2.2.2 Quantile-Quantile Plots

Quantile-quantile (Q-Q) plots are plots of sample order statistics against some "expected" quantiles from a standard normal distribution. The underlying theory behind a normal probability plot is that the plot will be linear except for random fluctuations in the data under the null hypothesis. Any systematic deviation from linearity in the probability plot indicates that the data are not normal.

The first step in making a Q-Q plot is to sort the observations to obtain the order statistics $x_{(1)} \leq x_{(2)} \leq \ldots \leq x_{(n)}$. These are the empirical quantiles used as the ordinates on the plot. Values for the abscissa of the empirical quantiles must now be chosen. Since a probability plot is a judgmental rather than formal testing procedure for assessing normality, the choice of plotting position is less critical than for regression tests (Section 2.3). The most common plotting positions for Q-Q plots are $p_i = (i - .5)/n$ and $p_i = i/(n + 1)$. Use of i/n as a plotting position prevents the use of the largest value in the plot. A discussion of the relative merits of these and other plotting positions is given in Section 2.2.1.

The pairs $(\Phi^{-1}(p_i), x_{(i)})$ are then plotted; ascertainment of the linearity of the plot is all that remains. Systematic deviation from linearity in a Q-Q plot is manifested according to the type of alternative the data comes from. Figure 2.4a is a normal Q-Q probability plot of 50 randomly generated normal variates using (2.4) as the plotting position.

A long-tailed density ($\beta_2 > 3$) is indicated when the lower tail turns downward and the upper tail curves upward. Figure 2.4b is a probability plot of a random sample from a double exponential distribution ($\beta_2 = 6$). Figure 2.4c shows a normal probability plot for a sample from a short-tailed symmetric (uniform) distribution. Short-tailed distributions are indicated by an S-shaped pattern. A normal Q-Q plot of any symmetric distribution is typically symmetric and linear in the center of the data.

Figure 2.4d shows a random sample from a right skewed (in this case, exponential) density; here the data show a U-shaped or "humped" pattern. If the data were left skewed, the plot would be concave downward.

Figure 2.4e shows a gap between two fairly linear components of the probability plot. This is the pattern for a mixture of two normal components with a fair amount of separation between components (Chapter 11).

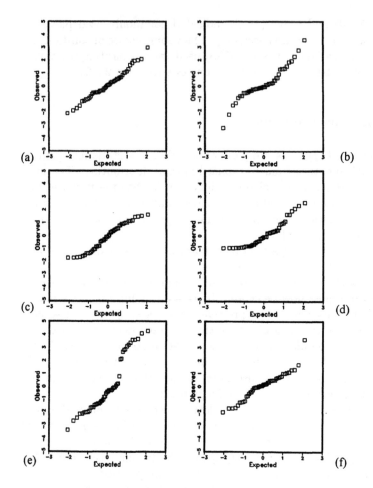

Figure 2.4 Normal probability Q-Q plots for 50 random numbers generated from (a) normal; (b) double exponential; (c) uniform; (d) exponential; (e) mixture of two normal components with same variance but different means; (f) normal distribution with a single outlier.

Finally, Figure 2.4f shows a normal sample with a single outlier: this is indicated by an approximately linear probability plot with a single outlying point.

　　　If the plot is acceptably linear then normal parameter and percentile estimates may be obtained from the plot. Under normality the transformation of normal values to standard normal values is given by $z = (x - \mu)/\sigma$ or $x = \mu + \sigma z$. From this it is obvious that if the data are sufficiently linear in the plot to accept normality, then a regression of the sample quantiles on the expected quantiles yields an equation where the slope is an estimate of σ and the intercept is an estimate of the mean μ. An eyeball estimate of μ and σ can be obtained based on a hand drawn best-fitting line, or the slope and intercept can be calculated using least squares or some other type of regression. Alternatively, if a fitted line is drawn, a line drawn perpendicularly from the abscissa value of 0.5 to the fitted line and then projected from this point to the y-axis gives an estimate of the median of the distribution; this value can also be used as an estimate of the mean. Similarly, by repeating this procedure at the values of $p = 0.025$ and 0.975, the 2.5th and 97.5th percentiles can be estimated. Since the distance between these two percentiles is approximately 4 standard deviations, an estimate of σ can be obtained. Finally, any percentile of the distribution can be estimated using the projection of lines from the corresponding p value on the x-axis to the line and then to the y-axis.

It is necessary to be cautious when estimating high or low percentiles or the standard deviation using percentiles since the farther out in the tails one goes, the larger the variance of the estimated percentile. Indeed, if the data are not linear in the plot, none of the estimates except the median are valid.

By investing in normal probability paper, a Q-Q plot can easily be produced by hand. The abscissa on probability paper is scaled in p proportionally to the expected quantiles of a standard normal distribution so that a plot of $(p, \Phi^{-1}(p))$ is linear. The abscissa limits typically run from 0.0001 to 0.9999. The vertical scale is linear and does not require that the data be standardized in any manner; also available is probability paper that is scaled logarithmically on the y-axis for use in determining whether data is lognormally distributed. On probability paper, the pairs $(p_i, x_{(i)})$ are plotted.

For plots done by hand, the advantage of Q-Q plots done on normal probability paper is that percentiles and cumulative probabilities can be directly estimated, and $\Phi^{-1}(p_i)$ need not be obtained to create the plot.

2.2.3 Percent-Percent Plots

Percent-percent (P-P) plots are usually limited to comparisons of two samples rather than between a sample and a theoretical distribution (Wilk and Gnanadesikan, 1968). Therefore, relatively little use has been made of P-P plots in the area of goodness of fit. However, Gan and Koehler (1990) suggested a standardized P-P plot for use with a single sample of data; their main purpose was to introduce some goodness of fit tests based on the correlation between sample and theoretical probabilities (Section 2.3.2). They

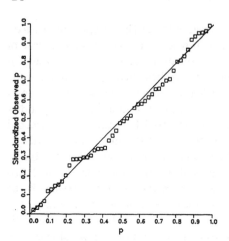

Figure 2.5 Standardized normal P-P plot for 50 normal random numbers (from Figure 2.4a).

presented results for a variety of null distributions, but we will limit our discussion to those for the normal case.

Their plotting method consists of plotting the pairs (p_i, Z_i), where $Z_i = \Phi((x - \hat{\mu})/\hat{\sigma})$. As with Q-Q plots, this is approximately a straight line if the sample comes from a normal distribution. They used the plotting position $p_i = i/(n + 1)$ for both their probability plots (Figure 2.5) and their tests.

In contrast to Q-Q plots, the baseline comparison of a P-P plot is always the 45° line between $(0, 0)$ and $(1, 1)$ on the probability plot; this eliminates the need to determine the best fitting straight line for comparison as in Q-Q plots. Non-normal data are indicated in the same fashion as Q-Q plots, e.g., S-shaped curves for symmetric short-tailed densities. However, parameter estimates and percentiles cannot be directly determined using a P-P plot.

2.2.4 Empirical Cumulative Distribution Function Plots

Empirical cumulative distribution function (ecdf) plots are plots of the sample order statistics with a plotting position (Figure 2.6); the choice of plotting position in an ecdf plot has not been the subject of as much scrutiny as it has in Q-Q plots or regression tests. The visual detection of non-normality in ecdf plots is more difficult than in Q-Q plots, especially for symmetric non-normal densities since all bell-shaped densities result in an S-shaped ecdf curve; ecdf plots are more effective at exposing outliers or

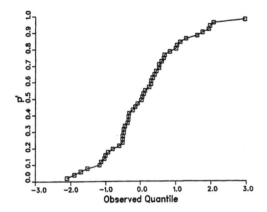

Figure 2.6 Empirical cumulative distribution function plot of 50 random normal variates (from Figure 2.4a).

skewness in a set of observations. Skewed densities, depending upon their degree of skewness, could have an asymmetric S-shape or could be concave or convex.

The advantages of ecdf plots in detecting normality are that the choice of plotting position is not of major concern; they are useful for detecting outliers, asymmetry and multimodality, although whether they are better than Q-Q plots is arguable; and, no special plotting paper or Φ^{-1} transformation is needed for the plot. As with Q-Q plots on probability paper, estimates of percentiles can be obtained directly. Empirical distribution function (EDF) tests for goodness of fit (Section 5.1) are based on the differences between the observed and theoretical ecdf.

2.2.5 Detrended Probability Plots

Another useful type of probability plot is the detrended Q-Q plot where $(x_{(i)} - \hat{\sigma}\Phi^{-1}(p_i))$ is plotted against the plotting position p_i or the expected quantile $\Phi^{-1}(p_i)$ for some estimate of the standard deviation $\hat{\sigma}$. If the observations come from a normal distribution, the result should be a straight line with zero slope.

These plots, like Q-Q plots, will be linear under a normal distribution regardless of the location and scale parameters; however, the line will be centered around the mean of the observed data. Percentiles can not be estimated. Non-normality is manifested in patterns similar to those observed in Q-Q plots. Figure 2.7a is a detrended Q-Q plot using p_i as the abscissa

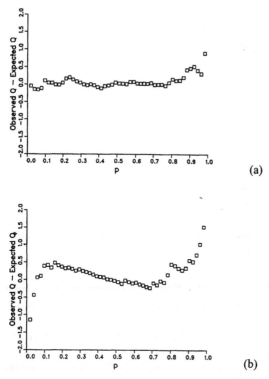

(a)

(b)

Figure 2.7 Detrended normal probability Q-Q plots for 50 random variates from (a) normal distribution (Figure 2.4a); (b) double exponential distribution (Figure 2.4b).

for a sample of 50 random normal variates; Figure 2.7b is the same plot for 50 random variates from a double exponential distribution.

2.3 Regression Tests

Although graphical methods are valuable for the identification of distributional characteristics, probability plots "can be sensitive to random occurrences in the data and sole reliance on them can lead to spurious conclusions" (D'Agostino, 1986). Therefore, it is also necessary to have more objective methods which can be used when it is necessary to verify distributional assumptions.

The expected relation between the $x_{(i)}$ and the $\Phi^{-1}(p_i)$ in a Q-Q plot intuitively suggests the use of a measure of the linear correlation in a probability plot as a location and scale free test for normality. Regression tests are all single-tailed omnibus tests of the form $T \leq c$ for the appropriate constant c, since by design the correlation in a probability plot is always positive with perfect correlation giving a value of unity.

2.3.1 Wilk-Shapiro Test

If a set of observations, x_i, come from a normal distribution, then on a normal probability plot

$$x_i = \mu + \sigma z_i.$$

If we denote the expected value of the ith order statistic by $\mathrm{E}(x_{(i)}) = w_i$ and $\mathbf{V}$ is the covariance matrix of the order statistics $x_{(1)}, \ldots, x_{(n)}$, then the best linear unbiased estimate of σ is obtained from the generalized least squares regression of the sample order statistics on their expected values, which is (up to a constant),

$$b = \mathbf{a}'\mathbf{x}$$

where $a = (\mathbf{w}'\mathbf{V}^{-1})/(\mathbf{w}'\mathbf{V}^{-1}\mathbf{V}^{-1}\mathbf{w})^{1/2}$, $\mathbf{a}'\mathbf{a} = 1$. In the vector $\mathbf{a}$, the entry $a_i = -a_{n-i+1}$ so that b can be calculated by

$$b = \sum_{i=1}^{[n/2]} a_{n-i+1}\left(x_{(n-i+1)} - x_{(i)}\right) \tag{2.5}$$

for the appropriate values a_i. The Wilk-Shapiro test statistic is defined as

$$W = b^2/((n-1)s^2).$$

where s^2 is the sample variance. Table B1 contains the coefficients a_i for samples up to size 50, and Table B2 contains selected percentiles for W.

One difficulty with W is that the elements of $\mathbf{V}$ (and hence the a_i) are known exactly only for samples up to size 20 (Sarhan and Greenberg, 1956). For sample sizes 21 up to 50, Shapiro and Wilk (1965) gave estimates of the a_i. A computational difficulty with W is that the calculation of b in (2.5) requires a different set of coefficients for each sample size, which decreases the portability of W. Extension of W to higher sample sizes would make this test even more cumbersome.

Royston (1982a, 1982c) presented a transformation of W to normality which was valid for sample sizes $7 \leq n \leq 2000$. The transformation was of the form

$$z = ((1 - W)^\lambda - \mu_y)/\sigma_y \tag{2.6}$$

Although his approximation still required expected values of normal order statistics and the covariance matrix, he used the Shapiro and Wilk (1965) approximations of $\mathbf{a}^* = \mathbf{w}'\mathbf{V}^{-1}$ for $\mathbf{a}$,

$$\hat{a}_i^* = \begin{cases} 2w_i & 1 < i < n \\ \left(\frac{\hat{a}_1^2}{1-2\hat{a}_1^2}\sum_{i=2}^{n-1}\hat{a}_i^2\right)^{1/2} & i=1,\ n \end{cases}$$

with

$$\hat{a}_1^2 = \hat{a}_n^2 = \begin{cases} g(n-1) & n \le 20 \\ g(n) & n > 20 \end{cases}$$

and

$$g(n) = \frac{\Gamma(n/2+.5)}{\sqrt{2}\Gamma(n/2+1)}.$$

Royston also used an approximation of $g(n)$,

$$g(n) = \left[\frac{6n+7}{6n+13}\right]\left[\frac{\exp(1)}{n+2}\left(\frac{n+1}{n+2}\right)^{n-2}\right]^{1/2}.$$

The parameters λ, μ_y and σ_y are estimated based on models dependent only on n, one model for $7 \le n \le 20$ and one for $21 \le n \le 2000$. These models are of the form

$$\sum \beta_i(\log(n)-d)^i$$

and are used to estimate λ, $\log(\mu_y)$ and $\log(\sigma_y)$. The estimates are used to transform W to a standard normal deviate (2.6), and normality is rejected if z is greater than the upper α critical value from a standard normal distribution. Model coefficients and values of d for each of the three parameters and two sets of sample sizes are given in Table 2.2.

Example 2.1. Darwin gave heights for 15 pairs of self- and cross-fertilized Zea mays plants grown in the same pots (Fisher, 1971); the paired differences in eighths of an inch (Data Set 2) are shown in Figure 2.8 as a Q-Q plot. The plot indicates a possible mixture of two normal components, or two outliers. Calculation of W using (2.5) results in a value of 0.9006, just under the $\alpha = 0.10$ critical value of 0.901 for a sample of size 15.

Estimates of the transformation parameters are $\lambda = 0.1079$, $\mu = 0.724$ and $\sigma = 0.00427$. Applying the transformation to the calculated W results in a value of $z = 1.293$, $\alpha = 0.098$.

Table 2.2 Model coefficients for estimation of parameters used to transform the Wilk-Shapiro test to normality.

parameter	λ	$\log(\mu_y)$	$\log(\sigma_y)$
$7 \leq n \leq 20$			
d	3	3	3
β_0	0.118898	-0.37542	-3.15805
β_1	0.133414	-0.492145	0.729399
β_2	0.327907	-1.124332	3.01855
β_3	-	-0.199422	1.558776
$21 \leq n \leq 2000$			
d	5	5	5
β_0	0.480385	-1.91487	-3.73538
β_1	0.318828	-1.37888	-1.015807
β_2	0	-0.04183209	-0.331885
β_3	-0.0241665	0.1066339	0.1773538
β_4	0.00879701	-0.03513666	-0.01638782
β_5	0.002989646	-0.01504614	-0.03215018
β_6	-	-	0.003852646

Royston (1986) gave a modified version of the Wilk-Shapiro test statistic for when there are ties in the data.

2.3.2 Probability Plot Correlation Tests

Probability plot correlation test statistics are the Pearson correlations between a set of expected normal order statistics and the sample order statistics. Different authors have investigated different sets of expected order statistics, based on a selected plotting position. Probability plot correlation tests can easily be extended to other null distributions although the distributions of the test statistics are different.

Shapiro-Francia Test

As an approximation to W, Shapiro and Francia (1972) suggested that for larger sample sizes the order statistics can be treated as independent. By circumventing the necessity of knowing $\mathbf{V}$, they obtained the least squares regression of the sample order statistics on the expected values,

$$W' = (\mathbf{a}^*\mathbf{x})^2 / ((n-1)s^2)$$

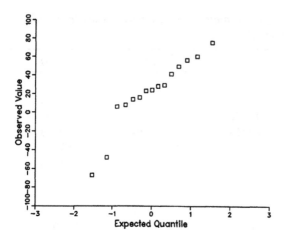

*Figure 2.8 Q-Q plot of Darwin's data on paired differences between self-
and cross-fertilized plants ($n = 15$).*

where

$$\mathbf{a}^* = \mathbf{w}'/\sqrt{(\mathbf{w}'\mathbf{w})}.$$

W' as given above is the squared Pearson correlation between the elements
of $\mathbf{a}^*$ and $\mathbf{x}$, i.e., the R^2 of the regression of $\mathbf{x}$ on $\mathbf{a}^*$.

The Shapiro-Francia test requires only that the expected values of
the order statistics, w_i, be known; these are given, for example, in Harter
(1961) for selected sample sizes up to 400. As with the Wilk-Shapiro test
statistic, the need for expected values makes this test unwieldy, although
the required values are available for more sample sizes. Royston (1982b)
gave a FORTRAN subroutine which calculates the expected normal order statistics, making storage of coefficients unnecessary to compute W'.
Shapiro and Francia (1972) and Weisberg (1974) gave empirical percentage
points for W'.

Weisberg and Bingham (1975) suggested using one of Blom's plotting
positions to obtain an approximation to w_i:

$$\tilde{w}_i = \Phi^{-1}\left(\frac{i - .375}{n + .25}\right)$$

and substituting $\tilde{w}$ into Shapiro and Francia's formula for W' to get an approximate (and easier to calculate) W test statistic. Although percentage
points for W' were given only for $n = 5$, 20 and 35, they suggested using
the critical values for W' for larger samples; using the critical values for
W for samples up to size 50 will result in only a slightly more conservative
test.

Filliben's Probability Plot Correlation

Filliben (1975) recommended the use of the correlation between the sample order statistics and estimated median values of the theoretical order statistics. For a sample of size n, Filliben used

$$m_i = \begin{cases} 1 - m_n & i = 1 \\ (i - .3175)/(n + .365) & 1 < i < n \\ (.5)^{1/n} & i = n \end{cases}$$

where the m_i were estimated order statistic medians from a uniform distribution. He then used the transformation $M_i = \Phi^{-1}(m_i)$ to obtain an estimate of the median value of the ith normal order statistic. The test statistic is

$$r = \frac{\sum_{i=1}^{n} x_{(i)} M_i}{\sqrt{\sum_{i=1}^{n} M_i^2} \sqrt{(n-1)s^2}}$$

since $\overline{M} = 0$. The advantages of this test over the Wilk-Shapiro test are that the expected values of the normal order statistics need not be calculated, and the coefficients from the generalized least squares regression do not need to be obtained.

Example 2.2. For the 65 TSP measurements in Data Set 1, the value of r is 0.934, which is less than 0.967, the 0.005-level critical value given by Filliben (1975).

Schlittgen's Percentiles

Using analysis of variance and exploratory data analysis techniques for two-way tables, Schlittgen (1988) determined models for obtaining the percentage points of the probability plot correlation test using Hazen's plotting position, $p_i = (i - .5)/n$. Two models were required, one for $5 \le n \le 10$ and another for $11 \le n \le 200$. These models were of the form

$$\left[ln \left(\frac{\rho_{n,\alpha}}{1 - \rho_{n,\alpha}} \right) \right]^{\lambda} = \beta_0 + \beta_1 n^{\delta} + \beta_2 n^{2\delta} + \beta_3 \mid ln(\alpha) \mid^{\eta} + \beta_4 \mid ln(\alpha) \mid^{2\eta}$$

where $\rho_{n,\alpha}$ is the lower-α percentage point of the correlation for a sample of size n. The parameter values for the two models are given in Table 2.3.

Table 2.3 Schlittgen's model parameters for obtaining critical values for probability plot correlation tests.

parameter	$5 \leq n \leq 10$	$11 \leq n \leq 200$
λ	1.23	1.44
δ	0.23	-0.12
η	0.55	0.50
β_0	2.59693	51.68485
β_1	1.71096	-97.92602
β_2	0.266912	53.91643
β_3	-2.25466	-3.23570
β_4	0.237798	0.257537

Example 2.3. For a sample of 25 observations and a correlation test at the $\alpha = 0.05$ level, Schlittgen's model gives a critical value of 0.959. By comparison, Filliben (1975) gave a critical value of 0.958 for his test, and the critical value is 0.959 for using either Cunnane's plotting position (Looney and Gulledge, 1984) or Blom's plotting position (Looney and Gulledge, 1985b). The critical value of the Wilk-Shapiro test for a sample of 25 is 0.918 (Shapiro and Wilk, 1965); however, this test statistic is analagous to a squared correlation test statistic, and the square root of the critical value is 0.958.

Schlittgen also gave the model parameters for the correlation between Hazen's plotting position and sample order statistics for the uniform null distribution.

Other Q-Q Correlation Tests

There is no theoretical underpinning which makes the Shapiro-Francia, Filliben or Hazen choices of plotting position superior to any other for correlation tests of fit. For these and other plotting positions Looney and Gulledge (1984, 1985a, 1985b) obtained null percentage points and compared the correlation tests in terms of power to see if any resulted in an improved test for normality, using the simple Pearson correlation as the test statistic for each plotting position. They compared the seven plotting positions

(VIII) $p_i = (i - .5)/n$ (VII) $p_i = (i - .375)/(n + .25)$
(IV) $p_i = i/(n + 1)$ (X) $p_i = (i - .3)/(n + .4)$
(I) $p_i = \Phi(E(x_{(i)}))$ (XI) $p_i = (i - .4)/(n + .2)$
(IX) $p_i = (i - 0.3175)/(n + 0.365)$

(Roman numerals correspond to those in Table 2.1). Their choice of plotting positions was based on the following criteria: VIII and IV are the most common Q-Q plotting positions; I is the expected value of the ith order statistic and is commonly used for testing purposes (Shapiro-Francia test); VII is Blom's estimate used to obtain the expected value; X and IX are suggested estimates of the median value of order statistics; Cunnane (1978) is the source for XI. There is virtually no difference in the lower percentile points of these seven test statistics; hence, Schlittgen's model should be useful for obtaining critical values for any plotting position. Also, the similarity between these tests and the Wilk-Shapiro W test would indicate that Royston's transformation to normality (Section 2.3.1) can be used to determine significance.

The results of their power study are discussed in Chapter 7.

Percent-Percent Plot Correlation

Gan and Koehler (1990) proposed two tests based on the correlation in a standardized P-P plot. Using the Weibull plotting position $p_i = i/(n + 1)$, both test statistics are essentially the squared correlations from the standardized probability plot defined by either:

$$k^2 = \frac{[\sum_{i=1}^{n}(Z_{(i)} - \overline{Z})(p_i - \overline{p})]^2}{\sum_{i=1}^{n}(Z_{(i)} - \overline{Z})^2 \sum_{i=1}^{n}(p_i - \overline{p})^2} \qquad (2.7)$$

or

$$k_0^2 = \frac{[\sum_{i=1}^{n}(Z_{(i)} - 0.5)(p_i - 0.5)]^2}{\sum_{i=1}^{n}(Z_{(i)} - 0.5)^2 \sum_{i=1}^{n}(p_i - 0.5)^2} \qquad (2.8)$$

where $Z_{(i)} = \Phi[(x_{(i)} - \hat{\mu})/\hat{\sigma}]$. Note that for (2.7), $\overline{p} = 0.5$ for their choice of plotting position. When $\overline{Z}$ is not close to 0.5 these two tests have different characteristics. Similar to correlation tests based on Q-Q plots, the distribution and critical values of these tests are dependent on the specified null distribution.

For all null distributions, they were able to estimate the αth lower tail percentage point for k_0^2 and k^2 using the formula

$$1 - k_\alpha^2 = (a_\alpha + b_\alpha pn)^{-1} \qquad (2.9)$$

Table 2.4 Model parameters used to obtain critical values for Gan and Koehler's percent-percent probability plot correlation test.

percentile	test	a_α	b_α
0.001	k^2	0.9536	0.3677
	k_0^2	0.5974	0.3245
0.005	k^2	0.5985	0.4785
	k_0^2	0.4178	0.4310
0.01	k^2	0.1898	0.5505
	k_0^2	0.4522	0.4941
0.05	k^2	0.7419	0.7733
	k_0^2	0.7420	0.7098
0.10	k^2	1.5880	0.9379
	k_0^2	1.0953	0.8633

although coefficients are different for each null distribution. The coefficients a_α and b_α for selected percentiles are presented in Table 2.4 for the normal null distribution.

Example 2.4. For the Darwin data, we obtain values of $k^2 = 0.938$ and $k_0^2 = 0.927$. For a sample of size 15, the 0.05-level critical values using (2.9) and the coefficients from Table 2.4 are 0.919 for k^2 and 0.912 for k_0^2, neither test rejecting the null hypothesis of normality.

2.3.3 LaBreque's Tests for Nonlinearity

LaBreque (1977) developed three tests designed to detect systematic nonlinear components in a probability plot. Under normality, $E(X_i) = \mu + \sigma E(Z_i)$ where Z_i is a standard normal random variable. If the data are not normal, then

$$E(X_i) = \mu + \sigma \xi_i + \alpha \phi_2(\xi_i) + \beta \phi_3(\xi_i) \tag{2.10}$$

where $\xi_i = E(Z_i)$ and ϕ_j is a jth degree polynomial. Using orthogonal polynomials, LaBreque obtained three test statistics F_1, F_2, F_3 corresponding to (2.10) and the models

$$E(X_i) = \mu + \sigma \xi_i + \alpha \phi_2(\xi_i) \tag{2.11}$$

Table 2.5 Model parameters for obtaining critical values for LaBreque's nonlinearity tests ($n \geq 12$).

percentile	test	a_0	a_1
0.01	F_1	4.618	1.067
	F_2	6.900	-2.948
	F_3	6.893	-1.087
0.05	F_1	2.992	-0.3591
	F_2	3.822	-0.1809
	F_3	3.815	-0.4944
0.10	F_1	2.331	-0.5660
	F_2	2.682	0.1450
	F_3	2.667	-0.1937

$$E(X_i) = \mu + \sigma\xi_i + \beta\phi_3(\xi_i) \tag{2.12}$$

The quadratic equation (2.11) seems the intuitive model for asymmetric alternatives, where the probability plot is U-shaped (or inverted U). The third degree polynomial (2.12) seems intuitive for detecting S-shaped probability plots, i.e., non-normal symmetric data. The combined (omnibus) test statistic is given by (2.10).

The test statistics are

$$F_j = D_j/ks^2$$

where D_j is the reduction in residual sum of squares for model j and k is the number of parameters used. These test statistics are calculated as:

$$F_1 = ((\hat{\alpha}/c_\alpha)^2 + (\hat{\beta}/c_\beta)^2)/2s^2$$
$$F_2 = (\hat{\alpha}/c_\alpha)^2/s^2$$
$$F_3 = (\hat{\beta}/c_\beta)^2/s^2$$

which are independent of $\bar{x}$ and s^2, but not of each other. In fact, the omnibus test statistic, F_1, is found by taking the average of F_2 and F_3.

Similar to the Wilk-Shapiro W, these test statistics are formed by linear combinations of the order statistics,

$$\hat{\alpha}/c_\alpha = \begin{cases} \sum_1^{n/2} a_i(x_{(i)} - x_{(n-i+1)}) & \text{n even} \\ \sum_1^{[n/2]} a_i(x_{(i)} - x_{(n-i+1)}) + a_{(n+1)/2}x_{(n+1)/2} & \text{n odd} \end{cases} \tag{2.13}$$

$$\hat{\beta}/c_\beta = \sum_{i=1}^{[n/2]} b_i(x_{(i)} - x_{(n-i+1)}) \tag{2.14}$$

                Cum.
     Freq.   Freq.

       1   1   7 I 2
       5   6   8 I 01467
       2   8   9 I 24
       3   11   10 I 367
       6   17   11 I 125689
      12   29   12 I 233456667888
       2   31   13 I 23
       1   32   14 I 2

Figure 2.9 Stem-and-leaf plot of newborn baby birthweights (n = 32).

Unfortunately, similar to the Wilk-Shapiro test, $[n/2 + 1]$ coefficients are needed for the calculation of F_2 and F_3, and twice that number for F_1. The coefficients for (2.13) and (2.14) are obtained from

$$\hat{\alpha} = c_\alpha^2 \Phi_2 V^{-1} x$$
$$\hat{\beta} = c_\beta^2 \Phi_3 V^{-1} x$$

where

$$c_\alpha^2 = (\Phi_2' V^{-1} \Phi_2)^{-1}$$
$$c_\beta^2 = (\Phi_3' V^{-1} \Phi_3)^{-1}$$
$$\Phi_2 = \xi^{(2)} - c_\mu^2 1' V^{-1} \xi^{(2)} 1$$
$$\Phi_3 = \xi^{(2)} - c_\sigma^2 1' V^{-1} \xi^{(3)} 1$$
$$c_\mu^2 = (1' V^{-1} 1)^{-1}$$
$$c_\sigma^2 = (\xi' V^{-1} \xi)^{-1}.$$

V is the variance-covariance matrix of the standard normal order statistics and $\xi^{(k)} = \{\xi_1^k, \xi_2^k, \ldots, \xi_n^k\}$. LaBreque gave the coefficients a_i and b_i for $n = 4(1)64$.

In contrast to regression tests, all three of these tests are significant for large values of the test statistic. LaBreque gave percentage points for F_1, F_2 and F_3 for $n = 4(1)12$, and models of the form $a_0 + a_1 n^{-1/2}$ for critical values for $n \geq 12$. The model parameters for 0.10, 0.05 and 0.01 tests are given here in Table 2.5; for other selected percentiles the model parameters can be found in LaBreque (1977).

Example 2.5. Armitage and Berry (1987) gave 32 newborn birth-weights (Data Set 3). Figure 2.9 is the stem-and-leaf plot for these data. The birth weights are skewed to the left, and may be bimodal. The values of LaBreque's test statistics for this data set are

$$F_1 = 12.1$$
$$F_2 = 23.4$$
$$F_3 = 0.72$$

with corresponding critical values obtained from the model of 2.93, 3.79 and 3.73. F_1 (the omnibus test) and F_2 (the skewness test) reject the normality of the sample, while F_3 does not.

Table B3 contains the coefficients needed to calculate Labreque's test statistics up to sample sizes of 64, and Table B4 contains critical values of the tests for sample sizes between 4 and 12.

2.4 Further Reading

Graphical methods for general exploratory examination of data can be found in a number of texts, such as Tukey (1977), Mosteller and Tukey (1977), and Hoaglin, Mosteller and Tukey (1983, 1985). FORTRAN and BASIC programs for some of these methods are provided in Velleman and Hoaglin (1981). Descriptions of exploratory plots and probability plotting are provided in Chambers et al. (1983). Probability plotting methods and regression/correlation tests (including those for non-normal null distributions) are described in the appropriate chapters of D'Agostino and Stephens (1986).

References

Armitage, P., and Berry, G. (1987). **Statistical Methods in Medical Research**. Blackwell Scientific Publications, Oxford, U.K.

Benard, A., and Bos-Levenbach, E.C. (1953). The plotting of observations on probability paper. Statistica Neerlandica 7, 163-173.

Blom, G. (1958). **Statistical Estimates and Transformed Beta Variables**. John Wiley and Sons, New York.

BMDP Statistical Software Manual, 1985 Reprinting (1983). University of California Press, Berkeley, CA.

Chambers, J.M., Cleveland, W.S., Kleiner, B., and Tukey, P.A. (1983). **Graphical Methods for Data Analysis**. Duxbury Press, Boston.

Cunnane, C. (1978). Unbiased plotting positions - a review. Journal of Hydrology 37, 205-222.

D'Agostino, R.B. (1986). Graphical analysis. In D'Agostino, R.B., and Stephens, M.A., eds., **Goodness-of-Fit Techniques**. Marcel Dekker, Inc., New York.

D'Agostino, R.B., and Stephens, M.A., eds. (1986). **Goodness-of-Fit Techniques**. Marcel Dekker, Inc., New York.

Filliben, J.J. (1975). The probability plot coefficient test for normality. Technometrics 17, 111-117.

Fisher, R.A. (1971). **The Design of Experiments**. Hafner Press, New York.

Gan, F.F., and Koehler, K.J. (1990). Goodness-of-fit tests based on P-P probability plots. Technometrics 32, 289-303.

Gringorten, I.I. (1963). A plotting rule for extreme probability paper. Journal of Geophysical Research 68, 813-814.

Harter, H.L. (1961). Expected values of normal order statistics. Biometrika 48, 151-165.

Hazen, A. (1930). **Flood Flows**. John Wiley and Sons, New York.

Kimball, B.F. (1960). On the choice of plotting positions on probability paper. Journal of the American Statistical Association 55, 546-560.

Hoaglin, D.C., Mosteller, F., and Tukey, J.W., eds. (1983). **Understanding Robust and Exploratory Data Analysis**. John Wiley and Sons, New York.

Hoaglin, D.C., Mosteller, F., and Tukey, J.W., eds. (1985). **Exploring Data Tables, Trends, and Shapes**. John Wiley and Sons, New York.

LaBreque, J. (1977). Goodness-of-fit tests based on nonlinearity in probability plots. Technometrics 19, 293-306.

Larsen, R.I., Curran, T.C., and Hunt, W.F., Jr. (1980). An air quality data analysis system for interrelating effects, standards, and needed source reductions: Part 6. Calculating concentration reductions needed to achieve the new national ozone standard. Journal of the Air Pollution Control Association 30, 662-669.

Looney, S.W., and Gulledge, Jr., T.R. (1984). Regression tests of fit and probability plotting positions. Journal of Statistical Computation and Simulation 20, 115-127.

Looney, S.W., and Gulledge, Jr., T.R. (1985a). Probability plotting positions and goodness of fit for the normal distribution. The Statistician 34, 297-303.

Looney, S.W., and Gulledge, Jr., T.R. (1985b). Use of the correlation coefficient with normal probability plots. American Statistician 39, 75-79.

Mage, D. (1982). An objective graphical method for testing normal distributional assumptions using probability plots. American Statistician 36, 116-120.

Mosteller, F., and Tukey, J.W. (1977). **Data Analysis and Regression**. Addison Wesley, Reading, Massachusetts.

Royston, J.P. (1982a). An extension of Shapiro and Wilks' W test for normality to large samples. Applied Statistics 31, 115-124.

Royston, J.P. (1982b). Algorithm AS 177. Expected normal order statistics (exact and approximate). Applied Statistics 31, 161-165.

Royston, J.P. (1982c). AS 181. The W test for normality. Applied Statistics 31, 176-180.

Royston, J.P. (1986). A remark on AS 181. The W test for normality. Applied Statistics 35, 232-234.

Sarhan, A.E., and Greenberg, B.G. (1956). Estimation of location and scale parameters by order statistics from singly and doubly censored samples, Part 1. Annals of Mathematical Statistics 27, 427-451.

Schlittgen, R. (1988). Goodness-of-fit tests for uniformity and normality based on the correlation coefficient. The Statistician 37, 375-381.

Shapiro, S.S., and Francia, R.S. (1972). Approximate analysis of variance test for normality. Journal of the American Statistical Association 67, 215-216.

Shapiro, S.S., and Wilk, M.B. (1965). An analysis of variance test for normality (complete samples). Biometrika 52, 591-611.

Tukey, J.W. (1977). **Exploratory Data Analysis**. Addison Wesley, Reading, Massachusetts.

Velleman, P.F., and Hoaglin, D.C. (1981). **Applications, Basics, and Computing of Exploratory Data Analysis**. Duxbury Press, Boston, Massachusetts.

Weisberg, S. (1974). An empirical comparison of the percentage points of W and W'. Biometrika 61, 644-646.

Weisberg, S., and Bingham, C. (1975). An approximate analysis of variance test for non-normality suitable for machine calculation. Technometrics 17, 133-134.

Wilk, M.B., and Gnanadesikan, R. (1968). Probability plotting methods for the analysis of data. Biometrika 55, 1-17.

CHAPTER 3

TESTS USING MOMENTS

"In the case of testing the hypothesis that a sample has been drawn from a normally distributed population, it seems likely that for large samples and when only small departures from normality are in question, the most efficient criteria will be based on the moment coefficients of the sample, e.g. on the values of $\sqrt{\beta_1}$ and β_2."

E.S. Pearson, 1935

Karl Pearson is credited with having been the first to recognize that deviations in distribution from the normal could, for the most part, be characterized by differences in the third and fourth standardized moments. It follows naturally that formal testing for normality could be accomplished by evaluating sample moments and comparing them to theoretical moments. This required knowledge of the distribution of the sample moments under normality.

Pearson (1930a), Fisher (1930), Hsu and Lawley (1939), and Geary and Worlledge (1947) contributed to the knowledge base of normal moments, calculating up to the seventh moment of the population kurtosis for normal samples. Attempts to approximate the distributions of sample skewness and kurtosis for normal samples, and hence obtain critical values for testing purposes, have continued since that time. One of the earliest attempts was Geary (1947), who fit the Gram-Charlier system to sample moments and absolute moments for large samples. D'Agostino (1970a) presented an approximately normal transformation of sample skewness to normality. Attempts to obtain similar results for kurtosis have been more difficult: small sample results have been mostly based on simulation. Shenton and Bowman (1975) have developed an adequate fit for kurtosis for moderate size samples.

Since the Wilk-Shapiro test was developed in 1965, other tests for normality have proliferated. However, none of these tests has proven to be better, in terms of having higher power, than moment tests under general circumstances.

In this chapter, we briefly review population moments (Section 3.1), and then describe moment tests for normality, with emphasis on skewness and kurtosis (Section 3.2). Absolute moment tests are discussed in Section 3.3. We conclude with Section 3.4, in which we present tests which have been shown to be similar to moment tests, either by mathematical form or by derivation.

3.1 Population Moments

One distinction which can be made between shapes of distributions is based on the comparison of the standardized central moments. The first two central moments of a probability density function $f(x)$ are defined as

$$\mu = \mu_1 = \int_{-\infty}^{\infty} x\, f(x) dx$$

$$\sigma^2 = \mu_2 = \int_{-\infty}^{\infty} (x - \mu)^2 f(x) dx$$

These are the location (mean) and scale (variance) parameters. They are by themselves usually of no interest when testing general distributional assumptions. Higher central moments are given by

$$\mu_k = \int_{-\infty}^{\infty} (x - \mu)^k f(x) dx$$

where k is an integer greater than 1. For k odd ($k > 2$), $\mu_k = 0$ when $f(x)$ is symmetric.

The standardized moments are the important values when comparing the shape of an empirical to that of a theoretical distribution, since they are location and scale invariant. For our purposes we may define the kth standardized moment as

$$\psi_k = \mu_k/\mu_2^{k/2}$$

so that the kth moment is adjusted for the population variance. Emphasis has been placed on the estimates of the third and fourth standardized moments in testing for normality. Classically, these parameters have been denoted as

$$\sqrt{\beta_1} = \mu_3/\mu_2^{3/2} = \mu_3/\sigma^{3/2}$$

and

$$\beta_2 = \mu_4/\mu_2^2 = \mu_4/\sigma^4.$$

For the normal distribution these values are 0 and 3, respectively. Figures 3.1a and 3.1b show densities skewed to the left ($\sqrt{\beta_1} < 0$) and to the right ($\sqrt{\beta_1} > 0$), respectively. Figures 3.1c and 3.1d show examples of "long-tailed" ($\beta_2 > 3$) and "short-tailed" ($\beta_2 < 3$) densities, respectively.

For the normal distribution the central moments are given by the formula

$$\mu_k = \begin{cases} \frac{k!\sigma^k}{(k/2)!2^{k/2}} & k \text{ even}, k \geq 2 \\\\ 0 & k \text{ odd}, k \geq 3 \end{cases}$$

The standardized moments for any normal distribution are the same as the moments for a N(0,1) density.

Since the variance of a standard normal distribution is unity and the mean is 0, the moments of a N(0, 1) density are defined as

$$\psi_k = \int_{-\infty}^{\infty} x^k \phi(x) dx$$

where $\phi(x)$ is the density of a standard normal random variable.

Comparison of the standardized moments estimated from a sample of data with the theoretical moments from a normal distribution seems a natural thing to do when testing distributional assumptions. Several problems arise, however:

(1) The population moments are not necessarily unique. For example, the odd moments for all symmetric distributions are 0, although there are skewed distributions with odd moments equal to 0 (Ord, 1968). Also, although most non-normal symmetric distributions have a fourth moment different from 3, there are some non-normal densities with a kurtosis value of 3 (e.g., Johnson, Tietjen and Beckman, 1980; Balanda and MacGillivray, 1988). Some examples of this phenomenon are shown in Figure 3.2.

(2) The different moments of a distribution indicate different characteristics. A test for skewness which does not reject normality for a given sample of data does not necessarily indicate that the data are acceptably normal. More than one moment test or a combination of tests may be

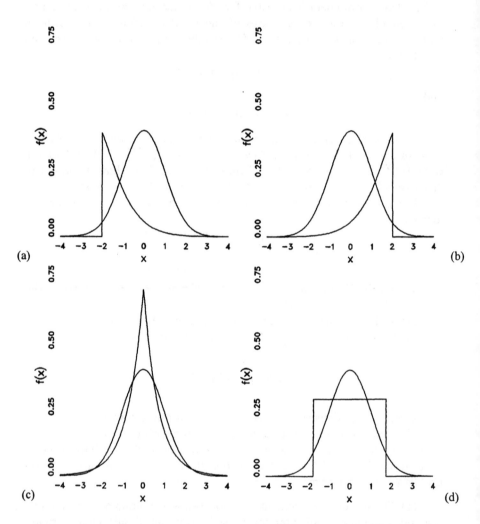

Figure 3.1 Probability densities which are skewed (a, b), "long-tailed" (c) and "short-tailed" (d) in comparison to the normal.

required unless some specific characteristic to be tested for is identified beforehand.

(3) The moment estimates are not independent (Bowman and Shenton, 1975); therefore, multiple testing procedures using a test for symmetry and a test for tail length with a simple reduction in the individual significance levels to obtain an overall alpha significance level is inappropriate to

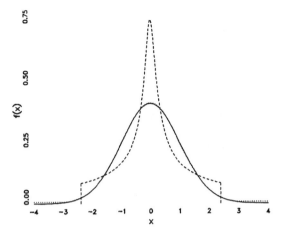

Figure 3.2 Symmetric distributions with $\beta_2 = 3$: — *normal;* - - - *Tukey* $\lambda(5.2)$; $\cdots$ *Tukey* $\lambda(.135)$.

obtain exact α-level critical values. Multiple testing requires adjustment for correlation between moments (e.g., Bowman and Shenton, 1975). Pearson, D'Agostino and Bowman (1977) found that the sample estimates of $\sqrt{\beta_1}$ and β_2 were nearly independent for sample sizes greater than 100.

(4) For $k > 4$, the variance of the moment estimates is usually considered too large to have an efficient test, although Thode (1985) found that moment tests with $k > 4$ had higher power in some instances.

The remainder of this chapter will be concerned with the use of estimated moments as tests for normality, and other moment-type tests. We identify moment-type tests by their mathematical similarity to sample moment estimates.

3.2 Moment Tests

The kth sample moment of a random variable X from a probability density $f(x)$ is defined as

$$m_k = \sum_{i=1}^{n}(x_i - \overline{x})^k / n$$

where x_i are the n observations of the random variable X, $\overline{x}$ is the sample mean and k is a positive integer greater than or equal to 2. Note that m_2 is similar to the sample standard deviation s^2 except for the denominators, which are n and $n-1$, respectively. The most commonly used moment-type test is the (standardized) moment test, based on the test statistic

$$g_k = m_k / m_2^{k/2}.$$

Odd moment tests are those based on a test statistic for which k is an odd integer. These tests are measures of asymmetry, the usual test being the skewness test, denoted $\sqrt{b_1}$. Even moment tests are measures of symmetric departures from normality, the most common test statistic being kurtosis, denoted b_2.

3.2.1 The Skewness Test

The third moment test for skewness is given by the test statistic

$$\sqrt{b_1} = m_3 / m_2^{3/2}.$$

When $\sqrt{b_1} > 0$ the data are skewed to the right (i.e., the long tail of the density goes off to the right), and when $\sqrt{b_1} < 0$, the data are skewed to the left (long tail is to the left). Under most circumstances, a two-tailed test is used, unless the direction of the long tail is assumed known. Table B5 contains 0.5%, 1%, 2.5% and 5% upper and lower tail critical values for $\sqrt{b_1}$.

Ferguson (1961) described outliers in a sample in terms of two models, one in which there is a shift in mean for some observations (model A) and the other in which there is a shift in variance (model B). He proved that for model A with the mean unknown, the locally best invariant one-sided test for outliers is $\sqrt{b_1}$, under the conditions that there are less than 50% outliers, and all outliers are shifted in the same direction. It is not required, via his proof, that each of the outliers have the same shift in mean. While $\sqrt{b_1}$ is one of the best tests for detecting skewed distributions, it is also a powerful test for symmetric long-tailed alternatives when the sample size is small (Shapiro, Wilk and Chen, 1968; Saniga and Miles, 1979; Thode, Smith and Finch, 1983).

Example 3.1. Fifty one measurements of the time (in minutes) it takes to assemble a mechanical part are given in Data Set 4 (Shapiro, 1980). Figure 3.3 is a plot of the frequency distribution of these data. The frequency distribution indicates possible

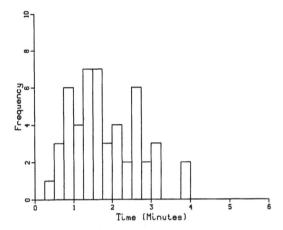

Figure 3.3 Frequency distribution of assembly times (n = 51).

bimodality, and there are two data values slightly separated from the remainder of the data. For this sample, the moments are

$$\bar{x} = 2.84$$

$$m_2 = 0.781$$

$$m_3 = 0.360$$

$$\sqrt{b_1} = g_3 = m_3/m_2^{3/2} = 0.522$$

Since the expected value of $\sqrt{b_1}$ under normality is 0, we reject the null hypothesis of normality if $\sqrt{b_1}$ is sufficiently far from 0. Using a two-sided test, the distribution of $\sqrt{b_1}$ being symmetric about 0, from Table B5 we find the 0.05 level critical value is 0.642. Since the absolute value of $\sqrt{b_1}$ is not greater than 0.642, we do not reject the hypothesis of normality, at least from the standpoint of skewness.

Under the null hypothesis of normality, $\sqrt{b_1}$ is asymptotically normal with mean 0 and variance $6/n$ (e.g., Kendall and Stuart, 1977). However, the sample size must be somewhat large before a simple normal approximation can be used to obtain critical values for testing purposes. For finite samples, the variance is more properly given by

$$\sigma^2(\sqrt{b_1}) = \frac{6(n-2)}{(n+1)(n+3)} \qquad (3.1).$$

Table 3.1 First four sampling moments of $\sqrt{b_1}$ for a normal distribution.

n	$\mu(\sqrt{b_1})$	$\sigma^2(\sqrt{b_1})$	$\sqrt{\beta_1}(\sqrt{b_1})$	$\beta_2(\sqrt{b_1})$
25	0	0.190	0	3.58
50	0	0.107	0	3.45
100	0	0.057	0	3.28
500	0	0.012	0	3.07
1000	0	0.006	0	3.04
5000	0	0.003	0	3.02
asymptotic	0	6/n	0	3.00

Pearson (1930a) obtained an estimate of the fourth moment of the distribution of $\sqrt{b_1}$ in an expansion to the order of n^{-3}, the mean and third moment of the distribution of $\sqrt{b_1}$ being 0. Shortly thereafter, Pearson (1930b) determined the exact fourth moment of the distribution, based on the results of Fisher (1930), who obtained exact expressions for the moments of ratios of k-statistics. The fourth moment of the distribution of $\sqrt{b_1}$ is

$$\beta_2(\sqrt{b_1}) = 3 + \frac{36(n-7)(n^2+2n-5)}{(n-2)(n+5)(n+7)(n+9)} \qquad (3.2).$$

Table 3.1 contains the first four moments of the distribution of $\sqrt{b_1}$ for selected sample sizes between 25 and 2000; it can be seen that using a simple normal approximation to obtain critical values even for samples of size 100 could give incorrect results, although for samples of size 500 $\beta_2(\sqrt{b_1})$ has sufficiently converged to the normal value of 3.

D'Agostino (1970a) obtained a transformation of the null distribution of $\sqrt{b_1}$ to normality which works well for small sample sizes, $n \geq 8$. This transformation is essentially a fit of a Johnson's unbounded (S_U) curve but bypasses the need for $sinh^{-1}(x)$ tables. Let

$$Y = \sqrt{b_1}\left\{\frac{(n+1)(n+3)}{6(n-2)}\right\}^{1/2}$$

and

$$B_2 = \frac{3(n^2+27n-70)(n+1)(n+3)}{(n-2)(n+5)(n+7)(n+9)}.$$

Then $X(\sqrt{b_1})$ is distributed N(0,1) where

$$X(\sqrt{b_1}) = \delta \log(Y/\alpha + \sqrt{(Y/\alpha)^2+1}) \qquad (3.3)$$

with α and δ defined by

$$W^2 = \sqrt{2(B_2 - 1)} - 1$$

$$\delta = 1/\sqrt{\log(W)}$$

$$\alpha = \sqrt{2/(W^2 - 1)}.$$

$X(\sqrt{b_1})$ can then be compared to a standard normal table to determine whether a sample should be rejected as coming from a normal distribution.

D'Agostino (1970a) determined the accuracy of this approximation by comparing the results with tabulated critical values of $\sqrt{b_1}$ (Pearson and Hartley, 1966) for $n \geq 25$, and with simulated critical values for $8 \leq n < 25$, showing only trivial differences. In addition, D'Agostino and Tietjen (1973) compared this approximation for $n \leq 35$ to the normal approximation of $\sqrt{b_1}$, and to results using Cornish-Fisher expansions, t or Pearson Type VII distributions and Monte Carlo simulations. For these sample sizes, all methods were essentially in agreement except for the normal approximation.

Example 3.2. Continuing with the 51 observations of Data Set 4 from Example 3.1, we obtain the transformation to normality described by D'Agostino (1970a).

$$Y = 0.522 \left\{ \frac{(52)(54)}{6(49)} \right\}^{1/2} = 1.613$$

$$B_2 = \frac{3(2601 + 1377 - 70)(52)(54)}{(49)(56)(58)(60)} = 3.448$$

$$W^2 = \sqrt{2(2.448)} - 1 = 1.213$$

$$\delta = 1/\sqrt{\log(1.101)} = 3.218$$

$$\alpha = \sqrt{2/0.213} = 3.064$$

Then

$$X(\sqrt{b_1}) = 3.218 \log(1.613/3.064 + \sqrt{(1.613/3.064)^2 + 1}) = 1.62$$

The two-tailed probability of this test can then be obtained from a standard normal table, giving the test a value of $\alpha = 0.105$

(probability of a greater absolute value, two-sided), as before not rejecting the possibility of normality.

The t-distribution approximation given by D'Agostino and Tietjen (1973) is that of a random variable T with ν degrees of freedom given by

$$\nu = \frac{4\beta_2(\sqrt{b_1}) - 6}{\beta_2(\sqrt{b_1}) - 3}$$

where

$$T = \sqrt{b_1}(\frac{\nu}{\nu - 2})^{\frac{1}{2}}/\sigma(\sqrt{b_1}),$$

$\sigma^2(\sqrt{b_1})$ is given by (3.1) and $\beta_2(\sqrt{b_1})$ is given by (3.2). This approximation requires that $\beta_2(\sqrt{b_1}) > 3$, which constraint is satisfied for $n \geq 8$.

3.2.2 The Kurtosis Test

The fourth moment test for testing symmetric departures from normality, is obtained by calculating

$$b_2 = g_4 = m_4/m_2^2.$$

The b_2 test is, along with $\sqrt{b_1}$, one of the oldest and most powerful tests for normality under a wide range of alternatives. Pearson (1935) and Geary (1947) expected that b_2 was the most powerful test for symmetric non-normality, especially for large sample sizes and small departures from normality. Ferguson (1961) showed that for model A (see Section 3.2.1) if there are less than 21% outliers with a shift in either direction, then an upper-tailed b_2 test is the locally best invariant test for detecting outliers in a normal sample. However, under model B, the upper tailed b_2 test is the locally best invariant test regardless of how many spurious observations there are.

Example 3.3. Using the same data as in Example 3.1, we observe that the value of the kurtosis test statistic for this data is

$$m_4 = 1.57$$

$$b_2 = m_4/m_2^2 = 2.57$$

*The upper and lower 2.5 percentiles for samples of size 51 (for
a two-sided 0.05 test) are 4.34 and 2.08, respectively (Table B6).
Since 2.08 < 2.57 < 4.34, we cannot reject the null hypothesis of
normality due to a symmetric departure based on these observations.*

Like $\sqrt{b_1}$, b_2 is asymptotically normal with mean 3 and variance $24/n$.
However, the sample size must be quite large before the distribution of b_2
becomes fairly well behaved. This is at least partially due to the one-tailed
nature of the test, it being constrained to being greater than 0 but unlimited
in the positive direction. In the same papers where he gave the moments
of the distribution of $\sqrt{b_1}$, Pearson also gave the first four moments of the
distribution of b_2 for finite samples of size n from a normal distribution,
both an approximation to order n^{-3} (1930a) and exact (1930b). The exact
moments are

$$\mu(b_2) = \frac{3(n-1)}{(n+1)} \tag{3.4}$$

$$\sigma^2(b_2) = \frac{24n(n-2)(n-3)}{(n+1)^2(n+3)(n+5)} \tag{3.5}$$

$$\sqrt{\beta_1}(b_2) = \sqrt{\frac{216}{n}} \left\{ \frac{(n+3)(n+5)}{(n-3)(n-2)} \right\}^{1/2} \frac{(n^2-5n+2)}{(n+7)(n+9)} \tag{3.6}$$

$$\beta_2(b_2) = 3 + \frac{36(15n^6 - 36n^5 - 628n^4 + 982n^3 + 5777n^2 - 6402n + 900)}{n(n-3)(n-2)(n+7)(n+9)(n+11)(n+13)}$$

As can be seen by the value of these moments in Table 3.2, even for
sample sizes up to 5000 a normal approximation may be inappropriate.
Although the first seven moments are known exactly, it has not been possible to obtain a good approximation. Pearson (1930a) obtained lower and
upper 5% and 1% critical values of b_2 for samples of size 50 and greater
using an approximation based on a Pearson Type IV curve. Geary (1947)
used the first four moments to fit a Gram-Charlier curve to b_2, although he
restricted his results to large samples ($n \geq 500$). D'Agostino and Tietjen
(1971) gave percentiles of the distribution of b_2 for samples between sizes
7 and 50, based on simulation. D'Agostino and Pearson (1973) performed

Table 3.2 First four sampling moments of b_2 for a normal distribution.

n	$\mu(b_2)$	$\sigma^2(b_2)$	$\sqrt{\beta_1}(b_2)$	$\beta_2(b_2)$
25	2.77	0.535	2.10	8.90
50	2.88	0.357	1.58	8.41
100	2.94	0.207	1.28	6.77
500	2.99	0.047	0.64	4.00
1000	2.99	0.024	0.46	3.52
5000	3.00	0.012	0.33	3.27
asymptotic	3.00	24/n	0	3.00

more extensive simulations to obtain percentiles to the 0.1% level for samples of sizes 20 through 200. Their results are presented as probability contour plots.

Bowman and Shenton (1975) developed a transformation of b_2 to a standard normal using Johnson S_U and S_B curves as a prelude to developing an omnibus test based on combining $\sqrt{b_1}$ and b_2. For $n \geq 25$, the transformation is

$$X_S(b_2) = \gamma_2 + \delta_2 \sinh^{-1}((b_2 - \eta)/\lambda_2). \tag{3.7}$$

while for $n < 25$,

$$X_S(b_2) = \gamma_2 + \delta_2 \log\left(\frac{b_2 - \eta}{\eta + \lambda_2 - b_2}\right). \tag{3.8}$$

The equation (3.7) is a Johnson S_U curve and (3.8) is a Johnson S_B curve. The S_U parameters can be obtained by equating the first four moments of b_2 with the appropriate curve and solving for the parameters. Obtaining the S_B is somewhat more difficult.

In their 1973 work, D'Agostino and Pearson also provided an empirical transformation to normality for b_2, $X_e(b_2)$. This involved calculating the value of b_2, obtaining an empirical p-value of b_2 from their contour plots, and using the standard normal z value corresponding to that p-value. This was shown to be somewhat closer to a normal transformation than $X_S(b_2)$, especially for small samples (Bowman and Shenton, 1975).

Anscombe and Glynn (1983) approximated D'Agostino and Pearson's (1973) results using the first three moments of b_2 (given by eqs. 3.4-3.6) for samples of size 20 or greater using a simpler transformation to normality. The test Z is distributed as a standard normal random variable under the null hypothesis where

$$Z = (2/9A)^{-\frac{1}{2}} \left\{ 1 - \frac{2}{9A} - \left(\frac{1 - 2/A}{1 + x\sqrt{2/(A-4)}} \right)^{\frac{1}{3}} \right\}$$
(3.9)

where x is the standardized b_2 using (3.4) and (3.5),

$$x = \frac{b_2 - \mu(b_2)}{\sigma(b_2)}$$

and

$$A = 6 + \frac{8}{\sqrt{\beta_1(b_2)}} \left(\frac{2}{\sqrt{\beta_1(b_2)}} + \sqrt{1 + 4/\beta_1(b_2)} \right)$$

Large positive values of Z indicate densities longer-tailed than normal, while large negative values would indicate densities shorter-tailed than normal.

Another transformation of b_2 to normality was given by Royston (1985) which is adequate for samples down to size 15. He also provided a computer program to perform the calculations. This transformation resulted in a normally distributed variable, q,

$$q = \log(\log(A + y + \sqrt{1 + y^2}))$$

where

$$y = \frac{b_2 - \epsilon}{\lambda}$$

$$\epsilon = 0.8 + \frac{2.2}{1 + \exp(-0.02687(n - 100))}$$

$$\lambda = \begin{cases} 0.17162 \exp(0.02253n) & 15 \leq n \leq 100 \\ \exp(-0.06174) + \frac{0.711}{1 + \exp(2.0581(X - 0.6374))} & n > 100 \end{cases}$$

$$A = \begin{cases} 0.3155 - 1.1555X(1 + 0.6143X + 0.3423X^2) & 15 \leq n \leq 100 \\ 0.3155 - 0.31287(\exp(-2.295X) - 1) & n > 100 \end{cases}$$

$$X = \log(n/100).$$

However, the difficulty with this transformation is that for samples less than 5000, estimates of the mean and variance for transforming q to a standard normal random variable require the use of cubic B-spline functions for interpolation, not something to be accomplished via hand calculations. For samples of 5000 and greater, the estimate of a standard normal variable

$$z = \frac{q - \eta}{\theta}$$

can be obtained by estimating the mean η and standard deviation θ by

$$H = A + Y + \sqrt{1 + Y^2}$$

$$\eta = E(q) = \log \log(H)$$

$$\theta = \sigma_q = \left(\frac{H - A}{(\lambda H \log(H))(H - A - Y)} \right) \sigma(b_2)$$

$$Y = \frac{E(b_2) - \epsilon}{\lambda}$$

where $E(b_2)$ and $\sigma^2(b_2)$ are the first two moments of b_2 as given above.

3.2.3 Bivariate Tests Using $\sqrt{b_1}$ and b_2

Since $\sqrt{b_1}$ and b_2 are tests for skewed and symmetric alternatives, respectively, an obvious choice for an omnibus test against all alternatives would be some combination of the two tests. If the two tests were independent and normally distributed, then the simple combination of the two standardized test statistics

$$\chi^2 = \left\{ \frac{\sqrt{b_1} - E(\sqrt{b_1})}{\sigma_1} \right\}^2 + \left\{ \frac{b_2 - E(b_2)}{\sigma_2} \right\}^2$$

would be χ^2 with two degrees of freedom. However, as noted previously, very large samples are needed for the asymptotic normality to be nearly true, especially for b_2. In addition, D'Agostino and Pearson (1974) noted that $\sqrt{b_1}$ and b_2 are not independent, although they are uncorrelated.

Bowman and Shenton (1975) developed combination tests using the Johnson S_U transformations for $\sqrt{b_1}$ and b_2,

$$K_S^2 = X_S^2(\sqrt{b_1}) + X_S^2(b_2)$$

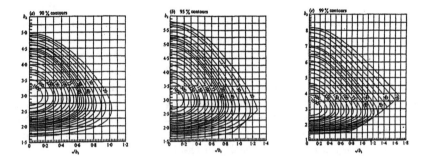

Figure 3.4 Critical value contours for the joint distribution of $\sqrt{b_1}$ and b_2 based on the bivariate moment test statistic K_S^2 (from Bowman and Shenton, 1975; used with permission of the Biometrika Trustees).

and, alternatively,

$$K_e^2 = X_S^2(\sqrt{b_1}) + X_e^2(b_2)$$

where $X_S(\sqrt{b_1})$ is given by (3.3) and $X_S(b_2)$ is given by (3.7). As a simpler alternative, use of the Anscombe and Glynn (1983) transformation Z (3.9) rather than $X_S(b_2)$ is suggested.

For the appropriate transformation, $X_S(\sqrt{b_1})$ and $X_S(b_2)$ are each approximately distributed as standard normal random variables; however, they are not independent. Therefore, Bowman and Shenton constructed critical value contours for sample sizes 20(5)65, 75 , 85, 100, 120, 150, 200, 250, 300, 500 and 1000 for the 90%, 95% and 99% levels (Figure 3.4). An alternative to using contours is a regression model for the critical points of K_S^2 for $30 \leq n \leq 1000$, given by

$$C_{90} = 4.50225 + 5.39716 \times 10^{-5}n + 8.59940 \times 10^{-8}n^2$$

$$C_{95} = 6.22848 + 1.52485 \times 10^{-3}n + 1.38568 \times 10^{-6}n^2$$

$$C_{99} = 10.8375 + 2.80482 \times 10^{-3}n + 4.97077 \times 10^{-7}n^2$$

(Bowman and Shenton, 1986). Since K_S^2 is nearly distributed as a χ_2^2 random variable, these equations are expected to be slightly conservative for larger sample sizes.

Shenton and Bowman (1977) noted, however, that for small samples the conditional density of $\sqrt{b_1}$ given b_2 was bimodal for large values of b_2. In large samples this feature was not too noticeable. The probability contours for K_S^2 were not in conformance with this characteristic of the joint density; although the transformation from a density skewed towards the higher values of kurtosis to a symmetric N(0,1) distribution must be taken into account for b_2, it would seem obvious from the quadratic form of K_S^2 that if $\sqrt{b_1}$ is near zero (and hence, $X_S(\sqrt{b_1})$ is near zero), a larger value of b_2 is necessary to attain a fixed critical value. In order to define a critical boundary curve that more closely resembles the shape of the joint null density, they used a joint density of the form

$$f(\sqrt{b_1}, b_2) = w(\sqrt{b_1})g(b_2 \mid \sqrt{b_1})$$

where w is the normal density of the S_U transformed $\sqrt{b_1}$ and g is the conditional density of b_2, which they determined to be similar to a gamma density. This conditional density is

$$g(b_2 \mid \sqrt{b_1}) = \frac{k^\theta (b_2 - 1 - b_1)^{\theta - 1}}{\Gamma(\theta)} \exp(-k(b_2 - 1 - b_1))$$

where

$$\theta = \theta(\sqrt{b_1}) = a + b\sqrt{b_1} + cb_1.$$

Under normality, the parameters k, a, b and c for g are given by

$$a = \frac{(n-2)(n+5)(n^2 + 27n - 70)}{6\Delta}$$

$$b = 0$$

$$c = \frac{(n-7)(n+5)(n+7)(n^2 + 2n - 5)}{6\Delta}$$

$$k = \frac{(n+5)(n+7)(n^3 + 37n^2 + 11n - 313)}{12\Delta}$$

with

$$\Delta = (n-3)(n+1)(n^2 + 15n - 4),$$

and it is necessary that $n > 7$. Figure 3.5 shows the comparison of the contours of critical values for K_S^2 and the joint density model for samples of size 50. The joint density follows more closely the shape of the density of the simulated samples under the null distribution. It can also be noted that K_S^2 is more likely to accept samples which show long tails but are symmetric and samples which are short-tailed and skewed, whereas the

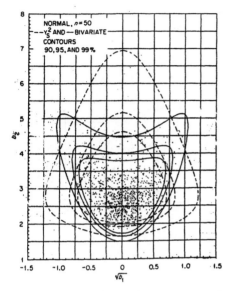

Figure 3.5 Critical value contours for K_S^2 and the joint density model $f(\sqrt{b_1}, b_2)$ for $n = 50$ (from Shenton and Bowman, 1977; reproduced with permission from the Journal of the American Statistical Association. Copyright 1977 by the American Statistical Association. All rights reserved.).

joint density method is somewhat more likely to accept samples which are long-tailed and skewed. Figure 3.6 gives critical value contours under the joint distribution for sample sizes ranging from 20 to 1000.

Example 3.4. The $(\sqrt{b_1}, b_2)$ values for the 51 assembly times are $(0.522, 2.57)$. This is within the 95% contours for both K_S^2 and $f(\sqrt{b_1}, b_2)$; the null hypothesis of normality can not be rejected.

Pearson, D'Agostino and Bowman (1977) also suggested a rectangular bivariate test using $\sqrt{b_1}$ and b_2, where each test is done separately at the

$$\alpha' = 1 - (1 - \alpha)^{\frac{1}{2}}$$

level, where α is the overall test level. It was shown that this test was conservative, in that it rejected normal samples slightly less often than desired. They provided corrected α' values which were dependent on the sample size.

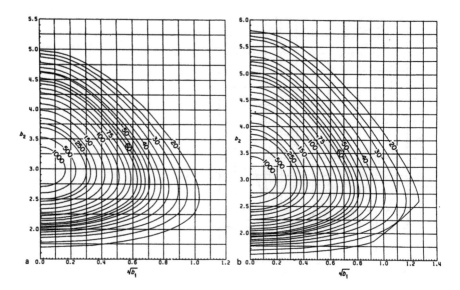

Figure 3.6 Critical value contours for $\sqrt{b_1}$ and b_2 based on the joint density $f(\sqrt{b_1}, b_2)$ (from Bowman and Shenton, 1986; reprinted by courtesy of Marcel Dekker, Inc.).

3.2.4 Higher Moment Tests

Moment tests with $k > 4$ are generally not used as tests for normality since the variance of the sample moments rapidly becomes large as k increases. Little has been done in determining sampling distributions or power comparisons for higher moment tests. Geary (1947) determined large sample distributional properties of absolute moment tests (Section 3.3), which include even moment tests, and claimed that kurtosis was the best test for detecting symmetric departures from normality. Thode, Smith and Finch (1983) included the fifth and sixth moment tests in a power study for detecting scale contaminated normal samples, and observed that the sixth moment test had essentially the same power as kurtosis, which was the most powerful test among all commonly used tests included in the study. Thode (1985) showed that the sixth moment test was also as good as kurtosis in detecting a uniform alternative and a Gram-Charlier alternative used by Geary (1947) in his study of large sample properties of absolute moment tests. Box (1953) implied that higher even moment tests are more powerful for detecting short-tailed alternatives. More details on higher even moment tests are included in the discussion of absolute moment tests (Section 3.3).

Higher order odd moment tests have also been virtually ignored as tests for normality. These tests would be used as measures of skewness. Only one study of the power of a higher order odd moment test is known to the authors (Thode, Smith and Finch, 1983) where the fifth moment test was included in a study to detect scale contaminated normal samples. Overall the fifth moment test had poor power at detecting these alternatives, but it should be noted that these are symmetric alternatives. However, the fifth moment test performed about as well as skewness, where the strong point for both tests was in rejecting small samples. This is most likely a result of the rejection of the sample because of outliers, which cause a sample to look skewed.

3.3 Absolute Moment Tests

Similar to moments, population absolute moments are defined as

$$\nu_c = \int_{-\infty}^{\infty} \mid x - \mu \mid^c f(x)dx, \quad c \neq 0.$$

where c can be any real number. The cth sample absolute moment is

$$v_c = \sum_{i=1}^{n} \mid x_i - \overline{x} \mid^c /n.$$

Absolute moment test statistics can then be defined analogously to those of moment tests,

$$a(c) = v_c/v_2^{c/2} = v_c/m_2^{c/2}$$

where c can be any real number except 0 or 2, usually a positive integer. Even moment tests are a subset of this general class of tests, which are most appropriate for detecting symmetric alternatives to normality.

Geary (1935a) laid the groundwork for using absolute moment tests as tests for normality by suggesting $a(1)$ (originally denoted w'_n, changed to a in Geary (1936), before he utilized the notation $a(c)$ for the general class of absolute moment tests in his 1947 work) as an alternative to kurtosis especially for small samples, due to the difficulty in obtaining small sample critical values for b_2. Pearson (1935) agreed with this sentiment, mostly for the two reasons (a) the test a had a more manageable sampling distribution, especially for small samples, and (b) a was easier to calculate than b_2; it must be remembered that this work was done before the advent of computers and hand calculators.

Geary (1947) claimed that the most powerful absolute moment test (at least, asymptotically) was $a(4)$ (kurtosis), although there would be little difference in power for c between 2 and 5. Thode, Liu and Finch (1988) showed that the optimal choice of c for moderate to very large samples was dependent on the alternative distribution. Small sample power of some absolute moment tests for selected alternatives are given in Thode, Smith and Finch (1983) and Thode (1985).

One theoretical justification for the use of absolute moment tests is given by Box (1953), who showed that they are "almost" likelihood ratio tests (Section 4.1) of the normal vs other members of the exponential power distribution (EPD) family, which is given by

$$f_c(x) = \frac{c}{2\beta\Gamma(1/c)} e^{\frac{-|x-\mu|^c}{\beta^c}}$$

where $f_c(x)$ is the EPD density function with exponent c and mean μ, and β is an appropriate constant. This family includes the normal distribution ($c = 2$), the Laplace or double exponential distribution ($c = 1$) and the uniform distribution as a limiting form ($c \to \infty$). More generally, EPD densities with $c < 2$ have long tails and those with $c > 2$ have short tails.

If the mean of the EPD were known, the likelihood that a sample comes from the EPD(c) distribution is given by

$$L_c = \sum_{i=1}^{n} |x_i - \mu|^c /n \qquad (3.10)$$

(ignoring constants). For unknown mean, the maximum likelihood estimate $\hat{\mu}$ of μ under EPD(c) would be substituted in (3.10). For the normal, Laplace and uniform distributions, the MLE's of the mean are $\bar{x}$, x_M and the midrange $(x_{(n)} + x_{(1)})/2$, respectively. MLE's of the mean for other values of c can not be obtained in closed form (Turner, 1960).

For known mean the Neyman-Pearson likelihood ratio test of the EPD with exponent c vs the EPD with exponent 2 (normal) is based on the statistic

$$LRT(c,2) = \frac{(\sum_{i=1}^{n} |x_i - \mu|^c /n)^{1/c}}{(\sum_{i=1}^{n} |x_i - \mu|^2 /n)^{1/2}}$$

so that $a(c) = [LRT(c,2)]^c$. By using $\bar{x}$ as the estimate of the mean in both numerator and denominator, we identify the $a(c)$ tests as "approximate" likelihood ratio tests under a normal null hypothesis.

In 1947, Geary described the asymptotic distribution theory of the general class of absolute moment tests. Although there were some errors in his calculations (Gastwirth and Owens, 1977; Shenton, Bowman and Lam, 1979) the adjustments were minor in regard to his results for detecting symmetric non-normal alternatives. Absolute moment tests are asymptotically normal when sampling from any symmetric distribution (Gastwirth and Owens, 1977), with parameters dependent on the true sampling distribution. The asymptotics are attained sooner for small c and require an increasingly larger n as c increases for approximate normality to hold (regardless of the sampling distribution). For example, Gastwirth and Owens (1977) compared Monte Carlo estimates of the power of $a(1)$, b_2 and $\sqrt{b_1}$ with calculated values based on asymptotic theory for samples of size 20, 50 and 100. Also included were the estimates of power when sampling from the normal distribution, which should equal the α level of the test. For b_2, the normal approximations were poor for all sample sizes, giving estimates of power that were too large. For $\sqrt{b_1}$, estimates of power were comparable between the two methods for n = 100, but were slightly discrepant for the other sample sizes. For $a(1)$, the normal values were good using the normal approximation, suggesting that it is a valid method for obtaining estimates of critical values for the test. Power estimates for $n = 20$ using the normal approximation were somewhat high, however.

Upper and lower critical values for selected $a(c)$ tests are given in Table B7.

3.3.1 Geary's Test

Geary (1935a) first suggested the test a (originally denoted w_n') as an alternative to b_2 because the small sample properties were more tractable for a. Geary's test statistic is the ratio of the mean deviation to standard deviation,

$$a = a(1) = \sum_{i=1}^{n} \mid x_i - \bar{x} \mid /(n\sqrt{m_2}).$$

Geary originally published the critical values of w_n, based on the mean deviation from the true rather than sample mean, suggesting that the critical values of w_n could be used. These critical values were obtained for samples down to size 6. In 1936, he obtained moments for a and published upper and lower critical values (10%, 5% and 1%) for samples down to size 11.

The negative correlation between a and b_2 (Geary, 1935b) indicates the utilization of a as a test for normality: for long-tailed alternatives, a lower tailed test is used; for short-tailed alternatives, an upper tailed test is used. For an unknown type of symmetric departure from normality, a two-tailed test is used.

The use of a as a test of normality has led to some conflicting results in determining the best circumstances for using a. Geary (1947) claimed that for large samples, b_2 was the best test for alternatives with longer tails than the normal based on using a moderately ($\beta_2 = 3.5$) long-tailed Gram-Charlier alternative. Box (1953) indicated that a was similar to the likelihood ratio test of a double exponential alternative to a normal ($\beta_2 = 6$), implying that it was therefore a good test for long-tailed alternatives. Further, it was shown that a is asymptotically equal to the likelihood ratio test and most powerful location and scale invariant test for a double exponential vs a normal (Section 4.2.3). Even for small samples, Geary's test was shown to be equivalent in power to the LR test (Thode, 1985) and better than kurtosis (Gastwith and Owens, 1977; Thode, 1985) for that alternative.

D'Agostino (1970b) published a simple transformation to standard normality to even further simplify the use of a. Using the asymptotic mean and standard deviation of a (0.7979 and $0.2123/\sqrt{n}$, respectively), he suggested that even for small samples

$$z = \frac{\sqrt{n}(a - 0.7979)}{0.2123}$$

is a sufficient approximation to a standard normal random variable. Indeed, for samples of $n \geq 41$ the actual α levels of this approximate test are near the presumed levels (for the 10%, 5%, 2% and 1% levels) and are somewhat conservative at all α levels for $11 \leq n \leq 31$.

Example 3.5. Using the data from Example 3.1, we find

$$v_c = 0.7148$$

which results in a value of

$$a = 0.826$$

Using D'Agostino's transformation,

$$z = \frac{\sqrt{35}(0.826 - 0.7979)}{0.2123} = 0.78.$$

Comparing z to a normal table results in non-rejection of the normality hypothesis.

Now, with the availability of small sample percentage points for b_2, easier computational ability via computer and calculator, and the failure of a to dominate kurtosis in power except in a limited number of circumstances, a has become less popular as a test for normality.

3.3.2 Higher Absolute Moment Tests

The behavior of absolute moment tests ($c > 2$) is similar to that of even moment tests which is not surprising since even moment tests are a subset of the absolute moment tests. As with moment tests, absolute moment tests with $c > 4$ are usually ignored because of their higher variability; $a(3)$ is also ignored in deference to $a(4)$. This is in part due to the fact that Geary (1947) showed that for large samples, $a(4)$ was the best absolute moment test for detecting non-normal symmetric alternatives. However, Box (1953) suggested that larger $a(c)$ tests would be somewhat more appropriate for short-tailed alternatives, and Thode (1985) showed this to be true. It was also shown (Thode, 1985; Thode, Liu and Finch, 1988) that, for longer-tailed distributions and large samples, the best value of c for testing purposes seemed to be between 2.5 and 3.5. However, these results were based on a limited set of alternatives; additional work needs to be conducted for other alternatives and smaller samples.

3.4 Moment-Type Tests

Two tests found in the literature, which were derived independently of moment tests, can be shown to be similar in derivation (Grubbs' outlier test) or in form (U test) to moment/absolute moment tests. In this section they are described in terms of their similarity to moment tests; elsewhere in this text they are covered in more detail from the standpoint of their original derivation and purpose.

3.4.1 Grubbs' Outlier Test

The numerator for an absolute moment test statistic is a power function of the absolute value of the deviations from the mean of a sample. As c increases the numerator becomes more dependent on the larger deviations, since these have a larger proportional increase. Finally, as $c \to \infty$, the numerator becomes essentially a function of the largest deviation in the sample. Then the cth root of the corresponding absolute moment test statistic is

$$a(\infty) = x_L/\sqrt{m_2} = \sqrt{\frac{n-1}{n}} T$$

where x_L is the largest deviation from the sample mean,

$$x_L = \max(x_{(n)} - \bar{x}, \bar{x} - x_{(1)})$$

where T is Grubbs' (1950) test statistic for a single outlier in either direction (Section 6.2.1), which has been used extensively for detecting large observations in the tails of a distribution. When testing for outliers, normality is rejected if T (or $a(\infty)$) is too large; however, in the context of likelihood ratio tests, normality is rejected in favor of a short-tailed distribution if the largest deviation from the mean is too small (i.e., a lower-tailed test is used).

> *Example 3.6. Twenty-eight brands of margarine were tested for sodium content (Data Set 5). For these data*
>
> $$\bar{x} = 101.96$$
> $$s = 9.16$$
>
> *The observation in the data set that is furthest from the mean is 120, resulting in the test statistic*
>
> $$T = |120 - 101.96|/9.16 = 1.97$$
>
> *Comparison of this value of T to the lower-tailed critical value (Table B8) indicates that the null hypothesis cannot be rejected; the most extreme value in this data set is not too close to the mean to be identified as a "short tail".*

By noting that the limiting form ($c \to \infty$) of the EPD distribution is the uniform density, it follows that $a(\infty)$ or, equivalently, T would be valuable in detecting short-tailed distributions. In this sense, the test $T < t$, for the appropriate constant t, is "like" the likelihood ratio test for a uniform vs a normal distribution, in that $\bar{x}$ is substituted for the uniform MLE (midrange) of μ. As an illustration, it was shown in Thode (1985) that for samples up to size 100, T was more powerful than all tests used except the respective likelihood ratio test for the uniform alternative. Surprisingly, it also performed well at detecting some long-tailed distributions; for example, it had 92% relative power to b_2 at detecting scale contaminated normal samples (Thode, Smith and Finch, 1983). Because of the ease of computation and good power properties shown for this test, we recommend this test as a good "quick and dirty" test for normality. We must state, however, that there are actually few power studies which include T and which test for general alternatives to normality; most comparisons with T have been to other outlier tests.

3.4.2 Uthoff's Likelihood Ratio Test

Uthoff's U test is the likelihood ratio test of a double exponential distribution against a normal distribution (Section 4.1.3), and is asymptotically equivalent to Geary's test (Uthoff, 1973). In fact, the test statistic is identical to a except that v_1, the average deviation from the mean used as the numerator in a, is replaced by the average deviation from the median, an asymptotically equivalent estimator of the scale parameter for a normal distribution,

$$U = \sum_{i=1}^{n} | x_i - x_M | /(n\sqrt{m_2})$$

where x_M is the sample median. As with Geary's a test, a lower-tailed test is used to detect a symmetric long-tailed alternative to normality. Selected percentiles for this test are given in Table B9.

This test suggests a whole new class of moment-type tests based on the median as the location estimate, and could even be extended to tests using the mode or other location esimate in the test. However, it seems unlikely that there would be any significant advantage to using these types of tests over other tests.

3.5 Further Reading

Except for the sample skewness and kurtosis tests, there is very little written with respect to moment tests. Theoretical issues regarding moments are covered in more or less detail in any number of texts, e.g., Kendall and Stuart (1977). Bowman and Shenton (1986) present an alternate summary of $\sqrt{b_1}$ and b_2 tests to the one presented in this chapter, including combinations of the two tests. The reader is directed towards the original publications for additional information on moments and moment tests (e.g., the Geary articles cited in the reference section of this chapter).

References

Anscombe, F.J., and Glynn, W.J. (1983). Distribution of the kurtosis statistic b_2 for normal statistics. Biometrika 70, 227-234.

Balanda, K.P., and MacGillivray, H.L. (1988). Kurtosis: a critical review. American Statistician 42, 111-119.

Bowman, K.O., and Shenton, L.R. (1975). Omnibus test contours for departures from normality based on $\sqrt{b_1}$ and b_2. Biometrika 62, 243-250.

Bowman, K.O., and Shenton, L.R. (1986). Moment $(\sqrt{b_1}, b_2)$ techniques. In D'Agostino, R.B., and Stephens, M.A., eds., **Goodness of Fit Techniques**, Marcel Dekker, New York, 279-329.

Box, G.E.P. (1953). A note on regions for tests of kurtosis. Biometrika 40, 465-468.

D'Agostino, R.B. (1970a). Transformation to normality of the null distribution of g_1. Biometrika 57, 679-681.

D'Agostino, R.B. (1970b). Simple compact portable test of normality: Geary's test revisited. Psychological Bulletin 74, 138-140.

D'Agostino, R.B., and Pearson, E.S. (1973). Tests for departure from normality. Empirical results for the distributions of b_2 and $\sqrt{b_1}$. Biometrika 60, 613-622.

D'Agostino, R.B., and Pearson, E.S. (1974). Correction to 'Tests for departure from normality. Empirical results for the distributions of b_2 and $\sqrt{b_1}$.' Biometrika 61, 647.

D'Agostino, R.B., and Tietjen, G.L. (1971). Simulation probability points of b_2 for small samples. Biometrika 58, 669-672.

D'Agostino, R.B., and Tietjen, G.L. (1973). Approaches to the null distribution of $\sqrt{b_1}$. Biometrika 60, 169-173.

Ferguson, T.S. (1961). On the rejection of outliers. Proceedings of the 4th Berkeley Symposium on Mathematical Statistics and Probability, University of California Press, Berkeley, CA.

Fisher, R.A. (1930). The moments of the distribution for normal samples of measures of departure from normality. Proceedings of the Royal Society of London 130, 16-28.

Gastwirth, J.L., and Owens, M.E.B. (1977). On classical tests of normality. Biometrika 64, 135-139.

Geary, R.C. (1935a). The ratio of the mean deviation to the standard deviation as a test of normality. Biometrika 27, 310-332.

Geary, R.C. (1935b). Note on the correlation between β_2 and w'. Biometrika 27, 353-355.

Geary, R.C. (1936). Moments of the ratio of the mean deviation to the standard deviation for normal samples. Biometrika 28, 295-305.

Geary, R.C. (1947). Testing for normality. Biometrika 34, 209-242.

Geary, R.C., and Worlledge, J.P.G. (1947). On the computation of universal moments of tests of statistical normality derived from samples drawn at random from a normal universe. Application to the calculation of the seventh moment of b_2. Biometrika 34, 98-110.

Grubbs, F.E. (1950). Sample criteria for testing outlying observations. Annals of Mathematical Statistics 21, 27-58.

Hsu, C.T., and Lawley, D.N. (1939). The derivation of the fifth and sixth moments of b_2 in samples from a normal population. Biometrika 31, 238-248.

Johnson, M.E., Tietjen, G.L., and Beckman, R.J. (1980). A new family of probability distributions with applications to Monte Carlo studies. Journal of the American Statistical Association 75, 276-279.

Kendall, M., and Stuart, A. (1977). **The Advanced Theory of Statistics, Vol. I**. MacMillan Publishing Co., New York.

Ord, J.K. (1968). The discrete Student's t distribution. Annals of Mathematical Statistics 39, 1513-1516.

Pearson, E.S. (1930a). A further development of tests for normality. Biometrika 22, 239-249.

Pearson, E.S. (1930b). Note on tests for normality. Biometrika 22, 423-424.

Pearson, E.S. (1935). A comparison of β_2 and Mr. Geary's w_n criteria. Biometrika 27, 333-352.

Pearson, E.S., D'Agostino, R.B., and Bowman, K.O. (1977). Tests for departure from normality: comparison of powers. Biometrika 64, 231-246.

Pearson, E.S., and Hartley, H.O. (1966). **Biometrika Tables for Statisticians, Vol. I**, 3rd ed., Cambridge University Press.

Royston, J.P. (1985). Algorithm AS 209. The distribution function of skewness and kurtosis. Applied Statistics 34, 87-94.

Saniga, E.M., and Miles, J.A. (1979). Power of some standard goodness of fit tests of normality against asymmetric stable alternatives. Journal of the American Statistical Association 74, 861-865.

Shapiro, S.S. (1980). **How to Test Normality and Other Distributional Assumptions**. American Society for Quality Control, Milwaukee, WI.

Shapiro, S.S., and Wilk, M.B. (1965). An analysis of variance test for normality (complete samples). Biometrika 52, 591-611.

Shapiro, S.S., Wilk, M.B., and Chen, H.J. (1968). A comparative study of various tests for normality. Journal of the American Statistical Association 62, 1343-1372.

Shenton, L.R., and Bowman, K.O. (1975). Johnson's S_U and the skewness and kurtosis statistics. Journal of the American Statistical Association 70, 220-228.

Shenton, L.R., and Bowman, K.O. (1977). A bivariate model for the distribution of $\sqrt{b_1}$ and b_2. Journal of the American Statistical Association 72, 206-211.

Shenton, L.R., Bowman, K.O., and Lam, H.K. (1979). Comments on a paper by R.C. Geary on standardized mean deviation. Biometrika 66, 400-401.

Thode, Jr., H.C. (1985). Power of absolute moment tests against symmetric non-normal alternatives. Ph.D. dissertation, University Microfilms, Ann Arbor, MI.

Thode, Jr., H.C., Liu, H.-K. and Finch, S.J. (1988). Large sample power of absolute moment tests. Communications in Statistics B17, 1453-1458.

Thode, Jr., H.C., Smith, L.A., and Finch, S.J. (1983). Power of tests of normality for detecting scale contaminated normal samples. Communications in Statistics - Simulation and Computation 12, 675-695.

Turner, M.E. (1960). On heuristic estimation methods. Biometrics 16, 299-301.

Uthoff, V.A. (1973). The most powerful scale and location invariant test of the normal versus the double exponential. Annals of Statistics 1, 170-174.

CHAPTER 4

OTHER TESTS FOR UNIVARIATE NORMALITY

"Although in many cases the techniques are more robust than the assumptions underlying them, still a knowledge that the underlying assumption is incorrect may temper the use and application of the methods."

S.S. Shapiro and M.B. Wilk, 1965

In addition to moment tests and regression tests, many other tests have been developed specifically for the purpose of detecting non-normal samples. Some, such as likelihood ratio tests and most powerful location and scale invariant tests, are derived from the desire to test for a specific alternative to normality; often, though, these tests are competitive with other tests for a larger set of alternatives. Other tests have as their alternatives a more general set of distributions, or are omnibus tests.

4.1 Likelihood Ratio Tests

Likelihood ratio (LR) tests are tests derived for a specific alternative against a null distribution, where the parameters may or may not be specified. Thus, there are as many LR tests as there are alternatives. The methods for development of LR tests will be discussed here.

    LR tests for normality are the ratio of the maximum likelihood function for the alternative to the maximum likelihood function for normality.

For any probability density function $f(x)$, the likelihood that a single observation x_i comes from that density is $f(x_i)$. For a sample of n observations, the likelihood that the sample comes from density $f(x)$ is

$$L = \prod_{i=1}^{n} f(x_i).$$

For distributions whose support is not a function of the parameters of the distribution, the maximum value of L with respect to the unknown parameters is calculated by taking the derivative of the likelihood with respect to each of the parameters, setting the derivative equal to zero, and solving for the unknown parameters. For ease in computation, $l = \log(L)$ is usually used to determine these maximum likelihood estimates (MLE's). For example, for the EPD distribution with fixed exponent c (Section 3.3), the log-likelihood, ignoring constants, is

$$\log(L_c) = -n \log[2\beta\Gamma(1/c)] + \beta^{-c} \sum_{i=1}^{n} |x_i - \mu|^c \qquad (4.1)$$

(Turner, 1960). The MLE's are then used in the likelihood equation to obtain the maximum likelihood. However, it is not always possible to obtain MLE's in closed form for a specific distribution.

The maximum likelihood, conditional on the observations, is found for both the alternative ($f_1(x)$) and null ($f_0(x)$) densities. The ratio of these likelihoods, L_1/L_0, is the LR test statistic. If the test statistic is greater than the appropriately chosen critical value, then the null hypothesis is rejected.

4.1.1 Normal Likelihood

The normal distribution is the EPD with $c = 2$. Ignoring constants, the likelihood is

$$L = \prod_{i=1}^{n} \sigma^{-1} \exp[-(x_i - \mu)^2/2\sigma^2]$$

and the log-likelihood from (4.1) is

$$\log(L) = -n \log(\sigma) - \sum_{i=1}^{n} \frac{(x_i - \mu)^2}{2\sigma^2}$$

where $\beta^2 = 2\sigma^2$. Taking the derivatives with respect to μ and σ and setting the results to 0 give the MLE's

$$\widehat{\mu} = \bar{x}$$

and

$$\hat{\sigma}^2 = n^{-1} \sum_{i=1}^{n} (x_i - \bar{x})^2$$

so that the maximum log-likelihood is given by

$$\log(L_{max}) = -n \log \left(\frac{\sum_{i=1}^{n} (x_i - \bar{x})^2}{n} \right)^{1/2} - \frac{n}{2}$$

showing that the maximum likelihood is a function of s^2.

Caveat to Likelihood Ratio Tests for Normality

As stated in Chapter 1, LR test statistics are of the form $L_1/L_0 > c$, which would indicate that the observations are more likely to have come from $f_1(x)$ than $f_0(x)$. However, note that the maximum likelihood, L_{max}, for the normal distribution is proportional to $1/s$, not s. Below, we will show that the likelihood for some alternative distributions is proportional to $1/g(x)$, where $g(x)$ is some simple function of the observations. Therefore, the LR test as described above would be $s/g(x) > c$ for the appropriate constant c. However, in their popular form, the most common tests are given as the inverse, i.e., $g(x)/s$. Therefore, when testing for the appropriate alternative using a single-tailed test, the lower tail percentage points are used.

4.1.2 Uniform Alternative

The uniform distribution is the limiting EPD as $c \to \infty$. The maximum likelihood estimates of μ and β are the midrange and the range, respectively. From (4.1) we can see that as $c \to \infty$ the second term becomes 0, and the first term is essentially the range. Therefore, for a uniform alternative the well known range test (David, Hartley and Pearson, 1954),

$$u = (x_{(n)} - x_{(1)})/s$$

is the LR test statistic. This test also seems to be a good test against other short-tailed symmetric alternatives; for these types of alternatives a lower-tailed test should be used. A two-tailed range test can be used as a

test for non-normal symmetric distributions. Critical values for this test
are given in Table B10.

> *Example 4.1. The sodium levels in margarine (Data Set 5) have*
> *variance $s = 9.16$ and a range of $120 - 90 = 30$. For these data,*
> *$u = 3.27$, which is less than the lower 2.5th percentile of the null*
> *distribution of the test statistic for a sample of size 28. This*
> *indicates that these data are too short-tailed to have come from a*
> *normal distribution. For this data set, the range test rejects the*
> *null hypothesis while the Grubbs' test for a short-tailed alternative*
> *(Example 3.6) does not.*

4.1.3 Double Exponential Alternative

The double exponential distribution is the EPD with $c = 1$. The MLE's of
μ and β in (4.1) are the median and the sum of absolute deviations from
the median, respectively, resulting in the LR test statistic (Uthoff, 1973)

$$U = \sum_{i=1}^{n} |x_i - x_M|/(n\hat{\sigma}).$$

For the double exponential alternative (or other long-tailed symmetric alternative), a single-tailed test using lower percentage points should be used;
d can also be used in a two-tailed fashion. This LR test is asymptotically
equivalent to Geary's $a(1)$ test (Section 3.3.1), and is in fact calculated
identically except for the use of the median rather than the mean in the
numerator. This test has had only limited use, but has been investigated
for its power as a test for normality (Uthoff, 1973; Hogg, 1972; Smith, 1975;
Thode, Smith and Finch, 1983). Critical values for selected sample sizes
are given in Table B9.

> *Example 4.2. Forty subjects underwent a standard physical fit-*
> *ness test and pulse rate was measured at termination of the test*
> *(Langley, 1971, p. 63). A Q-Q plot of these data (Data Set 6) is*
> *presented in Figure 4.1, which indicates that these data are nor-*
> *mal except possibly for an outlying observation. For these data*
> *$\hat{\sigma}^2 = 336.9$ and the median is 113. This results in a test statistic*
> *value of*
> $$U = 0.78$$

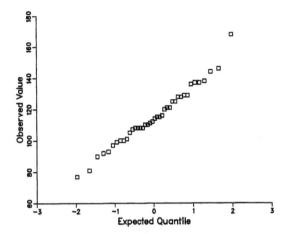

Figure 4.1 Q-Q plot of pulse rates measured after undergoing physical fitness test (n = 40).

which is not significant when compared to the lower 5% level critical value.

4.1.4 Other Likelihood Ratio Tests

Absolute moment tests have been shown to be "like" LR tests when the alternatives are from the exponential power family of distributions (Section 3.3).

The maximum likelihood estimates for some alternatives (e.g., normal mixtures, Section 11.3.2) cannot be evaluated in closed form, and require iterative procedures for calculation.

4.2 Most Powerful Location and Scale Invariant Tests

A test T is scale and location invariant provided

$$T(kx_1 - u, \ldots, kx_n - u) = T(x_1, \ldots, x_n)$$

for constants k and u. By using the likelihood ratio of the location and scale maximal invariant, we can obtain the most powerful location and scale invariant (MPLSI) test of a normal vs a specified alternative. Similar to LR tests, MPLSI tests are most powerful when used against the specific alternative they are designed for, although they may be comparable in power to other tests for a broader range of alternatives. Let $f(x_1, \ldots, x_n)$ be the continuous joint density of observations $x_1, \ldots, x_n$. Then

$$\lambda^* = \int_0^\infty \int_{-\infty}^\infty f(kx_1 - u, \ldots, kx_n - u)k^{n-2}dudk$$

$$= \int_0^\infty \int_{-\infty}^\infty \prod_{i=1}^n f(kx_i - u)k^{n-2}dudk$$

where the x_i are independent. Then the most powerful test of the density f_0 against f_1 which is invariant under changes in location and scale is given by

$$\lambda_1^* > C\lambda_0^*$$

which rejects the null hypothesis for the appropriate constant C (Hajek and Sidak, 1967).

4.2.1 Normal Maximal Invariant

For the normal distribution the location and scale maximal invariant is

$$\lambda_N^* = (1/2)n^{-1/2}\pi^{-n/2+1/2}\Gamma(n/2 - 1/2)[\sum(x_i - \overline{x})^2]^{-n/2+1/2}$$

(Uthoff, 1970). As discussed previously concerning LR tests (Section 4.1.1), this is a function of $1/s$, and the maximal invariants for the alternatives described below are also in the form of the inverse of some function of the observations; therefore, the common form of the test statistic, $g(x)/s$, where $g(x)$ is the function of observations for the alternative, should be compared to the lower tail of the distribution of the test statistic.

4.2.2 Uniform Alternative

For the uniform distribution the location and scale maximal invariant is

$$\lambda_U^* = (n^2 - n)^{-1}(x_{(n)} - x_{(1)})^{-n+1}$$

(Uthoff, 1970). As with the normal maximal invariant, the $-(n-1)$st root of λ_U^* is proportional to a basic function of the observations, in this case the range, so that the MPLSI test statistic for a uniform alternative is essentially the range test statistic

$$T_{U,N} = u = (x_{(n)} - x_{(1)})/s$$

(David, Hartley and Pearson, 1954). For this alternative, the MPLSI and LR tests (Section 4.1.2) are identical.

4.2.3 Double Exponential Alternative

Uthoff (1973) showed that the maximal invariant for the double exponential alternative is

$$\lambda_D^* = [\gamma(x_M)B_n]^{-n+1}$$

where x_M is the sample median,

$$\gamma(\hat{\theta}) = \sum_{i=1}^n |x_i - \hat{\theta}|$$

and (for n odd),

$$B_n = \left[\sum \frac{\gamma^{n-1}(x_M)\gamma^{-n+1}(x_{(i)})}{(2i-n)(n+2-2i)} \right]^{\frac{-1}{(n-1)}}.$$

For n even,

$$B_n = \left[\frac{(n-1)(x_{(n_2)} - x_{(n_1)})}{2\gamma(x_M)} + \frac{1}{2} + \sum_{i \neq n_1, n_2} \frac{\gamma^{n-1}(x_M)\gamma^{-n+1}(x_{(i)})}{(2i-n)(n+2-2i)} \right]^{\frac{-1}{(n-1)}}$$

where $n_1 = [(n+1)/2]$ and $n_2 = [n/2] + 1$. The $-(n-1)$st root of the MPLSI test statistic for this alternative is therefore proportional to

$$T_{D,N} = \gamma(x_M)B_n/s$$

and is asymptotically equivalent to Geary's test, $a(1) = \gamma(\bar{x})/(n\hat{\sigma})$, and the LR test for this alternative, $d = \gamma(x_M)/(n\hat{\sigma})$. A lower-tailed test is used to test for long-tailed symmetric alternatives.

4.2.4 Exponential Alternative

For the exponential alternative, the maximal invariant is given by

$$\lambda_E^* = n^{-1}\Gamma(n-1)(\sum(x_i - x_{(1)}))^{-n+1}$$
$$= n^{-n}\Gamma(n-1)(\bar{x} - x_{(1)})^{-n+1}$$

(Uthoff, 1970). The $-(n-1)$st root of λ_E^* is proportional to $(\bar{x} - x_{(1)})$. The MPLSI for this alternative reduces to

$$T_{E,N} = (\bar{x} - x_{(1)})/s$$

which is Grubbs' (1969) test statistic for a lower tail outlier from a normal sample. However, similar to the moment-type derivation of the two-tailed Grubbs' T as a test for short tails (Section 3.4.1), for this MPLSI test the normal hypothesis is rejected if the test statistic is too *small*, which would indicate that the smallest observation is too close to the mean, or is an "inlier". Percentiles for Grubbs' test are given in Table B8.

> *Example 4.3. As is typical of aerometric data, the TSP data of Data Set 1 are skewed to the right, indicating an exponential-like density. For these data, the mean and standard deviation are 39.3 and 23.5, respectively, and the minimum value is 7. The value of $T_{E,N}$ is calculated as 1.37, which is well below 1.67, the lower 0.01 level critical value for this test.*

If the alternative is an exponential distribution with the long tail to the left (right exponential distribution), the MPLSI is the Grubbs' test for an upper tail outlier

$$T_{R,N} = (x_{(n)} - \bar{x})/s$$

for which the null hypothesis is rejected if the test statistic is too small.

4.2.5 Cauchy Alternatives

Franck (1981) gives the maximal invariant for a Cauchy alternative

$$\lambda_C^* = (-1)^{(n-3)/2} \sum_{k=1}^{n} \left(\sum_{j<k} \frac{(x_j - x_k)^{n-1} \log|x_j - x_k|}{\prod_{l \neq j}(x_l - x_j) \prod_{l \neq k}(x_l - x_k)} \right)$$

for odd n; for even n,

$$\lambda_C^* = (-1)^{n/2} \sum_{k=1}^{n} \left(\sum_{j<k} \frac{|x_j - x_k|^{n-1}}{\prod_{l \neq j}(x_l - x_j) \prod_{l \neq k}(x_l - x_k)} \right)$$

The appropriate λ_C^* divided by λ_N^* gives the MPLSI for a Cauchy alternative to a normal, using a lower-tailed test for long-tailed symmetric alternatives. An appendix in Franck (1981) contains a FORTRAN program for calculating this test.

4.2.6 Combinations of MPLSI Tests

Using the logic that the combination of a good test for short tails and a good test for long tails would result in a good test for an unspecified symmetric alternative, Spiegelhalter (1977) used a combination of MPLSI tests to obtain a test that would be useful under more general conditions. He defined a test statistic against symmetric alternatives as

$$T_s = (\lambda_U^* + \lambda_D^*)/\lambda_N^*.$$

Due to the complexity of calculating λ_D^*, however, he examined a simplified approximation to T_s by substituting Geary's test for that component,

$$T_s' = [(c_n u)^{-n+1} + a(1)^{-n+1}]^{1/(n-1)}$$

where u is the range test statistic, $a(1)$ is Geary's test statistic, and

$$c_n = \frac{(n!)^{1/(n-1)}}{2n}.$$

Spiegelhalter (1980) extended T_s to an omnibus test by combining it with representative MPLSI tests for detecting skewed alternatives, namely the MPLSI tests for an exponential and right exponential alternative. He defined this omnibus test statistic to be

$$S_s = \left[\sum_{L \in \Lambda} \lambda_L^*/\lambda_N^* \right]^{-1}$$

where $\Lambda = (N, D, U, R, E)$.

Test critical values for T_s' and S_s are contained in Table B11. T_s and T_s' are upper-tailed tests for detecting symmetric alternatives; S_s is a lower-tailed omnibus test.

Example 4.4. For the pulse rate data (Data Set 6),

$$u = 4.90, \qquad a(1) = 0.79$$

and $c_{40} = 0.213$. Based on these results,

$$T_s' = 1.27$$

which does not exceed the 5% critical value for a sample of size 40.

4.3 U-Statistics as Tests for Normality

Several authors have presented tests for normality which consist of U-statistics (Hoeffding, 1948). U-statistics are asymptotically normal, and most of those presented below are adequately approximated by a normal distribution for small to moderate sample sizes as well.

4.3.1 Locke and Spurrier Tests

Locke and Spurrier (1976; 1977; 1981) considered using test statistics of the form U/s^p against nonsymmetric and symmetric alternatives where U is a location invariant U-statistic with kernel $\phi(x_1, \ldots, x_k)$ of order k, s^2 is the usual variance estimator and p is chosen to make the test location and scale invariant. The skewness test, $\sqrt{b_1}$, is a test of this type, its kernel being

$$\phi(x_1, x_2, x_3) = \sum_{i=1}^{3} \{x_i - \frac{1}{3}(x_1 + x_2 + x_3)\}^3.$$

The kernels of the U-statistics investigated by Locke and Spurrier have the general form

$$\phi(x_1, \ldots, x_k) = \sum_{i=1}^{k-1} c_i (x_{(i+1)} - x_{(i)})^p \qquad (4.2)$$

$k < n$, where the $x_{(i)}$ are the order statistics of $x_1, \ldots, x_k$, the c_i are real constants, and p is greater than zero.

Asymmetric Alternatives

When the alternative of interest is asymmetric, the constants in (4.2) are such that $c_i = -c_{k-i}$. Locke and Spurrier (1976) considered the U-statistics with kernel of order $k = 3$, $c_1 = -1$, $c_2 = 1$ and $p = 1$ or 2,

$$\phi(x_1, x_2, x_3) = (x_{(3)} - x_{(2)})^p - (x_{(2)} - x_{(1)})^p$$

and the respective test statistics $T_{pn} = U_{pn}/s^p$. The U_{pn} are calculated as

$$U_{pn} = \binom{n}{3}^{-1} \sum_{1 \leq i < j < k \leq n} \phi(x_i, x_j, x_k).$$

These statistics simplify to

$$U_{1n} = \binom{n}{3}^{-1} \sum w_i x_{(i)}$$

$$U_{2n} = \binom{n}{3}^{-1} \sum_{1 \leq i < j \leq n} v_{ij} (x_{(j)} - x_{(i)})^2$$

$$w_i = \binom{i-1}{2} - 2(n-i)(i-1) + \binom{n-i}{2}, \qquad \binom{b}{a} = 0 \text{ if } a > b$$

$$v_{ij} = i + j - n - 1.$$

As U-statistics, T_{1n} and T_{2n} are asymptotically normal and are symmetric about 0 for symmetric distributions. The asymptotic approximation is satisfactory when $n \geq 20$ for T_{1n} and $n \geq 50$ for T_{2n}. For T_{1n}, the variances for samples of size n under the normal approximation are

$$Var(T_{13}) = a_1$$

$$Var(T_{14}) = (a_1 + 3a_2)/4$$

and

$$Var(T_{1n}) = 3!(n-3)!\{a_1 + 3(n-3)a_2 + 1.5(n-3)(n-4)a_3\}/n!$$

for $n \geq 5$. The constants a_1, a_2 and a_3 are approximately 1.03803994, 0.23238211 and 0.05938718, respectively. The variances for T_{2n} are calculated identically, with values of 7.03804, 1.77417 and 0.49608 for a_1, a_2 and a_3, respectively.

Similar to $\sqrt{b_1}$, positive values of T_{pn} indicate skewness to the right while negative values indicate skewness to the left. Two-tailed tests can be used when the alternative is skewed in an unknown direction.

Locke and Spurrier found that T_{1n} and T_{2n} had good power properties compared to $\sqrt{b_1}$ but there was no advantage over skewness using T_{3n} and higher p.

Example 4.5. In a study of competing risks, Hoel (1972) presented data on the effects of environment on mice subject to radiation.

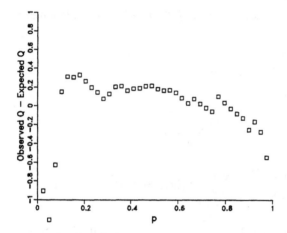

Figure 4.2 Detrended Q-Q plot of number of days until occurrence of cancer in mice subjected to radiation (n = 38).

The competing risks were from the various types of cancer that developed. In this example, we examine the number of days until cancer occurrence for 38 mice in a normal environment which develop reticulum sarcoma (Data Set 7). The detrended Q-Q plot for these data is shown in Figure 4.2 which shows that the data are highly skewed in nature.

Components of the test statistic T_{1n} are

$$U_{1,38} = -34.29$$

with the sample variance of the data $s^2 = 10480.68$, so that

$$T_{1,38} = -0.335.$$

Standardizing with $\mu = 0$ and $Var(T_{1,38}) = 0.0156$ results in a standard normal value of $z = -2.68$, indicating that the data are skewed to the left significantly more than would be expected from a normal sample.

Symmetric Alternatives

For symmetric departures from normality, the constants in (4.2) are such that $c_i = c_{k-i}$. Locke and Spurrier (1977; 1981) reported on two tests for symmetric alternatives, one test statistic being $T^* = U^*/s$ with $k = 4$, $c_1 = c_3 = 0$, $c_2 = 1$ and $p = 1$,

$$U^* = \binom{n}{4}^{-1} \sum_{i=1}^{n-3} \sum_{j=i+1}^{n-2} \sum_{k=j+1}^{n-1} \sum_{l=k+1}^{n} \phi(x_i, x_j, x_k, x_l)$$

where

$$\phi(x_1, x_2, x_3, x_4) = x_{(3)} - x_{(2)}.$$

This simplifies to

$$U^* = \binom{n}{4}^{-1} \sum w_i x_{(i)}$$

with

$$w_i = (i-1)(n-i)(2i-n-1)/2.$$

The other test investigated, T_D, was defined by the kernel (4.2) with $k = 2$, $c_1 = 1$ and $p = 1$. They determined that T_D was a constant times D'Agostino's D (Section 4.3.2).

   Although T^* is asymptotically normal, approximation of critical values for $n \geq 10$ is better using beta distributions (Locke and Spurrier, 1981). The lower tails of the distribution are used to test for heavy-tailed alternatives and upper tails are used to test for light-tailed alternatives. Critical values for selected sample sizes are given in Table B12.

   Example 4.6. For the margarine sodium level data (Data Set 5),

$$U^* = 6.18$$

which results in a value of

$$T^* = 0.675.$$

This value is just below the upper tail 5% level of 0.68 of the test statistic null distribution for a sample of size 28. Therefore, based on this test, we would not reject the null hypothesis of normality.

4.3.2 D'Agostino's D

D'Agostino (1971) proposed the D statistic as an extension of the Wilk-Shapiro test for moderate and large samples, since at the time the W test was only identified (via the required coefficients) for samples up to size 50.

Similar to the Wilk-Shapiro test statistic, D is the ratio of two estimates of σ (up to a constant): Downton's (1966) estimate of the standard deviation using order statistics, σ^*, to the maximum likelihood estimate, $\hat{\sigma}$. For a normal sample, Downton's estimate of σ is unbiased and nearly as efficient an estimator as s (D'Agostino, 1970), even for small samples.

The numerator for D'Agostino's test statistic is

$$T = \sum_{i=1}^{n} (i - \frac{1}{2}(n+1))x_{(i)}$$

$$= \sum_{i=1}^{[n/2]} (\frac{1}{2}(n+1) - i)(x_{(n-i+1)} - x_{(i)}).$$

T is related to σ^* by

$$\sigma^* = 2\sqrt{\pi}T/(n^2 - n).$$

Further, σ^* is equal to $\sqrt{\pi}g/2$ where

$$g = \frac{1}{n(n-1)} \sum_i \sum_j |x_i - x_j|$$

which is known as Gini's mean difference; like the tests by Locke and Spurrier, this is also a U-statistic (David, 1968), and in fact belongs to the family of U-statistics (4.2) derived by Locke and Spurrier (1977) for symmetric alternatives with kernel $\phi(x_1, x_2) = |x_1 - x_2|$, i.e., $k = 2$, $c_1 = 1$ and $p = 1$.

D'Agostino's test statistic is given by

$$D = \frac{T}{n^2\hat{\sigma}}.$$

Under normality, the asymptotic expected value and standard deviation are

$$E(D) = \frac{1}{2\sqrt{\pi}} = 0.28209479$$

$$\sigma_D = \frac{0.02998598}{\sqrt{n}}$$

For samples of size 50 and over, D'Agostino (1971) used Cornish-Fisher expansions to develop null percentage points for D and presented them in terms of its standardized version

$$Y = \frac{n^{1/2}(D - 0.28209479)}{0.02998598}$$

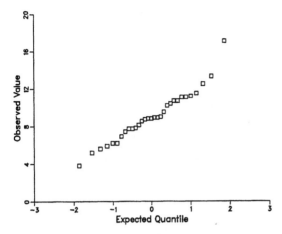

Figure 4.3 Q-Q plot of average daily wind speeds for July 1984 (n = 31).

D'Agostino (1972) later extended the D test to small samples and calculated small sample percentiles for Y using Pearson curves. These percentiles are given in Table B13. Although D is asymptotically normal, even for samples of size 1000 the percentiles of Y have not converged to those of a standard normal distribution, nor are they symmetric, indicating slow convergence to the asymptotic distribution.

Although D was originally derived as an omnibus test, Locke and Spurrier (1977) claimed that the kernel indicated that it might be weak against alternatives with considerable asymmetry. This was confirmed in many simulation studies which included D. D seems to be most powerful as a directional test for detecting heavy-tailed alternatives. For skewed alternatives or as an omnibus test, a two-tailed test should be used; as a directional test for symmetric non-normal alternatives, the lower tail should be used for heavy-tailed alternatives and the upper tail for light-tailed alternatives.

Example 4.7. Daily average wind speed at Long Island MacArthur airport for the month of July 1984 were obtained from NOAA (Data Set 8). These data are plotted in Figure 4.3 and appear normal with a possible upper tail outlier.

For these data, the test statistic components are

$$T = 694.80, \quad D = 0.273, \quad Y = -1.646$$

which is slightly above the lower 10% level of Y.

Table 4.1 Variance estimates of Oja's T_1 and T_2 test statistics under normality.

n	Var(T_1)	Var(T_2)	n	Var(T_1)	Var(T_2)
5	0.01512	0.01549	10	0.00358	0.00196
6	0.01045	0.00800	15	0.00218	0.00098
7	0.00740	0.00473	20	0.00144	0.00058
8	0.00575	0.00350	30	0.00089	0.00032
9	0.00457	0.00248	∞	$0.0214/n$	$0.0026/n$

4.3.3 Oja's Tests

Oja (1981) proposed two location and scale free U-statistics as goodness of fit tests and examined their value as tests of normality. The tests were intended to detect alternatives with different skewness and kurtosis from the null (not necessarily normal) distribution, and are based on ratios of combinations of sample order statistics. To detect an alternative more skewed than the null distribution, the proposed test statistic was

$$T_1 = \binom{n}{3}^{-1} \sum \frac{X_{(k)} - X_{(j)}}{X_{(k)} - X_{(i)}}$$

where the sum is over $1 \leq i < j < k \leq n$. For a symmetric alternative

$$T_2 = \binom{n}{4}^{-1} \sum \frac{X_{(k)} - X_{(j)}}{X_{(l)} - X_{(i)}}$$

was proposed, with the sum over $1 \leq i < j < k < l \leq n$.

The joint limiting distribution of T_1 and T_2 is bivariate normal and, under any symmetric null distribution with density function g, T_1 and T_2 are uncorrelated. Hence, the test statistic

$$S = \frac{(T_1 - E_g(T_1))^2}{Var_g(T_1)} + \frac{(T_2 - E_g(T_2))^2}{Var_g(T_2)}$$

can be used as an omnibus goodness of fit test, and is asymptotically χ^2 with two degrees of freedom where E_g and Var_g are the expected values and variances under the null distribution.

Davis and Quade (1978) gave the expected values under normality as $E(T_1) = 0.5$ and $E(T_2) = 0.298746$, and Oja (1981) gave $Var(T_1)$ and $Var(T_2)$ under normality for selected sample sizes up to size 30 using simulation methods (Table 4.1); the asymptotic variances are $0.0214/n$ and $0.0026/n$, respectively. Comparison with simulations indicated that the normal approximation to the distributions of T_1 and T_2 and the χ^2 approximation to S are good down to $n = 5$.

Example 4.8. For the July wind speed data, the results of testing for normality using Oja's three test statistics are

$$T_1 = 0.514$$

$$T_2 = 0.286.$$

Using the normal approximation with variances interpolated from Table 4.1, $z_1 = 0.47$ and $z_2 = -0.72$, both of which are well within the normal range. The combined test statistic is

$$S = (0.47)^2 + (-0.72)^2 = 0.74$$

which also does not reject normality based on a χ^2 distribution.

Oja (1983) also presented a family of U-statistics as tests of normality which have the form

$$T'_k = \sum_{i=1}^{n-1} \sum_{j=i+1}^{n} a_{ij} \log(X_{(j)} - X_{(i)}) \tag{4.3}$$

where $\sum a_{ij} = 0$. In particular, for testing against skewed alternatives the test statistic T'_1 is (4.3) with coefficients

$$a_{ij} = \frac{(i+j-n-1)}{3!(n-3)!}$$

and for symmetric alternatives T'_2 is (4.3) with coefficients

$$a_{ij} = \binom{n}{4}^{-1} \left[2(n-j)(i-1) - \binom{n-j}{2} - \binom{i-1}{2} \right].$$

As with T_1 and T_2, Oja claimed an omnibus test can be constructed using T'_1 and T'_2. Based on simulation results, $E(T'_1) = 0$ and $E(T'_2) = 0.4523$. Variances are given by

$$V(T'_1) = \binom{n}{3}^{-1} (0.11217n^2 - 0.44196n + 3.21425)$$

$$V(T_2') = \binom{n}{4}^{-1}(0.0874n^3 - 0.7431n^2 + 7.3099n - 14.0884).$$

Although these tests are asymptotically normal, small sample approximations using the normal distribution are not adequate. Oja did not provide any small sample distributional information or critical values for T_1' or T_2'.

For skewed alternatives, Oja (1983) also suggested the U-statistic obtained from (4.3) by using $a_{ij} = (i - n/2)\delta_{i,j-1}$,

$$U = \sum_{i=1}^{n-1}(i - n/2)\log(X_{(i+1)} - X_{(i)}).$$

No information on small sample or asymptotic distributions was provided for this test.

4.4 Other Tests for Normality

Some tests for normality have been designed to take advantage of the unique characteristics of the normal distribution, such as maximum entropy (Vasicek's test) and the independence of the mean and standard deviation (Lin and Mudholkar's test).

4.4.1 Vasicek's Sample Entropy Test

Vasicek (1976) developed a test for normality using an estimate of the sample entropy for $n \geq 3$. The entropy of a density $f(x)$ is

$$H(f) = \int_{-\infty}^{\infty} f(x)\log(f(x))dx.$$

An estimate of $H(f)$ can be calculated as

$$H_{mn} = n^{-1}\sum_{i=1}^{n}\log\left(\frac{n}{2m}(x_{(i+m)} - x_{(i-m)})\right)$$

where m is a positive integer, $m < n/2$ and $x_{(k)} = x_{(1)}$ for $k < 1$ and $x_{(k)} = x_{(n)}$ for $k > n$. Among all densities with a given variance σ^2, $H(f)$ is maximized by the normal density, with entropy

$$H(f) = \log\{\sqrt{2\pi e}\sigma\}$$

so that $\exp[H(f)]/\sigma \leq \sqrt{2\pi e}$ for all $f(x)$, equality being attained under normality. Therefore, an omnibus test for a sample of size n is defined by rejecting the null hypothesis if

$$K_{mn} \leq K^*$$

where K^* is the appropriate critical value for the test and

$$K_{mn} = \frac{n}{2m\hat{\sigma}} \{\prod_{i=1}^{n}(x_{(i+m)} - x_{(i-m)})\}^{1/n}.$$

This permits a number of tests to be defined for a sample of size n, i.e., for $m = 1, 2, \ldots, \{[n/2] - 1\}$. Vasicek did not propose a solution to those instances where $x_{(i+m)}$ and $x_{(i-m)}$ were tied in a sample for fixed m and some i, which would make K_{mn} identically 0.

Vasicek gave simulated 0.05 significance points of K_{mn} for $m = 1$, 2, 3, 4 and 5 for sample sizes ranging from 3 to 50 (Table B14). Based on simulation results Vasicek recommended that for $n = 10$, 20 and 50, optimal power was attained using $m = 2$, 3 and 4, respectively. In general, the best choice of m increases with n, while the ratio m/n goes to 0.

Example 4.9. The heights of the cross-fertilized Zea mays plants (Data Set 2) are plotted in Figure 4.4. From this plot we can observe two lower tail outliers, while the remainder of the data may be slightly skewed. We calculated two versions of Vasicek's test statistic, for m = 2 and 3,

$$K_{2,15} = 2.14$$

$$K_{3,15} = 1.96$$

both of which are below the 5% critical value for the respective test. Although we have identified the sample as being non-normal, the value of the test does not give an indication of the type of non-normality.

Similar to the Wilk-Shapiro W and D'Agostino's D, this test is also the ratio of two estimates of the standard deviation, since

$$K_{mn}/\sqrt{2\pi e} = \frac{\hat{\sigma}_e}{\hat{\sigma}}$$

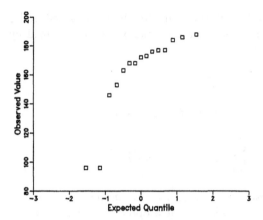

Figure 4.4 Q-Q plot of heights of cross-fertilized Zea mays *plants (n = 15).*

the term on the right being the ratio of the entropy estimate of the standard deviation to the maximum likelihood estimate.

4.4.2 Lin and Mudholkar's Test

Lin and Mudholkar (1980) developed a test for asymmetric alternatives based on the unique characteristic of the normal distribution that the mean and variance are independent. They suggest using the product moment correlation, r, of the n means obtained by deleting one observation at a time with the corresponding variance. Since the correlation is appropriate for a bivariate normal distribution, they use the transformation of the corresponding variance

$$y_i = (n^{-1}(\sum_{j\neq i} x_j^2 - (\sum_{j\neq i} x_j)^2/(n-1)))^{1/3}$$

to attain a more nearly normal marginal distribution for the variance. Although the n pairs $(\overline{x}_i, y_i)$ are not independent, Lin and Mudholkar used r to estimate the extent of the dependence. Equivalently, the product moment correlation of (x_i, y_i) can be used. High values of $|r|$ indicate non-normality.

Asymptotically r is $N(0, 3/n)$ under the null hypothesis; however, for moderate samples they suggest using Fisher's transformation

$$z = \log[(1+r)/(1-r)]/2.$$

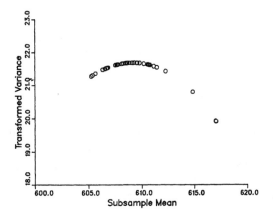

Figure 4.5 Plot of means and transformed variances, y_i, used for calculation of the Lin and Mudholkar (1980) test for the number of days until cancer occurrence data for mice (Data Set 7).

Under the null hypothesis, z is also asymptotically $N(0, 3/n)$; for small samples, z is nearly normal with mean 0 and variance

$$\sigma_n^2 = 3/n - 7.324/n^2 + 53.005/n^3.$$

An approximation of the percentile points of the distribution of z under normality was obtained using a Cornish-Fisher expansion. The upper α percentile of z is approximately

$$Z_\alpha = \sigma_n[U_\alpha + \gamma_{2n}(U_\alpha^3 - 3U_\alpha)/24] \tag{4.4}$$

where U_α is the upper α percentile of the standard normal distribution, $\Phi^{-1}(\alpha)$, and

$$\gamma_{2n} = -11.70/n + 55.06/n^2.$$

Example 4.10. For the number of days until cancer occurred in mice (Data Set 7) the value of Lin and Mudholkar's r is −0.655. Figure 4.5 is a plot of the means, $\bar{x}_i$, and transformed variances, y_i, for these data. The Fisher transformed value is

$$z = -0.784.$$

For a two-tailed $\alpha = 0.05$ level test, the lower and upper 2.5% critical values for z from (4.4) are ±0.53, indicating that the data are not from a normal distribution.

4.4.3 Quartile Means Test

In a comparison of tests of normality for detecting heavy-tailed alternatives, Smith (1975) presented a test for tail length based on the means of each quartile of a sample of data,

$$Z = (\bar{x}_4 - \bar{x}_1)/(\bar{x}_3 - \bar{x}_2)$$

where $\bar{x}_i$ is the mean of the ith quartile of the ordered data. Upper tail critical values for Z were obtained through simulation only for sample sizes of $n = 20, 50$ and 100; 0.05 values are 6.268, 4.999 and 4.720, respectively, where the test for heavy-tailed alternatives is an upper-tailed test. For short-tailed alternatives, a lower-tailed test should be used, although these critical values were not given.

4.5 Transformations to Normality

One can also use an approach consisting of a Box-Cox transformation (Box and Cox, 1964) and LR test for the transformation parameter as a test for normality. The Box-Cox transformation is a power transformation used to convert a sample of observations x_i to more near normal values y_i by

$$y_i = \begin{cases} (x_i^\lambda - 1)/\lambda & \lambda \neq 0 \\ \ln(x_i) & \lambda = 0 \end{cases}$$

under the constraint that all x_i are positive. The best transformation to normality is obtained by determining the maximum likelihood estimate of λ, where the log-likelihood is

$$L = -\frac{n}{2}\ln(\sigma_\lambda^2) + (\lambda - 1)\sum \ln(x_i) \qquad (4.5)$$

and σ_y^2 is the variance under the transformation. In the context of testing for normality, if the likelihood ratio test for $\lambda = 1$ does not reject or, equivalently, a confidence interval for λ contains 1, then the sample cannot be rejected as being from a normal distribution.

The precise maximum likelihood estimate of λ can be obtained by iterative methods using (4.5) and a computer optimization routine; however, in practice, (4.5) is usually evaluated using a grid of λ values (for example, Draper and Smith, 1981, suggest an initial grid within the range of (-2, 2). This range can be widened subsequently as necessary). A value of λ is then

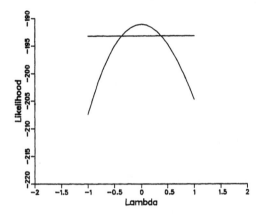

Figure 4.6 Likelihood for Box-Cox transformation of TSP data (Data Set 1).

chosen which is near to some common power transformation (e.g., square root when the MLE of λ is 0.4).

A $100(1-\alpha)\%$ confidence interval is obtained by finding the two values λ_L and λ_U, $\lambda_L < \lambda_U$ such that

$$2(L_\lambda - L_1) \leq \chi_1^2(\alpha)$$

where L_t is the likelihood when $\lambda = t$. In the region of $\hat{\lambda}_{max}$, it has been our experience that the likelihood function (4.5) is nearly symmetric, so that the confidence interval is of the form $\hat{\lambda}_{max} \pm k$ for the appropriate k.

Example 4.11. For the TSP data of Data Set 1, Figure 4.6 contains the plot of the likelihood for values of λ from -1(0.1)1. The maximum is attained at $\lambda = 0$, corresponding to a logarithmic transformation. The 95% confidence interval for λ is $(-0.4, 0.4)$, so as a test for normality the transformation method rejects normality for these data.

As a comparative result, for the raw data the Filliben probability plot correlation (Section 2.3.2) has a value of 0.934, rejecting the null hypothesis (Example 2.2); however, for the log transformed data a value of 0.994 is obtained, indicating a satisfactory transformation to normality.

The disadvantage of this method is the necessary condition that all observations be positive. In those instances where the condition does not hold, the observations can be "started" with a parameter λ_2 such that $\lambda_2 > \min(x_i)$, i.e.,

$$y_i = \begin{cases} ((x_i + \lambda_2)^{\lambda_1} - 1)/\lambda_1 & \lambda_1 \neq 0 \\ \ln(x_i + \lambda_2) & \lambda_1 = 0. \end{cases}$$

For this transformation, the log-likelihood is

$$L = -\frac{n}{2}\ln(\sigma_\lambda^2) + (\lambda_1 - 1)\sum \ln(x_i + \lambda_2)$$

and the $100(1-\alpha)\%$ confidence interval is obtained by

$$2(L_\lambda - L_1) \leq \chi_2^2(\alpha).$$

For the started transformation, a grid of (λ_1, λ_2) values must be inspected, or iterative methods used.

4.6 Measures of Skewness and Tail Length

In this section we present statistics which have been offered as measures of asymmetry and tail length by several authors as alternatives to moments $\sqrt{\beta_1}$ and β_2. Some of these measures are functions of the observations of a sample (e.g., Uthoff's T_1 and T_2); others have been presented as measures of distributional characteristics based on parameters and quantiles of the density (e.g., Groeneveld and Meeden's B_1 and Filliben's ρ), but could be used as tests of normality by substituting estimates of the measures based on the observations. Although some of the following measures have been suggested for use as tests for normality, they have not generally been studied with respect to their small sample characteristics, their power in comparison to other tests or, in the case of tail-length measures, whether they would be better at detecting short- or long-tailed alternatives.

4.6.1 Skewness Measures

Uthoff's T_2 and S_2

The MPLSI test statistic for an exponential (skewed) alternative to a uniform (symmetric) distribution is

$$T_{E,U} = (\bar{x} - x_{(1)})/(x_{(n)} - x_{(1)})$$

(Uthoff, 1968). Let x_α be the αth quantile of the observations: then, if in $T_{E,U}$ we substitute the median for the mean and x_α and $x_{1-\alpha}$ for $x_{(1)}$ and $x_{(n)}$, respectively, letting α be small we have the measure of skewness

$$T_2 = (x_{0.5} - x_\alpha)/(x_{1-\alpha} - x_\alpha).$$

Uthoff (1968) suggests the use of $\alpha \approx 0.01$.

As another measure of skewness, define the function $h_m(x_{(i)} : x_{(j)})$ to be Gini's mean difference using only the order statistics of the sample $x_{(i)}$ through $x_{(j)}$, $i < j, i + j = m$; for the full sample,

$$h_n(x_{(1)} : x_{(n)}) = \frac{2}{n(n-1)} \sum_{i=1}^{[n/2]} (n + 1 - 2i)(x_{(n+1-i)} - x_{(i)}).$$

Then another measure of skewness can be obtained by

$$S_2 = \frac{h_{n-n_3+1}(x_{(n_3)} : x_{(n)})}{h_{n_3}(x_{(1)} : x_{(n_3)}) + h_{n-n_3+1}(x_{(n_3)} : x_{(n)})}$$

where $n_3 = n/2$. Both T_2 and S_2 are U-statistics and therefore asymptotically normal.

Groeneveld and Meeden's B_1, $B_2(\alpha)$ and B_3

Groeneveld and Meeden (1984) suggested several statistics as measures of skewness. One general measure for a specified distribution F is

$$B_2(\alpha) = [F^{-1}(1 - \alpha) + F^{-1}(\alpha) - 2\nu_x]/[F^{-1}(1 - \alpha) + F^{-1}(\alpha)] \quad (4.6)$$

where ν_x is the median of the distribution and $F^{-1}(\alpha)$ is the αth quantile. For a sample of observations, this suggests the substitution of sample order statistics into (4.6) as a measure of skewness in a sample. The Bowley coefficient of skewness (Bowley, 1920) is a measure of this form,

$$B_1 = (Q_3 + Q_1 - 2Q_2)/(Q_3 - Q_1)$$

where the Q_i are the ith quartiles of the random variable X. A third measure of skewness is

$$B_3 = (\mu_x - \nu_x)/E(|X - \nu_x|)$$

where μ_x is the sample mean. B_1 can also be estimated using the sample observations. B_3 is closely related to the Pearson measure of skewness

$$q = (\mu_x - \nu_x)/\sigma$$

where ν_x is the mode. Both B_3 and q are equal to 0 for any symmetric distribution, and are positive for right skewed and negative for left skewed densities.

Filliben's ρ

As a measure of the skewness of the densities he was using as alternatives in a power study of the probability plot correlation test, Filliben (1975) used, in addition to $\sqrt{\beta_1}$,

$$\rho = [F^{-1}(0.975) + F^{-1}(0.5)]/[F^{-1}(0.975) + F^{-1}(0.025)].$$

Note that this is the distributional equivalent to Uthoff's T_2 statistic (Section 4.6.1) with $\alpha = 0.025$, except the numerator measures the upper tail of the distribution. For symmetric distributions $\rho = 0.5$, while for asymmetric distributions skewed to the right $0.5 < \rho < 1$. In general, ρ can range within the interval $[0, 1]$.

4.6.2 Measures of Symmetric Tail Heaviness

Uthoff's T_1 and S_1

Uthoff's (1968) test statistic T_1 for measuring tail length is

$$T_1 = (x_{1-\beta} - x_\beta)/(x_{1-\alpha} - x_\alpha)$$

where he recommended using α and β of about 0.01 and 0.25, respectively; the expected value of T_1 under normality is 0.289 for this configuration. T_1 was suggested by substituting $(x_{1-\alpha} - x_\alpha)$ for the range in the numerator of the range test $T_{U,N}$ and substituting the interquartile range in the denominator for the standard deviation. The inverse of the result was taken to make the range of T_1 between 0 and 1. Uthoff also demonstrated aspects of T_1 when $\beta = 0.32$ and $\alpha = 0.04$, for which the expected value is 0.269 under normality.

A second tail length measure was given by

$$S_1 = \frac{h_{n_2-n_1+1}(x_{(n_1)} : x_{(n_2)})}{h_{n_1}(x_{(1)} : x_{(n_1)}) + h_{n_2-n_1+1}(x_{(n_1)} : x_{(n_2)}) + h_{n-n_2+1}(x_{(n_2)} : x_{(n)})}$$

where n_1 and n_2 are about βn and $(1 - \beta)n$, respectively, and β is about 0.25; the h_n are defined in Section 4.6.1. Under these conditions the expected value of S_1 is 0.290 for the normal distribution.

Both T_1 and S_1 are U-statistics and are asymptotically normal, under any distribution.

Hogg's Q

In a study of robust estimation, Hogg (1967) considered the kurtosis statistic, b_2, as a selector function for determining the best estimator of a distributional parameter (see Section 12.2.7). In a study of goodness of fit, Hogg (1972) showed that over a variety of symmetric alternatives and null distributions the statistic V, which was essentially the range divided by the mean absolute deviation from the sample median (i.e., the LR test statistic for a uniform vs a double exponential distribution), seemed to be the best test. Noting that $\sum |x_i - x_M|$ can be written as $U_{0.5} - L_{0.5}$ where U_β and L_β are the sum of the $n\beta$ largest and smallest order statistics, respectively, and that V was the ratio of two linear functions of order statistics, he suggested that statistics of the form

$$Q = (U_\alpha - L_\alpha)/(U_\beta - L_\beta)$$

might be very useful as selector functions (equivalently, the averages of the U's and L's may be used). In particular, he reported that $\alpha = 0.05$ and $\beta = 0.50$ gave a satisfactory selector function. (Note that for small samples, Q is for all intents and purposes V).

Groeneveld and Meeden's B_2^* and B_3^*

For a symmetric distribution, assuming symmetry about 0, define a new distribution function $F^*(x) = 2F(x) - 1$ for $x > 0$. Groeneveld and Meeden (1984) defined two tail heaviness measures $B_2(F^*)$ and $B_3(F^*)$.

These measures are essentially the same as calculating the skewness measures B_2 and B_3 for the upper half of a symmetric distribution. Values which are more skewed than those for a half-normal distribution would indicate a distribution with heavier tails than a normal.

Filliben's τ_2

Filliben (1975) presented a measure of distributional tail length similar to Uthoff's T_1, although the α and β values were much smaller than those proposed by Uthoff. If

$$\tau_1 = [F^{-1}(0.9975) - F^{-1}(0.0025)]/[F^{-1}(0.975) - F^{-1}(0.025)]$$

then τ_2 is defined as

$$\tau_2 = (1 - 1/\tau_1)^{0.57854}.$$

Since τ_1 can range from 1 to ∞, τ_2 will always be a measure on $[0, 1]$.

References

Bowley, A.L. (1920). **Elements of Statistics**, 4th ed. Charles Scribner's Sons, New York.

Box, G.E.P., and Cox, D.R. (1964). An analysis of transformations. Journal of the Royal Statistical Society B 226, 211-252.

D'Agostino, R.B. (1970). Linear estimation of the normal distribution standard deviation. American Statistician 24, 14-15.

D'Agostino, R.B. (1971). An omnibus test of normality for moderate and large size samples. Biometrika 58, 341-348.

D'Agostino, R.B. (1972). Small sample probability points for the D test of normality. Biometrika 59, 219-221.

David, H.A. (1968). Gini's mean difference rediscovered. Biometrika 55, 573-575.

David, H.A., Hartley, H.O., and Pearson, E.S. (1954). The distribution of the ratio, in a single normal sample, of the range to the standard deviation. Biometrika 41, 482-493.

Davis, C.E., and Quade, D. (1978). U-statistics for skewness or symmetry. Communications in Statistics - Theory and Methods A7, 413-418.

Downton, F. (1966). Linear estimates with polynomial coefficients. Biometrika 53, 129-141.

Draper, N.R., and Smith, H. (1981). **Applied Regression Analysis**, 2nd ed. John Wiley and Sons, New York.

Filliben, J.J. (1975). The probability plot correlation coefficient test for normality. Technometrics 17, 111-117.

Franck, W.E. (1981). The most powerful invariant test of normal versus Cauchy with applications to stable alternatives. Journal of the American Statistical Association 76, 1002-1005.

Groeneveld, R.A., and Meeden, G. (1984). Measuring skewness and kurtosis. The Statistician 33, 391-399.

Grubbs, F.E. (1969). Procedures for detecting outlying observations in a sample. Technometrics 11, 1-21.

Hajek, J., and Sidak, Z.S. (1967). **Theory of Rank Tests**. Academic Press, New York.

Hoeffding, W. (1948). A class of statistics with asymptotically normal distributions. Annals of Mathematical Statistics 19, 293-325.

Hoel, D.G. (1972). A representation of mortality data by competing risks. Biometrics 28, 475-488.

Hogg, R.V. (1967). Some observations on robust estimation. Journal of the American Statistical Association 62, 1179-1186.

Hogg. R.V. (1972). More light on the kurtosis and related statistics. Journal of the American Statistical Association 67, 422-424.

Langley, R. (1971). **Practical Statistics**. Dover Publications, New York.

Lin, C.-C., and Mudholkar, G.S. (1980). A simple test for normality against asymmetric alternatives. Biometrika 67, 455-461.

Locke, C., and Spurrier, J.D. (1976). The use of U-statistics for testing normality against non-symmetric alternatives. Biometrika 63, 143-147.

Locke, C., and Spurrier, J.D. (1977). The use of U-statistics for testing normality against alternatives with both tails heavy or both tails light. Biometrika 64, 638-640.

Locke, C., and Spurrier, J.D. (1981). On the distribution of a new measure of heaviness of tails. Communications in Statistics - Theory and Methods A10, 1967-1980.

Oja, H. (1981). Two location and scale free goodness of fit tests. Biometrika 68, 637-640.

Oja, H. (1983). New tests for normality. Biometrika 70, 297-299.

Shapiro, S.S., and Wilk, M.B. (1965). An analysis of variance test for normality (complete samples). Biometrika 52, 591-611.

Smith, V.K. (1975). A simulation analysis of the power of several tests for detecting heavy-tailed distributions. Journal of the American Statistical Association 70, 662-665.

Spiegelhalter, D.J. (1977). A test for normality against symmetric alternatives. Biometrika 64, 415-418.

Spiegelhalter, D.J. (1980). An omnibus test for normality for small samples. Biometrika 67, 493-496.

Thode, H.C., Jr., Smith, L.A., and Finch, S.J. (1983). Power of tests of normality for detecting scale contaminated normal samples. Communications in Statistics - Simulation and Computation 12, 675-695.

Turner, M.E. (1960). On heuristic estimation methods. Biometrics 16, 299-301.

Uthoff, V.A. (1968). Some scale and origin invariant tests for distributional assumptions. Ph.D. thesis, University Microfilms, Ann Arbor, MI.

Uthoff, V.A. (1970). An optimum test property of two well-known statistics. Journal of the American Statistical Association 65, 1597-1600.

Uthoff, V.A. (1973). The most powerful scale and location invariant test of the normal versus the double exponential. Annals of Statistics 1, 170-174.

Vasicek, O. (1976). A test for normality based on sample entropy. Journal of the Royal Statistical Society B 38, 54-59.

CHAPTER 5

GOODNESS OF FIT TESTS

"...it is important that the particular goodness of fit test used be selected without consideration of the sample at hand, at least if the calculated significance level is to be meaningful. This is because a measure of discrepancy chosen in the light of an observed sample anomaly will tend to be inordinately large."

H.T. David, 1978

Prior to the interest in tests specific to identifying non-normality, initiated at least in part by the test of Shapiro and Wilk (1965), distributional assumptions were verified most often by more general goodness of fit tests which could be used for any simple null hypothesis $F_0(x)$. These tests are unique from the tests for normality described in the preceding chapters in that, under a simple hypothesis, their distribution is independent of the null distribution and therefore tables of critical values are identical regardless of the hypothesized null distribution. However, prior knowledge of distributional parameters is relatively rare in actual practice.

In general, the goodness of fit tests described here rely on the relation of the empirical distribution function of the observations to the hypothesized distribution function in some manner: e.g., by direct comparison, as in the Kolmogorov-Smirnov test; by comparing the number of actual with the expected number of observations occurring within bins defined by the null distribution, as in the χ^2 test; or by comparing spacings within the null distribution function.

It was not until relatively recently that null distributions or critical values of many of these tests were obtained for use with a composite null hypothesis. If $F_0(x)$ is completely specified, then $p_i = F_0(x_i)$ are uniform random variables; however, if one or more parameters are unknown, then the p_i obtained using estimated parameters are no longer uniform. If $F_0(x)$ depends only on a location and scale parameter, then the distribution of $F_0(x|\hat{\theta})$ under the null hypothesis depends on the functional form of $F_0(x)$, but not the parameters (David and Johnson, 1948; Stephens, 1986a).

In this chapter we will focus on general goodness of fit tests and their application to testing for normality when parameters are unknown. While we attempt to mention as broad a range of tests as possible, we will be concerned mainly with those tests that have been shown to have decent power at detecting normality; less powerful tests will be described in less detail, except for those with historical interest.

5.1 Tests Based on the Empirical Distribution Function

Empirical distribution function (EDF) tests are those goodness of fit tests based on a comparison of the empirical and hypothetical distribution functions. There are two general types of EDF test: those based on the maximum distance of empirical to null distribution function (Kolmogorov-Smirnov test and Kuiper's V), and quadratic tests (Anderson-Darling and Cramer-von Mises tests). As originally derived, these tests require a simple rather than compound hypothesis. More recently (Lilliefors, 1967; Stephens, 1974) they were expanded to include those circumstances where the mean and variance were not specified, and these have been shown to have power which was comparable to the Wilk-Shapiro test for some alternatives to normality.

The EDF of a sample, designated $F_n(x)$, is a step function defined as

$$F_n(x) = \begin{cases} 0 & x < x_{(1)} \\ i/n & x_{(i)} \leq x < x_{(i+1)} \quad i = 1, \ldots, n-1 \\ 1 & x_{(n)} \leq x. \end{cases}$$

As with any distribution function, $F_n(x)$ is the proportion (probability) of observations with a value less than or equal to x, with increasing steps of $1/n$ at each observation. Figure 5.1 is a plot of the EDF for the newborn baby birthweight data (Data Set 3). This differs from an ecdf plot (Section 2.2.4) in that steps are shown here; also, i/n is conventionally the plotting

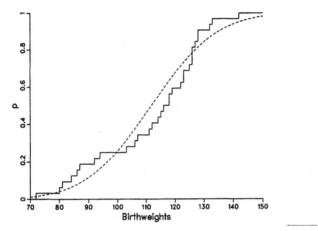

Figure 5.1 EDF plot of newborn birthweights (n = 32).

position used in EDF plots. EDF tests are based on the differences between the EDF and the distribution function based on the null hypothesis, in the normal case $p_{(i)} = \Phi([x_{(i)} - \hat{\mu}]/\hat{\sigma})$. EDF tests reject the null hypothesis (normality) when the discrepancies between the EDF and the hypothesized cumulative distribution function are too large; hence, these tests are all upper-tailed tests. There has been some concern that test values that are too small might indicate a lack of randomness in the data, and therefore lower-tailed or two-tailed tests should be used (Stephens, 1986a); however, this problem of *superuniformity* will not be considered here.

Although these tests can be used for $p_{(i)} = F_0(x_{(i)})$ for any distribution F_0, critical values for these tests are dependent upon the null distribution. Modifications to the EDF tests in this section have been derived (Stephens, 1974) so that the critical values for each test are independent of the sample size.

5.1.1 The Kolmogorov-Smirnov Test

The Kolmogorov-Smirnov statistics are based on the maximum differences between the EDF and the p_i. These are given by

$$D^+ = \max_{i=1,\ldots,n} [i/n - p_{(i)}]$$

$$D^- = \max_{i=1,\ldots,n} [p_{(i)} - (i - 1)/n]$$

$$D = \max[D^+, D^-].$$

Lilliefors (1967) was the first to address the issue of using EDF tests for composite hypotheses, when he investigated the null distribution of D under a normal composite null hypothesis and compared it to the χ^2 test in terms of power. Lilliefors gave a table of critical values for D based on simulation. However, for simplification, the modification

$$D^* = (\sqrt{n} - 0.01 + 0.85/\sqrt{n})D$$

(Stephens, 1974) can be compared to the critical value of 0.895 for an 0.05 level test for all sample sizes; critical values of the null distribution of D^* for other levels of significance are given in Table B15.

> *Example 5.1. For the birthweight data (Data Set 3), the value of D^+ is found at the sixth observation and D^- is found at the fourteenth observation. These values are 0.1006 and 0.1431, respectively. Therefore, $D = 0.1431$ and*
>
> $$D^* = (5.66 - 0.01 + 0.15)D = 0.830.$$
>
> *This value does not exceed the 0.05 level critical value of 0.895.*

5.1.2 Kuiper's V

The V test (Kuiper, 1960) is also based on a combination of D^+ and D^-, but is obtained using the sum rather than the maximum,

$$V = D^+ + D^-$$

with

$$V^* = (\sqrt{n} + 0.05 + 0.82/\sqrt{n})V$$

having fixed distribution under the null hypothesis for all sample sizes (Stephens, 1974). V can also be used for testing goodness of fit for distributions on a circle. Critical values for selected levels of significance for V^* are given in Table B15.

> *Example 5.2. From the values of D^- and D^+ in Example 5.1, for the birthweight data we calculate*
>
> $$V = 0.1431 + 0.1006 = 0.2437$$

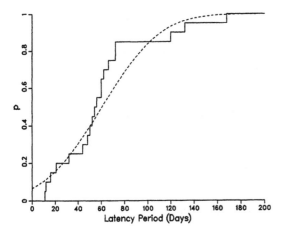

Figure 5.2 EDF plot of leukemia latency period (n = 20).

and

$$V^* = (5.66 + 0.05 + 0.14)V = 1.426$$

which does not attain significance.

5.1.3 Cramer-von Mises Test

A class of EDF goodness of fit tests was proposed by Anderson and Darling (1952), defined by

$$n \int_{-\infty}^{\infty} [F_n(x) - F(x)]^2 \psi(F(x)) dF(x) \tag{5.1}$$

where $\psi(F(x))$ is a weighting function. In particular, the weighting function $\psi(F(x)) = 1$ gives the Cramer-von Mises test statistic

$$W^2 = \frac{1}{12n} + \sum \left(p_{(i)} - \frac{2i - 1}{2n} \right)^2$$

with the modification

$$W^{2*} = (1.0 + 0.5/n)W^2$$

accounting for differences in sample size when using critical values.

Example 5.3. Figure 5.2 shows the EDF plot for the leukemia latency data (Data Set 9). For these data, the value of W^2 is 0.179, *resulting in the value of 0.184 for W^{2*}. From Table B15 we see that this value is between the 0.01 and 0.005 level critical values, 0.179 and 0.201, respectively, indicating a significant deviation from the normal distribution.*

Critical values for W^{2*} are given in Table B15. For distributions on a circle, Watson (1961, 1962) proposed

$$U^2 = \sum_{i=1}^{n} \left(p_{(i)} - \bar{p} + 0.5 - \frac{2i-1}{2n} \right)^2 + \frac{1}{12n}.$$

The relationship between U^2 and W^2 is given by

$$U^2 = W^2 - n(\bar{p} - 0.5)^2.$$

$$U^{2*} = (1.0 + 0.5/n)U^2$$

with critical values given in Table B15.

5.1.4 Anderson-Darling Test

Anderson and Darling (1954) used $\psi(p) = [p(1-p)]^{-1}$ as the weighting function in (5.1), resulting in the test statistic

$$A^2 = -n - n^{-1} \sum_{i=1}^{n} [2i - 1][\log(p_{(i)}) + \log(1 - p_{(n-i+1)})]$$

This weighting scheme gives more weight to the tails of the distribution than does W^2. Stephens (1986a) proposed the modification

$$A^{2*} = (1.0 + 0.75/n + 2.25/n^2)A^2$$

to obtain a set of critical values for all sample sizes; these are provided in Table B15.

Example 5.4. The observed daily July wind speed data (Data Set 8) indicate a slightly heavy-tailed distribution ($b_2 = 4.1$, see Appendix 1), so the Anderson-Darling test might be more appropriate for testing these data than other EDF tests. For these data,

$$A^2 = 0.301$$

Table 5.1 Upper 5% percentage points of the Green-Hegazy modified EDF tests for selected sample sizes (from Green and Hegazy, 1976).

Test	5	10	20	40	80	160
D_2	0.692	0.973	1.327	1.844	2.617	3.674
D_{22}	0.660	0.950	1.323	1.848	2.615	3.678
A_{21}	0.287	0.410	0.534	0.610	0.677	0.712
A_{22}	0.600	0.727	0.765	0.764	0.764	0.765

$$A^{2*} = 0.309$$

which is well below the 10% critical value.

5.1.5 Modified EDF Tests

Green and Hegazy (1976) presented modified EDF tests and compared their power under a variety of null distributions. Those that they suggested for a composite normal null were two tests based on a Kolmogorov-Smirnov (distance) type criterion,

$$D_2 = \sum_{i=1}^{n} |p_{(i)} - i/n|$$

$$D_{22} = \sum_{i=1}^{n} |p_{(i)} - (i+2)/(n+1)|$$

and two quadratic tests related to the Anderson-Darling test

$$A_{21} = -n - n/(n+1)^2 \sum_{i=1}^{n} \Big([(2i-1)\log p_{(i)} + (2i+1)\log(1 - p_{(n-i+1)}]$$
$$- [(2n+1)\log p_{(n)} - \log(1 - p_{(n)})] \Big)$$

$$A_{22} = -n - 2n/(n+1)^2 \sum_{i=1}^{n} i[\log p_{(i)} + \log(1 - p_{(n-i+1)})]$$
$$- n/(n+1)^2 \{0.25[\log p_{(1)} + \log(1 - p_{(n)})]$$
$$+ (n+0.75)(\log p_{(n)} + \log(1 - p_{(1)})\}$$

Critical values for 0.05-level significance tests under a normal null hypothesis are given in Table 5.1.

5.2 The χ^2 Test

In 1900, Karl Pearson presented his χ^2 goodness of fit test. Until relatively recently, this test was among the most useful of all tests for testing goodness of fit, irrespective of the null hypothesis. However, with the proliferation of tests of normality that has occurred since Shapiro and Wilk (1965) provided their test specifically for the composite normal null, the χ^2 test has fallen into disfavor as a test for normality, mostly because of its lack of power relative to other tests for normality; Moore (1986) recommended that the χ^2 test not be considered "... for testing fit to standard distributions for which special-purpose tests are available...". Here we present a review of the χ^2 test for the purpose of historical interest and completeness.

5.2.1 Development of the χ^2 Test

The well known χ^2 goodness of fit test is given by

$$\mathbf{X}^2 = \sum_{i=1}^{k} \frac{(n_i - np_i)^2}{np_i} \tag{5.2}$$

and the null hypothesis is rejected for values of $\mathbf{X}^2$ that are too large. To apply this test, the range of the n observations is divided into k mutually exclusive classes; n_i is the number of observations that fall into class i; and p_i is the probability that an observation will fall into class i under the null hypothesis. Then np_i is the number of observations which would be expected to occur in class i. The resulting test is the familiar calculation "sum of the observed minus expected squared over the expected".

Pearson (1900) derived the test using the following reasoning: if $\mathbf{x} = (x_1, \ldots, x_{k-1})$ has a nonsingular $(k-1)$-variate normal distribution with mean μ and covariance matrix Σ, then the quadratic form $(\mathbf{x}-\mu)'\Sigma^{-1}(\mathbf{x}-\mu)$ has a χ^2_{k-1} distribution. For large samples $(n_i - np_i)$ is approximately multivariate normal, and is nonsingular if only $k-1$ of the k classes are considered. Therefore, if $x_i = n_i - np_i$, then the quadratic form reduces to (5.2). This result holds under any null distribution.

For a simple hypothesis, this result is very straightforward, although the number k and size p_i of classes to use need to be considered. In addition, another important issue arises in the use of this test as a test for composite normality: whether the distribution of the test statistic is still χ^2. When parameters are estimated, (5.2) becomes

$$\mathbf{X}^2 = \sum_{i=1}^{k} \frac{(n_i - np_i(t))^2}{np_i(t)} \tag{5.3}$$

where t is the m-vector of parameter estimates. Since the $p_i(t)$ are estimates of the p_i, and are themselves random variables, the asymptotic χ^2_{k-1} distribution no longer applies. Common practice is to use the estimates of the parameters to obtain the p_i, calculate the test statistic and compare it to a χ^2_{k-m-1}, adjusting the degrees of freedom by m. If the multinomial maximum likelihood estimators are used to obtain the $p_i(t)$, then the asymptotic distribution is χ^2_{k-m-1} (Fisher, 1924). These estimates are the solutions to the m equations

$$\sum_{i=1}^{k} \frac{n_i}{p_i(t)} \frac{\delta p_i(t)}{\delta t_j} = 0 \quad j = 1, \ldots, m.$$

Estimators that are asymptotically equivalent to the multinomial maximum likelihood estimators are the minimum χ^2 estimators (Fisher, 1924), which are the solution to the equations

$$\sum_{i=1}^{k} \left(\frac{n_i}{p_i(t)} \right)^2 \frac{\delta p_i(t)}{\delta t_j} = 0 \quad j = 1, \ldots, m$$

and the minimum modified χ^2 estimators (Neyman, 1949), based on the solution to

$$\sum_{i=1}^{k} \frac{p_i(t)}{n_i} \frac{\delta p_i(t)}{\delta t_j} = 0 \quad j = 1, \ldots, m.$$

Since these estimators are asymptotically equivalent to the multinomial maximum likelihood estimates, the asymptotic distribution of $\mathbf{X}^2$ under these estimators is still χ^2_{k-m-1}. Unfortunately, under the normal null distribution none of the above estimators are able to be obtained in closed form, requiring the use of numerical optimization methods to obtain solutions.

When more efficient estimates are used (e.g., maximum likelihood estimates based on the n observations rather than the multinomial estimates based on the k classes), the asymptotic distribution of (5.2) is no longer χ^2_{k-m-1} (Fisher, 1928; Chernoff and Lehmann, 1954). In this instance there is a partial recovery of the m degrees of freedom lost by multinomial estimation and so the distribution is bounded between a χ^2_{k-1} and a χ^2_{k-m-1} distribution. For large k the difference may be ignored; however, for small k the use of a χ^2_{k-m-1} may lead to significant error in the results. Even for k as high as twenty, which is rare in usual practice, the true significance level could be as high as 0.09 for a 5% test when estimating the two parameters of the normal distribution (Table 5.2).

Table 5.2 Probability levels of the χ^2_{k-1} distribution for the 5th percentile of the χ^2_{k-3} distribution.

k	$\{x \mid P(\chi^2_{k-3} > x) = 0.05\}$	$P(\chi^2_{k-1} > x)$
5	5.99	0.20
6	7.81	0.17
7	9.49	0.15
8	11.07	0.14
9	12.59	0.13
15	21.02	0.10
20	27.59	0.09

5.2.2 Number of Cells

One disadvantage of the χ^2 test is that, for a given sample, the results obtained from the test are affected (substantially, in some cases) by the number and size of the k classes chosen.

For the simple hypothesis, Mann and Wald (1942) recommended that the k cells have equal probability under the null distribution, i.e., $p_i = 1/k$. This results in a more accurate approximation to the χ^2 distribution. While the general χ^2 test is consistent, it is unbiased only for the equiprobable cell case. The calculation (5.2) in the equiprobable cell case becomes simply

$$\mathbf{X}^2 = \frac{k}{n} \sum_{i=1}^{k} n_i^2 - n.$$

Kendall and Stuart (1973) showed that the best k to choose based on maximizing the power when the power is β is

$$k = b \left(\frac{\sqrt{2}n^2}{z_\alpha^2 + \Phi^{-1}(\beta)} \right)^{2/5} \tag{5.4}$$

where z_α is the upper α percentage point of the standard normal distribution. Mann and Wald (1942) considered the case when $\beta = 0.5$ ($\Phi^{-1}(\beta) = 0$), and suggested using $b = 4$. Note that (5.4) suggests that the "best" k decreases in a region with higher power for fixed sample size, i.e., as the discrepancies $(n_i - np_i)$ get larger. However, Kendall and Stuart (1973) recommended that, for $n \geq 200$, k can be reduced by as much as one half without serious loss of power. Also, k cannot be too large since the normal approximation will not be adequate if the expected frequencies are too small. They did not recommend the use of (5.4) when $n < 200$, in which case a smaller number of classes should be used. In contrast, Koehler and Larntz (1980) claimed that the Pearson χ^2 test is adequate at both the 0.05 and 0.01 levels for expected frequencies as low as 0.25 when $k \geq 3$, $n \geq 10$ and $n^2/k \geq 10$.

Example 5.5. Fifty-eight wells were tested for a variety of water quality parameters in Suffolk County, N.Y., in 1990; Data Set 10 contains the observed average alkalinity of these wells. Using the Koehler-Larntz criteria, the number of cells that can be used for these data is 232 (expected value = 0.25, $n^2/k = 14.5$). Although not recommended for $n < 200$, when using (5.4) to obtain k we find that $k = 79$ when $b = 4$, and $k = 40$ when using $b = 2$. Based on the recommendation of equiprobable cells and that np_i be greater than or equal to 5 yields $k = 11$, and using $np_i \geq 10$ (as an arbitrary alternative) yields $k = 5$.

Significance testing results vary widely when $k = 5$ and $k = 11$. When $k = 5$, the value of Pearson's test is $\mathbf{X}^2 = 10.5$, which exceeds both the χ_2^2 and the χ_4^2 critical values at the 0.05 levels (5.99 and 9.49, respectively). However, when 11 equiprobable cells are used, Pearson's test is only $\mathbf{X}^2 = 7.2$, which is not nearly significant (for either 8 or 10 degrees of freedom) at even the 0.10 level.

5.2.3 Other χ^2 Tests

Under a simple hypothesis, the likelihood ratio test statistic is

$$G^2 = 2L = 2 \sum_{i=1}^{k} n_i \log(n_i/np_i).$$

It has been shown that

$$L = \sum_{i=1}^{k} \left(\frac{1}{2} \frac{(n_i - np_i)^2}{np_i} + \frac{1}{6} \frac{(n_i - np_i)^3}{np_i^2} + \ldots + (j^2 + j)^{-1} \frac{(n_i - np_i)^{j+1}}{np_i^j} + \ldots \right)$$

(Fisher, 1924). Ignoring all but the first term in the summation, we see that $2L = \mathbf{X}^2$ and so the likelihood ratio test statistic is asymptotically equivalent to (5.2) and has an asymptotic χ_{k-1}^2 distribution. This test is subject to the same criticisms as Pearson's χ^2 test when parameters are estimated. Moore (1986) recommended the use of χ^2 over G^2 in situations where the hypothesis is simple or minimum χ^2 estimators are used.

A third type of χ^2 test is based on the Freeman-Tukey (1950) deviates,

$$FT^2 = \sum_{i=1}^{k} [\sqrt{n_i} + \sqrt{n_i + 1} - \sqrt{4np_i + 1}]^2$$

which, if terms of order $1/n$ are omitted, reduces to

$$FT^2 = 4 \sum_{i=1}^{k} (\sqrt{n_i} - \sqrt{np_i})^2$$

(Bishop, Fienberg and Holland, 1975). FT^2 is also asymptotically χ^2 distributed. Larntz (1978) indicated that $\mathbf{X}^2$ more accurately approximates the distribution of χ^2 under the null hypothesis than either G^2 or FT^2.

5.2.4 Recommendations on the Use of the χ^2 Test

The use of the χ^2 test is not recommended as a test for univariate normality, mostly because of its lack of power. However, the flexibility of the test is such that it is useful for testing multivariate normality (Chapter 9) rather than using other tests which are much more difficult to implement, and for censored data (Chapter 8). A further advantage, at least over EDF tests, is the need for only one set of tables for determining significance regardless of the form of the null distribution $F_0(x)$.

5.3 Other Methods of Testing for Composite Goodness of Fit

Here we present other methods of testing goodness of fit which have been derived specifically for composite null hypotheses. These include Neyman's smooth goodness of fit test modified for the composite case; using normalized spacings to reduce the data to (non-uniform) ordered values on the interval $(0, 1)$; and half- and multi-sample methods wherein psuedo-independent estimates of the parameters are used as the true values.

5.3.1 Neyman's Smooth Goodness of Fit Tests

Thomas and Pierce (1979) presented modified smooth goodness of fit tests (Neyman, 1937) which are applicable when the distribution parameters are

Table 5.3 *Coefficients a_{ms} for calculating modified Neyman smooth goodness of fit test statistics.*

m/s	1	2	3	4
1	16.3172	-	-	-
2	-27.3809	27.3809	-	-
3	89.7593	-156.6001	104.4001	-
4	-118.2638	436.8700	-637.2124	318.6062

unknown. The smooth tests are based on a sequence of sums of individual single degree of freedom χ^2 tests, so that the jth test statistic is (at least asymptotically) χ_j^2. Letting $y_i = \Phi[(x_i - \hat{\mu})/\hat{\sigma}]$, define

$$u_i = \sum_{j=1}^{n} (y_j^i - (i+1)^{-1})$$

$$t_m = n^{-1/2} \sum_{s=1}^{m} a_{ms} u_s, \qquad m = 1, 2, \dots.$$

Then the kth modified smooth test statistic, W_k, is

$$W_k = \sum_{m=1}^{k} t_m^2$$

so that $W_k = W_{k-1} + t_k^2$. W_k can be compared to a χ^2 distribution with k degrees of freedom to determine significance, achieving good agreement with the asymptotic distribution for samples as small as 20. Thomas and Pierce (1979) gave coefficients for calculating the W_k under composite normal, exponential and Weibull null hypotheses for $k = 1, 2, 3$, and 4; Table 5.3 contains the appropriate coefficients for the normal case.

Example 5.6. *The smooth goodness of fit test statistics for $k = 1, 2, 3, 4$ for the July wind speed data (Data Set 8) are calculated as*

$$W_1 = 1.12$$

$$W_2 = 1.99$$

$$W_3 = 3.43$$

$$W_4 = 4.29.$$

*None of these tests is significant, all of them having a test level
between 0.50 and 0.25.*

Neyman (1937) suggested the use of W_3 or W_4, and Quesenberry
(1986) preferred W_4. [It should be noted that their preferences were based
on the tests under a *simple* null hypothesis.]

5.3.2 Half-Sample and Multi-Sample Methods

Stephens (1978) suggested the half-sample method for testing for normality
and exponentiality. In this method, half of the sample is selected randomly
and parameter estimates are obtained. Then, using the whole sample the
$p_i = \Phi((x_i - \hat{\mu})/\hat{\sigma})$ are obtained and an EDF test is performed. The
asymptotic distribution of the EDF tests under this procedure is that of
the EDF test under the simple hypothesis; for the quadratic tests (A^2, W^2
and U^2), the finite distribution converges to the asymptotic for relatively
small sample sizes ($n \geq 20$). This procedure results in a substantial loss of
power, however, and is not invariant to the half sample initially selected to
obtain the parameter estimates.

Braun (1980) suggested a multi-sample method in which the parameters are first estimated using the entire sample. These estimates are then
used as the parameter values. For an overall α level test, the sample is
divided into m equal size (m/n) subsamples, and a chosen goodness of fit
test is performed on each subsample at the α/m level, using the simple hypothesis; if any of the test statistics exceeds the appropriate critical value
then the null hypothesis is rejected.

This procedure naturally requires a large sample in order to be effective; questions arise concerning the number of subsamples to be used
and/or the size of each subsample for a given n. Braun stated that if the
Cramer-von Mises test is to be used, then subsample size should be no
more than $n^{1/2}$; for the Kolmogorov-Smirnov test, subsample size should
be between 10% and 15% of n. Braun also suggested that the multi-sample
method should be used when there are natural groupings of the observations, making the subsample selection less arbitrary.

5.3.3 Tests Based on Normalized Spacings

Lockhart, O'Reilly and Stephens (1986) investigated the use of normalized
spacings for tests of composite hypotheses of normality. In particular,

normalized spacings are used to obtain z values which are on the interval $(0,1)$. These tests are also useful for testing goodness of fit with data censored at one or both ends.

For an ordered sample of n observations, let the normalized spacings be defined as

$$y_i = (x_{(i+1)} - x_{(i)})/(m_{i+1} - m_i), \qquad i = 1, \ldots, n-1$$

where m_i is the ith standard normal order statistic. Calculating

$$T_j = \sum_{i=1}^{j} y_i$$

leads to the $n-2$ ordered values

$$z_{(i)} = T_i/T_{n-1} \qquad i = 1, \ldots, n-2.$$

Lockhart, O'Reilly and Stephens (1986) considered the use of three tests with the normalized spacings. The first test of interest used the Anderson-Darling statistic,

$$A_{NS}^2 = -(n-2) - (n-2)^{-1} \sum_{i=1}^{n-2} (2i-1)[\log(z_{(i)}) + \log(1 - z_{(n-i-1)})].$$

Convergence to its asymptotic distribution is fairly rapid for this test; use of the asymptotic critical values with finite samples results in a test with an α level slightly greater than that selected. Critical values for A_{NS}^2 are given in Table B16.

Example 5.7. The 18 ordered $z_{(i)}$ calculated from the leukemia latency data (Data Set 9) are given in Table 5.4. The Anderson-Darling test statistic is

$$A_{NS}^2 = 2.50$$

which has a significance level of $0.01 < \alpha < 0.025$. In comparison, these same data had a significance level of $\alpha < 0.01$ using the Cramer-von Mises test of Section 5.1.3.

Table 5.4 Normalized spacings $z_{(i)}$ for the leukemia latency data (Data Set 9), used to calculate the Anderson-Darling test statistic in Example 5.7.

0.0030	0.0226	0.0550	0.1402	0.2454
0.2837	0.3040	0.3253	0.3471	0.3690
0.4125	0.4125	0.4329	0.4712	0.5238
0.5238	0.8346	0.8936		

Two other tests based on the normalized spacings were also investigated; these tests are similar to tests used to test goodness of fit for the extreme value and Weibull distributions (Mann, Scheuer and Fertig, 1973; Tiku and Singh, 1981).

$$Z_1 = \begin{cases} \sqrt{n-2}[z_{(\frac{n-1}{2})} - 0.5], & n \text{ odd}; \\ \sqrt{n-2}[z_{(\frac{n}{2})} - \frac{n}{2n-2}] & n \text{ even}, \end{cases}$$

and

$$Z_2 = \sqrt{n-2}(\bar{z} - 0.5)$$

where $\bar{z}$ is the mean of the $z_{(i)}$. Asymptotically under normality, Z_1 and Z_2 each have mean 0, and variance 0.1875 and 0.056084, respectively; convergence is fairly rapid (Lockhart, O'Reilly and Stephens, 1986).

5.4 Transformations to Uniformity

The tests described in this chapter have been, for the most part, based on tests originally designed to test for uniformity, obtained by the probability integral transform (PIT) $\Phi((x - \mu)/\sigma)$ under a simple null hypothesis. Modifications of the test statistic distributions were made in order to accomodate those cases where parameters needed to be estimated, since the PIT in the composite case does not result in uniform observations.

Methods of transforming normal variates to uniform observations without knowledge of the true mean and variance have been derived. These methods allow the use of EDF and other tests for uniformity on the transformed variables, using the tests under the simple hypothesis. Use of transformation methods has little advantage over tests for uniformity modified for unknown parameters; however, the transformations can be utilized as a basis for tests which have not been so modified, such as tests based on spacings.

Like the χ^2 test and the half-sample method, test statistic values obtained from a sample using the transformations are not unique. In particular, the transformation methods are not invariant under permutations of the observations. The values of the transformed variates, and hence of the test used, are affected by the sequence of observations. Transformation of normal variates to uniformity results in $n - 2$ uniform variates, equivalent to losing two degrees of freedom from parameter estimation.

5.4.1 Characterization of Normality

Characterizing properties of a distribution F are those where a statistic $t(x|F)$ has a known distribution if and only if F is true. Csorgo, Seshadri and Yalovsky (1973) presented two transformations of normal variates to uniformity based on such a characterization and compare the power of EDF tests using the transformed variates with the Wilk-Shapiro test.

For the first transformation, if a sample $x_i, i = 1, \ldots, n$ are independent and identically distributed from a normal distribution with unknown mean μ and variance σ^2, then

$$Z_j = (\sum_{i=1}^{j} x_i - jx_{j+1})/\sqrt{j(j+1)}, \qquad j = 1, \ldots, n-1$$

are iid normal with mean 0 and variance σ^2. From Kotlarski (1966), if $n \geq 4$, and using

$$Y_k = \sqrt{k}Z_{k+1}/\sqrt{Z_1^2 + Z_2^2 + \ldots + Z_k^2}, \qquad k = 2, \ldots, n-2,$$

then $u_i = G_i(Y_i), i = 1, \ldots, n$ are independent $U(0,1)$ random variables, where G_i is the Student's t distribution function with i degrees of freedom. This characterization did not yield goodness of fit tests which had power comparable to the Wilk-Shapiro test.

A second characterization, useful for odd sample sizes ($n = 2k+3, k \geq 2$), uses the Z_j as described above, and defines

$$Y_i = Z_{2n-1}^2 + Z_{2n}^2, i = 1, \ldots, k+1 = (n-1)/2.$$

Further define

$$S_{k+1} = \sum_{i=1}^{k+1} Y_i$$

and

$$\eta_r = \sum_{i=1}^{r} Y_i/S_{k+1} \qquad r = 1, \ldots, k.$$

Then the η_r act like the order statistics of k independent uniform random variables if and only if the x_i are normal. This characteriation yielded EDF tests that were comparable in power to the Wilk-Shapiro test when the alternative distribution was symmetric with $\beta_2 > 3$.

5.4.2 The Conditional Probability Integral Transformation

O'Reilly and Quesenberry (1973) discussed the conditional probability integral transformation (CPIT), which is a characterizing transformation which conditions on sufficient statistics for the unknown parameters in order to obtain $U(0,1)$ random variates. Quesenberry (1986) gave the transformation for normal random variables: for a sample from a normal distribution, given complete and sufficient statistics for μ and σ^2, define $\tilde{F}_n(x_1, \ldots, x_n)$ to be the distribution function of the sample given the statistics. Then the $n - 2$ random variables

$$U_{j-2} = \tilde{F}_j(x_j) \qquad j = 3, \ldots, n$$

are iid $U(0,1)$ random variables. In particular,

$$U_{j-2} = G_{j-2}(\sqrt{(j-1)/(j)}(x_j - \overline{x}_{j-1})/s_{j-1}) \qquad j = 3, \ldots, n$$

are iid uniform random variables, where

$$\overline{x}_k = k^{-1} \sum_{i=1}^{k} x_i$$

and

$$s_k = (k-1)^{-1} \sum_{i=1}^{k} (x_i - \overline{x}_k)^2$$

and G_j is the Student's t distribution function with j degrees of freedom.

5.5 Other Tests for Uniformity

Although the main focus of this text is testing for normality under a composite hypothesis, the methods given in the previous section allow the use of simple hypothesis tests when parameters are unknown.

However, we put less emphasis on these methods because (1) in general they seem to result in tests that are less powerful than many tests for normality given in Chapters 2 through 4 or the composite EDF tests given in Section 5.1; and (2) they are not invariant under permutations of the observations. Therefore, we mention only briefly some common goodness of fit tests used to ascertain the uniformity of a sample. In addition to those described below, of course, simple EDF tests could be used.

5.5.1 Tests Based on Spacings

The spacings of a sample are defined as the $n+1$ intervals on the hypothesized distribution function. Using Stephens (1986a) notation, the spacings are defined as

$$D_i = p_{(i)} - p_{(i-1)} \qquad i = 2, \ldots, n+1$$

with $D_1 = p_{(1)}$ and $D_{n+1} = 1 - p_{(n)}$. For uniform $p_{(i)}$, these spacings are exponentially distributed with mean $1/(n+1)$. Perhaps the most widely known test based on spacings is the Greenwood (1946) test,

$$G = \sum_{i=1}^{n+1} D_i^2$$

originally introduced as a method of ascertaining whether the incidence of contagious disease could be described as a Poisson process. Percentage points for G are given in Hill (1979), Burrows (1979), Stephens (1981) and Currie (1981).

Kendall's test (Kendall, 1946; Sherman, 1950) uses the sum of the absolute differences between the spacings and their expected values,

$$K = \sum_{i=1}^{n+1} |D_i - (n+1)^{-1}|.$$

Tests based on spacings of ordered uniform takens m $(m > 1)$ at a time have also been investigated (e.g., Hartley and Pfaffenberger, 1972; Del Pino, 1979; Rao and Kuo, 1984).

5.5.2 Pearson's Probability Product Test

In conjunction with their transformation, one test Csorgo, Seshadri and Yalovsky (1973) proposed is Pearson's probability product test (Pearson 1933) for testing the uniformity hypothesis. For this test,

$$PP_k = -2 \sum_{i=1}^{k} \log(u_i)$$

which is, under the null hypothesis, a χ^2 random variable with $2k$ degrees of freedom.

5.6 Further Reading

D'Agostino and Stephens (1986) contains a number of chapters which cover many of the topics described here in more detail. Stephens (1986a) describes EDF tests, including the case of the simple hypothesis and many non-normal null distributions. Moore (1986) provides more detail on χ^2 tests. Stephens (1986b) and Quesenberry (1986) discuss transformation methods.

References

Anderson, T.W., and Darling, D.A. (1952). Asymptotic theory of certain "goodness of fit" criteria based on stochastic processes. Annals of Mathematical Statistics 23, 193-212.

Anderson, T.W., and Darling, D.A. (1954). A test of goodness of fit. Journal of the American Statistical Association 49, 765-769.

Bishop, Y.M.M., Fienberg, S.E., and Holland, P.W. (1975). **Discrete Multivariate Analysis**. MIT Press, Cambridge, Mass.

Braun, H. (1980). A simple method for testing goodness of fit in the presence of nuisance parameters. Journal of the Royal Statistical Society B 42, 53-63.

Burrows, P.M. (1979). Selected percentage points of Greenwood's statistic. Journal of the Royal Statistical Society A 142, 256-258.

Chernoff, H., and Lehmann, E.L. (1954). The use of maximum likelihood estimates in χ^2 tests for goodness of fit. Annals of Mathematical Statistics 25, 579-586.

Csorgo, M., Seshadri, V., and Yalovsky, M. (1973). Some exact tests for normality in the presence of unknown parameters. Journal of the Royal Statistical Society B 35, 507-522.

Currie, I.D. (1981). Further percentage points of Greenwood's statistic. Journal of the Royal Statistical Society A 144, 360-363.

D'Agostino, R.B., and Stephens, M.A. (1986). **Goodness of Fit Techniques**. Marcel Dekker, New York.

David, H.T. (1978). Goodness of fit. In Kruskal, W.H., and Tanur, J.M., eds., **International Encyclopedia of Statistics**, The Free Press, New York.

David, F.N., and Johnson, N.L. (1948). The probability integral transformation when parameters are estimated from the sample. Biometrika 35, 182-192.

Del Pino, G.E. (1979). On the asymptotic distribution of k-spacings with applications to goodness of fit tests. Annals of Statistics 7, 1058-1065.

Fisher, R.A. (1924). The conditions under which χ^2 measures the discrepancy between observation and hypothesis. Journal of the Royal Statistical Society 87, 442-450.

Fisher, R.A. (1928). On a property connecting the χ^2 measure of discrepancy with the method of maximum likelihood. Atti Congresso Internazionale dei Mathematici 6, 95-100.

Freeman, M.F., and Tukey, J.W. (1950). Transformations related to the angular and square root. Annals of Mathematical Statistics 21, 607-611.

Green, J.R., and Hegazy, Y.A.S. (1976). Powerful modified EDF goodness of fit tests. Journal of the American Statistical Association 71, 204-209.

Greenwood, M. (1946). The statistical study of infectious disease. Journal of the Royal Statistical Society A 109, 85-110.

Hartley, H.O., and Pfaffenberger, R.C. (1972). Quadratic forms in order statistics used as goodness of fit criteria. Biometrika 59, 605-611.

Hill, I.D. (1979). Approximating the distribution of Greenwood's statistic with Johnson distributions. Journal of the Royal Statistical Society A 142, 378-380.

Kendall, M.G. (1946). Discussion of Professor Greenwood's paper. Journal of the Royal Statistical Society A 109, 103-105.

Kendall, M.G., and Stuart, A. (1973). **The Advanced Theory of Statistics**. Hafner Publishing Co., New York.

Koehler, K.J., and Larntz, K. (1980). An empirical investigation of goodness of fit statistics for sparse multinomials. Journal of the American Statistical Association 75, 336-444.

Kotlarski, I. (1966). On characterizing the normal distribution by Student's law. Biometrika 58, 641-645.

Kuiper, N.H. (1960). Tests concerning random points on a circle. Proceedings, Akademie van Wetenschappen A 63, 38-47.

Larntz, K. (1978). Small sample comparisons of exact levels for chi-squared goodness of fit statistics. Journal of the American Statistical Association 73, 253-263.

Lilliefors, H.W. (1967). On the Kolmogorov-Smirnov test for normality with mean and variance unknown. Journal of the American Statistical Association 62, 399-402.

Lockhart, R.A., O'Reilly, F.J., and Stephens, M.A. (1986). Tests of fit based on normalized spacings. Journal of the Royal Statistical Society B 48, 344-352.

Mann, N.R., Scheuer, E.M., and Fertig, K.W. (1973). A new goodness of fit test for the two parameter Weibull or extreme value distribution with unknown parameters. Communications in Statistics 2, 838-900.

Mann, H.B., and Wald, A. (1942). On the choice of the number of intervalsin the application of the chi-square test. Annals of Mathematical Statistics 13, 306-317.

Moore, D.S. (1986). Tests of chi-squared type. In D'Agostino, R.B, and Stephens, M.A., eds., **Goodness of Fit Techniques**, Marcel Dekker, New York.

Neyman, J. (1937). 'Smooth test' for goodness of fit. Skandinaviske Aktuarietiddkrift 20, 150-199.

Neyman, J. (1949). Contributions to the theory of the χ^2 test. In **Proceedings of the First Berkeley Symposium on Mathematical Statistics and Probability**, University of California Press, 239-273.

O'Reilly, F.J., and Quesenberry, C.P. (1973). The conditional probability integral transformation and applications to obtain composite chi-square goodness of fit tests. Annals of Statistics 1, 74-83.

Pearson, K. (1900). On the criterion that a given system of deviations from the probable in the case of a correlated system of variables is such that it can be reasonably supposed to have arisen from random sampling. Philosophical Magazine 50, 157-175.

Pearson, K. (1933). On a method of determining whether a sample of size n supposed to have been drawn from a parent population having a known probability integral has probably been drawn at random. Biometrika 25, 379-410.

Quesenberry, C.P. (1986). Some transformation methods in goodenss of fit. In D'Agostino, R.B, and Stephens, M.A., eds., **Goodness of Fit Techniques**, Marcel Dekker, New York.

Rao, J.S., and Kuo, M. (1984). Asymptotic results on the Greenwood statistic and some of its generalizations. Journal of the Royal Statistical Society B 46, 228-237.

Shapiro, S.S., and Wilk, M.B. (1965). An analysis of variance test for normality (complete samples). Biometrika 52, 591-611.

Sherman (1950). A random variable related to the spacings of sample values. Annals of Mathematical Statistics 21, 339-361.

Stephens, M.A. (1974). EDF statistics for goodness of fit and some comparisons. Journal of the American Statistical Association 69, 730-737.

Stephens, M.A. (1978). On the half-sample method for goodness of fit. Journal of the Royal Statistical Society B 40, 64-70.

Stephens. M.A. (1981). Further percentage points for Greenwood's statistic. Journal of the Royal Statistical Society A 144, 364-366.

Stephens, M.A. (1986a). Tests based on EDF Statistics. In D'Agostino and Stephens, eds., **Goodness of Fit Techniques**, Marcel Dekker, New York.

Stephens, M.A. (1986b). Tests for the uniform distribution. In D'Agostino and Stephens, eds., **Goodness of Fit Techniques**, Marcel Dekker, New York.

Thomas, D.R., and Pierce, D.A. (1979). Neyman's smooth goodness of fit test when the hypothesis is composite. Journal of the American Statistical Association 74, 441-445.

Tiku, M.L., and Singh, M. (1981). Testing for the two parameter Weibull distribution. Communications in Statistics - Theory and Methods 10, 907-918.

Watson, G.S. (1961). Goodness of fit tests on a circle. Biometrika 48, 109-114.

Watson, G.S. (1962). Goodness of fit tests on a circle. II. Biometrika 49, 57-63.

CHAPTER 6

TESTS FOR OUTLIERS

"Actually, the rejection of "outlying" observations may be just as much a practical (or common sense) problem as a statistical one and sometimes the practical or experimental viewpoint may naturally outweigh any statistical contributions."

F.E. Grubbs, 1950

There is perhaps as much or more interest in identifying outliers in a sample than there is for determining whether a sample is normal. Outlier tests are seldom used as tests for normality *per se.* Their usual function is the identification of an observation(s) which may have undue influence on an estimation or testing procedure; what to do with them once we find them is sometimes open to debate.

6.1 On the Need to Identify Outliers

Usually when performing a statistical test or estimation procedure, we assume that the data are all observations of iid random variables, often from a normal distribution. Sometimes, however, we notice in a sample one or more observations that stand out from the crowd. These observations are commonly called outliers.

Where do outliers come from, why do we care about them, and what can we do if we have them?

Outliers can stem from a variety of sources. Some of the most common sources are:

1. Purely by chance. In normal samples, about one out of every 20 observations can be expected to be greater than 2 standard deviations from the mean, about one in a hundred can be 2.5 standard deviations from the mean. Presence of an outlier does not necessarily mean that there is a problem.

2. Failure of the data generating process. In this instance, outlier identification may be of extreme importance in isolating a breakdown in an experimental process.

3. A subject being measured may not be homogeneous with the other subjects. In this case, we would say the sample comes from a contaminated distribution, i.e., the sample we are observing has been contaminated with observations which do not belong.

4. Failure of the measuring instrument, whether mechanical or human.

5. Error in recording the measurement, for example, on the data sheet or log, or in computer entry.

David and Paulson (1965) summarized why outliers should be identified:

(1) to screen the data for further analysis;

(2) to flag problems in the data generation process (see (2) above);

(3) the extreme points may be the only points of real interest.

What do we do if we have outliers? If possible, we should try to identify the reason for being so extreme. It is of utmost importance that the first thing to be done when identifying an observation as an outlier that *the data be checked as to their accuracy in recording and data entry.* This is usually, but not always, the easiest data verification that can be made. For example, Gibbons and McDonald (1980a, 1980b) performed some diagnostics on the regression of mortality rates on socioeconomic and air pollution data for some Standard Metropolitan Statistical Areas (SMSA's), as previously reported by Lave and Seskin (1970, 1977). SMSA's which had high influence on one or more coefficients in the resulting regression were identified. A check of the socioeconomic data from an easily obtainable source (U.S. Bureau of the Census, 1973) showed that some of the data for one of the outlier SMSA's (Providence, Rhode Island) had had the decimal point transposed one place to the left.

If an identified outlier(s) has been caused by data recording error, the value should be corrected and analysis can proceed. If it is determined that the data have been recorded correctly, investigation into other reasons why an observation is extreme should be done if at all possible. If the data have been investigated thoroughly and other sources of error have been identified, it may not be possible to recover the true data value (e.g., a scale was found to be incorrectly calibrated). In this instance, the only alternative is to throw out the contaminated data. However, all data which have been subject to the same source of error should also be eliminated, not just the extreme values.

If no source of error is discovered, several alternatives are available: for example, robust methods of analysis can be used (see Chapter 12). Other ways of accommodating outliers include (Tietjen, 1986): (1) removing the outliers and proceeding with analysis; (2) removing the outliers and treating the reduced sample as a censored sample; (3) Winsorize the outliers, that is, replace them with the nearest non-outlier value; (4) replace the outliers with new observations; (5) use standard methods for analyzing the data, both including and excluding the outliers, and reporting both results. Interpretation of results becomes difficult if the two results differ drastically.

Subjective or ad hoc methods of outlier identification are often used, such as graphical analysis or standardized observations. For exploratory or descriptive analyses, these are usually sufficient. Histograms, stem-and-leaf plots or box plots can be used to pinpoint extreme values. Using standardized values and flagging those values of 2 or greater as possible outliers is common.

Outlier tests are more formal procedures which have been developed for detecting outliers when a sample comes from a normal distribution and there is reason to believe a "contaminated" observation or "slippage" has occurred. Outlier tests are also useful as general tests of normality in the detection of skewed distributions with small samples (David, 1981). In addition, it has been shown (Thode, Smith and Finch, 1983) that certain outlier tests are nearly as powerful or more powerful than general tests of normality when sampling from scale mixtures of normal distributions. However, comparison of outlier tests (except the range test) with tests of normality for general alternatives and comparison of outlier tests with each other for outlier testing has been relatively rare (e.g., Ferguson, 1961; Johnson and Hunt, 1979; Thode, Smith and Finch, 1983; Thode, 1985).

Since an outlier test is often used as an "objective" method of identifying an outlier after already having examined the data themselves or are sometimes used sequentially without regard to the true α level of the sequential test, Grubbs recommended that tests for outliers always be done at a more conservative α level than the common 0.05 level.

Barnett and Lewis (1994) have presented a summary of tests for outliers and their critical values, many of which are specific to the detection of outliers in normal samples. They include tables of critical values for many variations of the same test, including those where one or both of the parameters are known. Here we will only consider those tests where both mean and variance are unknown.

Several outlier tests also turn out to be optimal tests in some way for detecting a non-normal alternative other than those with slippage. For example, outliers may indicate a contaminated distribution with small contaminating fraction p (relative to n), or may be indicative of a heavy-tailed or skewed distribution in a small sample. Therefore, outlier tests can also be used as tests for alternative distributions rather than simply as tests for aberrant observations.

6.2 Tests for k Specified Outliers

For certain outlier tests, it is known (or assumed) that the number of possible outliers in a sample is known *a priori*; in addition, some of the tests also assume knowledge of the direction of the outlying observation(s). In this section we describe outlier tests of these two types, which consist mainly of Grubbs' and Dixon's tests and extensions thereof.

When two or more outliers occur in a data sample, a test for a single outlier may not detect one outlier since the other outlier(s) inflates the variance of the sample (in the case of tests like Grubbs' test or the range test) or the numerator (in the case of Dixon's test), thereby "masking" the detection of the largest outlier. For this reason, it is also necessary to consider testing for more than one outlier. On the other hand, if $k > 1$ outliers are tested for and there are less than k outliers, one of two things may occur:

(1) The test may reject the null hypothesis of no outliers because of a large influence the true outliers have on the test statistic, thereby identifying more outliers than there really are. This is an effect known as *swamping*.

(2) The true extreme values may not have enough influence to attain a significant test statistic, so no outliers are identified. This can be thought of as reverse masking.

6.2.1 Grubbs' Test for a Single Outlier

Grubbs' (1950) test is perhaps the most widely known and intuitively obvious test for identifying a single outlier in a sample of n observations. It also seems to have high potential as a test for normality against other types of alternatives. As a test for an outlying observation, regardless of direction, Grubbs' test is given by the maximum deviation from the sample mean,

$$T = \max(T_1, T_n)$$

where

$$T_1 = \frac{(\bar{x} - x_{(1)})}{s}$$

$$T_n = \frac{(x_{(n)} - \bar{x})}{s}$$

where $\bar{x}$ and s are the sample mean and standard deviation, respectively. In the event that the direction of the possible outlier is known, T_1 would be used for identifying an extreme low observation, and T_n would be used to identify an extreme high observation (Grubbs, 1950; 1969). In each case the test statistic is compared to the upper α percentile of the distribution, with large values of the test statistic indicating the presence of an outlier.

Grubbs (1950) showed that T_n was identical to the ratio of the squared sum of differences with and without the suspected outlier,

$$S_n^2/S^2 = \frac{\sum_{i=1}^{n-1}(x_{(i)} - \bar{x}_n)^2}{\sum_{i=1}^{n}(x_{(i)} - \bar{x})^2} = 1 - \frac{T_n^2}{n-1}$$

where $\bar{x}_n$ is the mean of the sample excluding $x_{(n)}$. The analogous test statistic for testing for a lower-tail outlier,

$$S_1^2/S^2 = \frac{\sum_{i=2}^{n}(x_{(i)} - \bar{x}_1)^2}{\sum_{i=1}^{n}(x_{(i)} - \bar{x})^2} = 1 - \frac{T_1^2}{n-1}$$

has a similar relation to T_1.

Example 6.1. *Thirteen average annual erosion rates (m/year) were estimated for the East Coast states (Data Set 11). For these data, negative values indicate erosion and positive values indicate accretion. Virginia seems to have a very high erosion rate compared to the other states (Figure 6.1). For these data,*

$$T_1 = \frac{-0.83 - (-4.2)}{1.23} = 2.74$$

$$T_n = \frac{0.7 - (-.83)}{1.23} = 1.24$$

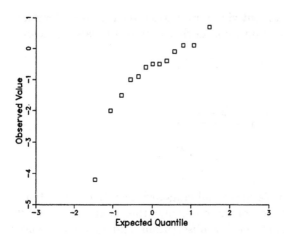

Figure 6.1 Q-Q plot of average annual erosion rates (m/yr) for 13 East Coast states.

so that $T = 2.74$, which is just significant at the 0.01 level.

Uthoff (1970) showed that T_1 and T_n were the MPLSI goodness of fit tests for left and right exponential alternatives to the normal distribution, respectively (Section 4.2.4). Surprisingly, Box (1953) suggested that T is "like" the LRT for a uniform alternative to a normal distribution, implying that it is also a powerful test against short-tailed distributions. Whereas the test statistic is compared to the upper critical value of the null distribution (either T_1, T_n or T, as appropriate) when testing for an outlier, testing for a short-tailed or skewed distributions requires the use of the lower tail of the null distribution of the test statistic.

Critical values for T, T_1 and T_n (Grubbs and Beck, 1972) are given in Table B8.

6.2.2 Dixon Tests for a Single Outlier

Dixon's test statistic for a single upper-tail outlier (Dixon, 1950; 1951) is the ratio of the distance between the two largest observations to the range,

$$r_{10} = \frac{x_{(n)} - x_{(n-1)}}{x_{(n)} - x_{(1)}}.$$

Similarly, to test for a single lower-tail outlier,

$$r'_{10} = \frac{x_{(2)} - x_{(1)}}{x_{(n)} - x_{(1)}}$$

is used. A test for a single outlier in an unknown direction is obtained by using $r = \max(r_{10}, r'_{10})$. This is the simplest of the outlier tests computationally.

> *Example 6.2. For the 15 height differences between cross- and self-fertilized Zea mays plants (Data Set 2), two lower values may be outliers. As a test for a single lower-tail outlier,*
>
> $$r'_{10} = \frac{-48 - (-67)}{75 - (-67)} = 0.134$$
>
> *which does not exceed the 5% critical value of 0.339.*

Upon inspection, it would not be expected that r'_{10} would identify an outlier in Example 6.2 because of the masking effect of $x_{(2)}$. To avoid masking, Dixon also proposed alternative tests of a type similar to r_{10} which would bypass masking effects of other observations. The general form of these test statistics (for detecting an upper-tail outlier) is defined by

$$r_{jk} = \frac{x_{(n)} - x_{(n-j)}}{x_{(n)} - x_{(k+1)}}$$

which eliminates the influence of $j - 1$ other large observations and k extreme small observations. Dixon suggested tests using $j = 1, 2$ and $k = 0, 1, 2$. To avoid a masking effect due to two upper outliers one should use $j = 2$; $k = 1$ or 2 should be considered for robustness against lower-tail outliers or long tails. Similarly, test statistics r'_{jk} are used for single lower-tail outliers. In general, Dixon suggested using r_{10} for very small samples, r_{21} for sample sizes 8 to 13, and r_{22} for samples sizes greater than 15. Critical values for r_{jk} are given in Table B17.

> *Example 6.3. To test for a lower outlier while guarding against the masking effect of $x_{(2)}$ for the plant height differences,*
>
> $$r'_{20} = \frac{14 - (-67)}{75 - (-67)} = 0.570.$$
>
> *which well exceeds the 1% critical value of 0.523; therefore, we have identified $x_{(1)} = -67$ as a lower-tail outlier.*

The sequential use of r'_{10} and r'_{20} in Examples 6.2 and 6.3 is indicative of a masking effect, and hence there is the possibility of two outliers in this sample. However, more formal sequential procedures for identifying multiple outliers are given below (Section 6.4).

6.2.3 Range Test

The range test has been discussed earlier (Section 4.1.2, 4.2.2) as the LR and MPLSI test for a uniform alternative. The range test statistic is

$$u = \frac{x_{(n)} - x_{(1)}}{s}.$$

One of the earlier suggested uses for u was as a test for outliers and as an alternative to b_2 (David, Hartley and Pearson, 1954), where for some examples they showed that u was identifying samples with a single outlier (in either direction) where b_2 was not rejecting normality. Alternatively, Barnett and Lewis (1994) suggested the use of the range test as a test "of a lower and upper outlier-pair $x_{(1)}$, $x_{(n)}$ in a normal sample".

While the use of u for detecting short-tailed alternatives requires comparison to the lower percentiles of the test statistic distribution, testing for an outlier or upper/lower outlier pair requires comparison of u to the upper α percentile of the distribution (Table B10).

6.2.4 Grubbs Test for k Outliers in a Specified Tail

Grubbs (1950) presented extensions of his test for a single outlier to a test for exactly two outliers in a specified tail ($S_{n-1,n}^2/S^2$ and $S_{1,2}^2/S^2$ for two upper and two lower outliers, respectively); this was subsequently extended to a test for a specified number k, $1 < k < n$, of outliers in a sample (Tietjen and Moore, 1972). Here it is assumed that we know (or suspect we know) how many observations are outliers as well as which of the tails contains the outlying observations. The test, denoted L_k (and its opposite tail analog L_k^*), is the ratio of the sum of squared deviations from the sample mean of the non-outliers to the sum of squared deviations for the entire sample,

$$L_k = \frac{\sum_{i=1}^{n-k}(x_{(i)} - \bar{x}_{n-k})^2}{(n-1)s^2}$$

where $\bar{x}_{n-k}$ is the mean of the lowest $n-k$ order statistics. This is the test for k upper-tail outliers. Similarly,

$$L_k^* = \frac{\sum_{i=k+1}^{n}(x_{(i)} - \bar{x}_{n-k}^*)^2}{(n-1)s^2}$$

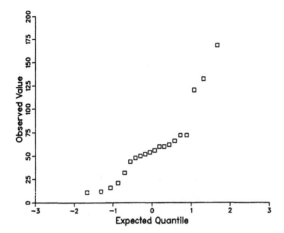

Figure 6.2 Q-Q plot of leukemia latency period in months for 20 patients following chemotherapy.

is the test statistic for k lower-tail outliers, where $\bar{x}^*_{n-k}$ is the mean of the highest $n - k$ order statistics. Obviously, L_k and L^*_k are always less than 1; if the variance of the subsample is sufficiently smaller than that of the full sample, then there is an indication that the suspected observations are indeed outliers. Therefore, the test statistic is compared to the lower tail of the null test distribution, and the null hypothesis of no outliers is rejected if L_k (L^*_k) is smaller than the appropriate critical value (Table B18).

Example 6.4. Data on the latency period in months of leukemia for 20 patients following chemotherapy (Data Set 9) may contain 3 large values (over 100 months), whereas the remaining cases have no latency period greater than 72 (Figure 6.2). For this data set,

$$L_k = S^2_{18,19,20}/S^2 = 6724/30335 = 0.222$$

which is well below the 0.01 significance level of 0.300 for $k = 3$ and $n = 20$.

6.2.5 Grubbs Test for One Outlier in Each Tail

Grubbs (1950) also presented a test for identifying one outlier in each tail, using

$$S^2_{1,n}/S^2 = \frac{\sum_{i=2}^{n-1}(x_{(i)} - \bar{x}^2_{1,n})}{\sum_{i=1}^{n}(x_{(i)} - \bar{x})^2}.$$

This is essentially L_k where the numerator is based on the $n-2$ observations $x_{(2)}$ through $x_{(n-1)}$. When k outliers are specified, with $m < k$ in the lower tail and $k - m$ in the upper tail, a similar test can be defined using the appropriately trimmed numerator sum of squares. Critical values for this test are given in Table B19.

6.2.6 Tietjen-Moore Test for k Outliers in One or Both Tails

Tietjen and Moore (1972) presented an outlier test statistic, denoted E_k, which is similar to L_k but which examines the sample for the k largest deviations from the sample mean, without regard to which tail they come from, i.e., the outliers can be from either or both tails. To compute E_k, first it is necessary to obtain r_i, the absolute deviation from the sample mean of each of the observations,

$$r_i = |x_i - \bar{x}|$$

These residuals are sorted to obtain the order statistics of the r_i, and the ordered observations $z_{(i)}$ are defined as the signed value of $r_{(i)}$. The test statistic for k outliers is then computed similarly to L_k, using the $z_{(i)}$,

$$E_k = \frac{\sum_{i=1}^{n-k}(z_{(i)} - \bar{z}_{n-k})^2}{\sum_{i=1}^{n}(z_{(i)} - \bar{z})^2}$$

where $z_{(i)}$ are the deviations from the mean, ordered irrespective of sign, $\bar{z}$ is the mean of all of the $z_{(i)}$ and $\bar{z}_{n-k}$ is the mean of $z_{(1)}, z_{(2)}, \ldots, z_{(n-k)}$.

Since this test statistic is the ratio of a reduced sum of squares to a full sum of squares, the comparison to determine significance is whether E_k is less than the appropriate critical value (Table B20).

Example 6.5. For the 31 observations in the July wind data (Data Set 8) there is one observation below 5 mph (3.8) and one above 14 mph (17.1). These two observations happen to be the largest deviations (in absolute value) from the sample mean of 9.01, with $z_{(31)} = 8.09$ and $z_{(30)} = -5.21$. The full sample sum of squares of z is 217.05, while the sum of squares based on the reduced sample of 29 observations is 124.19. This results in the test statistic

$$E_2 = 124.19/217.05 = 0.572.$$

Therefore, the reduction in the sum of squares is not quite suffi-cient, at the 0.05 level, to identify the two largest observations as simultaneous outliers, in comparison to the stated critical value of 0.568.

6.3 Tests for an Unspecified Number of Outliers

In practice it is relatively rare that the number or the direction of the outliers is known *a priori*. Suppose that there is more than one outlier in a sample; then testing for a single outlier may give misleading results because of a masking effect. Tietjen and Moore's tests require that the number of outliers be specified in advance. These tests give misleading results if there are more (or fewer) outliers than the number specified.

Certain tests have been suggested for outlier detection when the number of outliers is unknown. Use of outlier labeling based on box plots can identify a number of outliers, although this is an exploratory data analysis technique and therefore cannot be used when a significance level for the test is desired. Skewness and kurtosis have been suggested as outlier tests, but other types of departure from normality can also lead to significant results; further, the number of outliers is not identified when using those tests.

6.3.1 EDA - Outlier Labeling

Resistant outlier labeling is an exploratory data analysis technique (Tukey, 1977) related to the description of data using box plots (Section 2.1). Hoaglin, Iglewicz and Tukey (1986) discussed the use of the *inner fences* as a method of labeling possible outliers in an otherwise normal sample.

They used the definition of the lower and upper *fourths* (approximate quartiles, see Section 12.3.2) of the data to be those order statistics of a sample defined by $h_l = x_{(f)}$ and $h_u = x_{(n+1-f)}$, where $f = [(n+3)/2]/2$, brackets denoting the greatest integer less than or equal to the value. These fourths are the medians of the half-sample defined by splitting the data on the median, including the median in each half-sample if n is odd. In the case where f is noninteger, the fourths are taken as the average of the two adjacent order statistics. The f-spread is then defined as $d_f = h_u - h_l$.

The identification of possible outliers is accomplished by labeling all observation outside of the inner fences

$$[h_l - 1.5d_f, h_u + 1.5d_f]$$

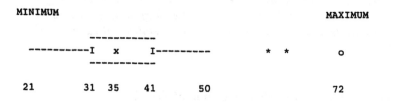

Figure 6.3 Box plot of average water well alkalinity for 58 wells.

as "outside", or possible outliers. Observations outside of the interval

$$[h_l - 3d_f, h_u + 3d_f]$$

are denoted "far outside".

In a normal population, Hoaglin, Iglewicz and Tukey (1986) showed that for samples sizes between 20 and 75, 1-2% of the observations would be expected to be outside, purely by chance; for samples of size 100 and over, less than 1% of the observations would be outside, with the expected proportion under an infinite population being 0.7%.

The proportion of far outside values in a normal samples was shown to be less than 1% for sample sizes between 6 and 19, and less than 0.1% for sample sizes 20 and greater (Table B21). The population proportion of far outside values is 0.00023%.

> *Example 6.6. A box plot of the 58 well average alkalinity values (Data Set 10) is shown in Figure 6.3. The fourths of these data are 31 and 41, so the inner fences are 31 − 15 = 16 and 41 + 15 = 56 (truncated to the nearest inner observations, 21 and 50). The body of the data, both within the fourths and within the inner fences, is symmetric. There are three outside observations (5.2%) including one (1.7%) far outside observation. Under an assumed normal population, then, there is an overabundance of outside and far outside values compared to what would be expected purely by chance.*

Outlier labeling is resistant to masking, since the cutoff values are based on centrally located order statistics, rather than on the variance.

Unlike the outlier detection methods discussed in the previous section or in the sequential tests discussed below, neither the number nor the maximum number of outlying observations need to be specified in advance of conducting the procedure. One other advantage of outlier labeling is that the number of outliers identified are not based on a preliminary inspection of the data.

Other fence lengths are briefly discussed in Hoaglin, Iglewicz and Tukey (1986), e.g., those based on $1.0d_f$ and $2.0d_f$.

6.3.2 Skewness Test

Ferguson (1961) showed that when a normal sample with unknown mean contains some observations which have a shift in mean (also unknown), the locally best invariant single-sided test for outliers is the skewness test (Section 3.2.1). It is not necessary that all of the shifted observations have the same mean, although it is required that they are all shifted in the same direction. Note that for this test the number of outliers is not specified beforehand, although the direction of the outliers must be known.

6.3.3 Kurtosis Test

Ferguson (1961) also showed that if less than 21% of the observations in a normal sample have a shift in mean, regardless of direction, then an upper-tailed kurtosis test (Section 3.2.2) is the locally best invariant test for detecting outliers. Here also it is not necessary that the shifted observations have the same mean.

In the case of a normal sample where all observations have the same mean but some have a shift in variance, then the upper-tailed kurtosis test is the locally best invariant test for identifying outliers regardless of how many spurious observations there are.

6.3.4 Normal Mixtures

The likelihood ratio test for mixtures of two normal components has been used to detect a small number of outliers (Aitkin and Wilson, 1980), i.e., when the mixing proportion p is small relative to the sample size. The test does not require a pre-specification of the number of outliers, and classification methods can be used to separate the outlier group from the remainder of the sample; furthermore, the overall significance level is based on the single test.

Similar to sequential tests, however, tests for normal mixtures are not as sensitive to outliers as tests with the correct number of outliers specified. Normal mixtures and tests are described more fully in Chapter 11.

6.4 Sequential Procedures for Detecting Outliers

Sequential or "many outlier" procedures can be used when the number of outliers in a sample is unknown, but the maximum number of outliers can be specified beforehand. If there are more outliers than the maximum specified, then the techniques may not work due to masking.

The sequential procedures described below are all performed in the same manner: after specifying k, the maximum number of outliers expected, the test is performed (step 1). Then, after deleting the observation farthest from the sample mean, the test is repeated for the resulting subsample. This process is repeated k times. The appropriate critical values are compared to determine if any of the steps gave a significant result.

Because it is usually desired that an overall rather than individual α level be attained, critical values are obtained by using the marginal distributions of the test statistics, say $t_1, \ldots, t_k$, to determine critical values $\lambda(\beta)$ such that

$$P[t_i > \lambda_i(\beta)] = \beta, \qquad i = 1, \ldots, k$$

so that

$$P\left(\bigcup_{i=1}^{k} [t_i > \lambda_i(\beta)]\right) = \alpha.$$

If all $t_i < \lambda_i(\beta)$, then no outliers are declared; if any of the $t_i > \lambda_i(\beta)$ then m outliers are declared, where m is the largest value of i such that $t_i > \lambda_i(\beta)$.

Since sequential methods are defined to give an overall α significance level over every step in the procedure, sequential tests will not be as sensitive at picking out m outliers as a test designed for detecting exactly m outliers. Prescott (1979) suggested that a sequential procedure for no more than three outliers is sufficient, since if "there are more than a few outliers present the problem ceases to be one of outlier detection and perhaps the underlying structure of the data should be examined in some other way".

6.4.1 Extreme Studentized Deviate Procedure

The extreme studentized deviate (ESD) procedure is essentially a stepwise Grubbs T test for a single outlier in either tail. At each step, the ESD is calculated

$$ESD_i = \max_{j=1,\ldots,n-i+1} | x_j - \bar{x}_i | / s_i, \qquad i = 1,\ldots,k$$

where $\bar{x}_i$ and s_i^2 are the mean and variance, respectively, of the subsample remaining after the first $i - 1$ steps and observation deletions. Jain (1981) gave critical values for the ESD procedure for k up to 5 (Table B21).

6.4.2 Studentized Range Procedure

The studentized range (STR) procedure is a stepwise test-and-deletion method using the range statistic,

$$STR_i = (x^i_{(n-i+1)} - x^i_1)/s_i, \qquad i = 1,\ldots,k.$$

Here the $x^i_{()}$ are the order statistics of the subsample at step i, and s_i^2 is the variance of the subsample. This method has been found to have poor power properties (Jain, 1981).

6.4.3 Sequential Kurtosis

A sequential kurtosis (KUR) procedure was discussed by Rosner (1975) and Jain (1981); critical values were given for k up to 5 by Jain (Table B23). This procedure consists of obtaining the kurtosis of the subsample at step i after the previous $i - 1$ deletions,

$$KUR_i = m_4^i/(m_2^i)^2, \qquad i = 1,\ldots,k.$$

where the m^i are the sample moments for subsample i at step i.

6.4.4 Prescott's Sequential Procedure

Prescott (1979) suggested the sequential use of Grubbs-type sum of squares ratios, using the ratios of sums of squares obtained by deleting the k observations one at a time. The steps in Prescott's procedure are conducted by using the criteria

$$D_j = S^2_{(j)}/S^2_{(j-1)} < \lambda_j(\beta) \qquad j = 1,\ldots,k$$

where $S^2_{(j)}$ is the sum of squares of the sample obtained by deleting the j observations furthest from the original sample mean (in either direction). The number of outliers identified by this test is m, where m is the maximum j such that $D_j < \lambda_j$.

Prescott also gave the formula for calculating $S^2_{(j)}$ from the full sample sum of squares, using only functions of the residuals. By letting

$$r^i = x_i - \bar{x},$$

i.e., the signed residual from the full sample mean, then

$$S^2_{(j)} = S^2 - \sum_{i=1}^{j}(r^i)^2 - \left(\sum_{i=1}^{j} r^i\right)^2 /(n-j)$$

where the summations are over the j most extreme deviations from the sample mean (note that the sign of r^i is retained in this calculation).

Example 6.7. In the leukemia latency data (Figure 6.2), as many as 3 outliers in the sample of 20 cases can be identified. The three most extreme observations from the sample mean of 60.4 are all in the upper tail, with values of 168, 132 and 120 (Data Set 9). The sums of squares for the full sample is $S^2 = 30334.8$ and for the reduced samples are

$$S^2_{(1)} = 18147.7$$

$$S^2_{(2)} = 11846.4$$

$$S^2_{(3)} = 6723.9$$

These result in the sequential test statistics

$$D_1 = 18147.7/30334.4 = 0.598$$

$$D_2 = 11846.4/18147.7 = 0.653$$

$$D_3 = 6723.9/11846.4 = 0.568$$

From the table of critical values for a sample size of 20 with a maximum of 3 outliers, the critical values for comparison to D_1, D_2 and D_3 are 0.545, 0.605 and 0.630, respectively, for an overall 0.05 significance level. Although D_1 and D_2 are not less than their respective critical values, D_3 is less than 0.630, indicating that *these three observations are outliers relative to the remainder of the sample. Note that these outliers would not have been detected had a significance level of 0.01 been used, while using L_3 resulted in a significance level well below 0.01 (Example 6.4).*

In keeping with his suggestion, Prescott gave critical values for his procedure only for $k = 2$ and 3 (Table B24).

6.4.5 Rosner's RST Procedure

Rosner (1975, 1977) suggested a modified sequence of Grubbs-type T statistics, R_i, for use as a sequential procedure. The method he suggested was different than the previously discussed sequential procedures in that he used a trimmed mean and variance which did not have to be recalculated after each step.

As with all sequential procedures, to perform the RST (R-statistic) procedure the maximum number of outliers, k, is determined. Then a trimmed mean and variance, based on the original sample trimmed on each side by k observations, are obtained

$$a = \sum_{i=k+1}^{n-k} x_{(i)}/(n - 2k)$$

$$b^2 = \sum_{i=k+1}^{n-k} (x_{(i)} - a)/(n - 2k - 1).$$

Then, if I_0 is the full sample,

$$R_1 = \max_{I_0} \mid x_i - a \mid /b = \mid x^0 - a \mid /b$$

$$I_1 = I_0 - x^0$$

$$R_2 = \max_{I_1} \mid x_i - a \mid /b = \mid x^1 - a \mid /b \ldots$$

so that the R_i are sequentially the largest standardized residuals from the trimmed mean. These test statistics are then compared to the critical values corresponding to the specified (α, n, k) to determine whether any outliers exist in the original sample.

Example 6.8. For the July wind data (Data Set 8), two outliers will again be specified, as in Example 6.5. The trimmed mean and standard deviation based on the 27 middle observations are

$$a = 8.89$$

$$b = 1.872$$

which result in test statistic values of

$$R_1 = (17.1 - 8.89)/1.872 = 4.386$$

$$R_2 = (8.89 - 3.8)/1.872 = 2.719.$$

From the table of critical values for a sample size of 31, using $\alpha = 0.05$ and $k = 2$ (Table B25), the critical values for comparison to R_1 and R_2 are about 4.60 and 3.55, respectively, indicating that there neither of the observations should be considered outliers. However, note that Grubbs' $T = 3.01$ would have just identified a single outlier at the 0.05 level, had that been the model specified.

Rosner (1977) gave critical values for selected sample sizes up to 100 for $\alpha = 0.10, 0.05$ and 0.01 and $k = 2, 3$ and 4; Jain (1981) gave similar values for k up to 5 (Table B25).

6.5 Further Reading

The most comprehensive (and recent) discourse on the subject of outliers is that of Barnett and Lewis (1994). In addition to the subject matter covered in this chapter, they discuss robust methods of estimation and testing in the presence of outliers; multivariate outliers (see Chapter 10); outlier detection in samples when one or more parameters are known; outliers when the null distribution is other than normal; and outliers in regression, time series and contingency tables. Hawkins (1980) and Tietjen (1986) are also useful, but more limited, references on this topic.

References

Aitken, M., and Wilson, G.T. (1980). Mixture models, outliers and the EM algorithm. Technometrics 22, 325-331.

Barnett, V., and Lewis, T. (1994). **Outliers in Statistical Data**, 3rd ed. John Wiley and Sons, New York.

Box, G.E.P. (1953). A note on regions for tests of kurtosis. Biometrika 40, 465-468.

David, H.A. (1981). **Order Statistics** 2nd ed. John Wiley and Sons, New York.

David, H.A., Hartley, H.O. and Pearson, E.S. (1954). The distribution of the ratio, in a single normal sample, of the range to the standard deviation. Biometrika 41, 482-493.

David, H.A. and Paulson, A.S. (1965). The performance of several tests for outliers. Biometrika 52, 429-436.

Dixon, W. (1950). Analysis of extreme values. Annals of Mathematical Statistics 21, 488-505.

Dixon, W. (1951). Ratios involving extreme values. Annals of Mathematical Statistics 22, 68-78.

Ferguson, T.S. (1961). On the rejection of outliers. Proceedings, Fourth Berkeley Symposium on Mathematical Statistics and Probability, University of California Press, Berkeley, 253-287.

Gibbons, D.I., and McDonald, G.C. (1980a). Examining regression relationships between air pollution and mortality. GMR-3278, General Motors Research Laboratories, Warren, Michigan.

Gibbons, D.I., and McDonald, G.C. (1980b). Identification of influential geographical regions in an air pollution and mortality analysis. GMR-3455, General Motors Research Laboratories, Warren, Michigan.

Grubbs, F. (1950). Sample criteria for testing outlying observations. Annals of Mathematical Statistics 21, 27-58.

Grubbs, F. (1969). Procedures for detecting outlying observations in samples. Technometrics 11, 1-19.

Grubbs, F., and Beck, G. (1972). Extension of sample sizes and percentage points for significance tests of outlying observations. Technometrics 14, 847-859.

Hawkins, D.M. (1980). **Identification of Outliers**. Chapman and Hall, New York.

Hoaglin, D.C., Iglewicz, B., and Tukey, J.W. (1986). Performance of some resistant rules for outlier labeling. Journal of the American Statistical Association 81, 991-999.

Jain, R.B. (1981). Percentage points of many-outlier detection procedures. Technometrics 23, 71-76.

Johnson, B.A., and Hunt, H.H. (1979). Performance characteristics for certain tests to detect outliers. Proceedings of the Statistical Computing Section, Annual Meeting of the American Statistical Association, Washington, D.C.

Lave, L.B. and Seskin, E.P. (1970). Air pollution and human health. Science 169, 723-733.

Lave, L.B. and Seskin, E.P. (1977). **Air Pollution and Human Health.** John Hopkins University Press, Baltimore.

Prescott, P. (1979). Critical values for a sequential test for many outliers. Applied Statistics 28, 36-39.

Rosner, B. (1975). On the detection of many outliers. Technometrics 17, 221-227.

Rosner, B. (1977). Percentage points for the RST many outlier procedure. Technometrics 19, 307-312.

Thode, Jr., H.C. (1985). Power of absolute moment tests against symmetric non-normal alternatives. Ph.D. dissertation, University Microfilms, Ann Arbor, Michigan.

Thode, Jr., H.C., Smith, L.A. and Finch, S.J. (1983). Power of tests of normality for detecting scale contaminated normal samples. Communications in Statistics - Simulation and Computation 12, 675-695.

Tietjen G.L. (1986). The analysis and detection of outliers. In D'Agostino, R.B., and Stephens, M.A., eds., **Goodness-of-Fit Techniques,** Marcel Dekker, New York.

Tietjen, G.L., and Moore, R.H. (1972). Some Grubbs-type statistics for the detection of outliers. Technometrics 14, 583-597.

Tukey, J.W. (1977). **Exploratory Data Analysis.** Addison-Wesley, Reading, MA.

U.S. Bureau of the Census (1972). **City and County Data Book, 1972.** U.S. Government Printing Office, Washington, D.C.

Uthoff, V.A. (1970). An optimum test property of two well-known statistics. Journal of the American Statistical Association 65, 1597-1600.

CHAPTER 7

POWER COMPARISONS FOR UNIVARIATE TESTS
FOR NORMALITY

"Depending on the nature of the alternative distribution and on the sample size, the various procedures show to better or worse advantage."

Shapiro, Wilk and Chen, 1968

The most frequent measure of the value of a test for normality is its power, the ability to detect when a sample comes from a non-normal distribution. All else being equal (which decidedly never happens) the test of choice is the most powerful. However, in addition to power which depends on both the alternative distribution and sample size, choice of test when assessing normality can be based on a variety of other reasons, including ease of computation and availability of critical values. Ideally, one would prefer the most powerful test for all situations, while in reality no such test exists.

7.1 Power of Tests for Univariate Normality

Often, while the specific alternative is not known, some general characteristics of the data may be known in advance (e.g., skewness). If not, there may be limited concerns about the types of departures from normality. For example, regression residuals which are symmetric but have short tails are

Table 7.1 Univariate tests for normality discussed in this chapter.

Test Symbol	Test Name	Reference Section
a	Geary's test	Section 3.3.1
A^2	Anderson-Darling test	Section 5.1.4
$\sqrt{b_1}$	skewness	Section 3.2.1
b_2	kurtosis	Section 3.2.2
χ^2	chi-squared test	Section 5.2
D	D'Agostino's D	Section 4.3.2
D^*	Kolmogorov-Smirnov	Section 5.1.1
E_k	Tietjen-Moore test for > 1 outlier	Section 6.2.6
EDF tests		Section 5.1
F_1, F_2	LaBreque's tests	Section 2.3.3
k^2	P-P correlation test	Section 2.3.2
K_{mn}	sample entropy test	Section 4.4.1
K_s^2	joint kurtosis/skewness test	Section 3.2.3
L_k	Grubbs' test for > 1 outlier	Section 6.2.4
MPLSI tests		Section 4.2
r	probability plot correlation	Section 2.3.2
R	rectangular skewness/kurtosis test	Section 3.2.3
S_s	omnibus MPLSI test	Section 4.2.6
T	Grubbs' outlier test	Section 3.4.1, 6.2.1
T_{1n}, T_{2n}	Locke and Spurrier tests	Section 4.3.1
T_2	Oja's test	Section 4.3.3
T^*	Locke and Spurrier test	Section 4.3.1
u	range test	Section 4.1.2, 6.2.3
U	Uthoff's test	Section 3.4.2, 4.1.3
U^2	Watson's test	Section 5.1.3
V	Kuiper's V	Section 5.1.2
W	Wilk-Shapiro test	Section 2.3.1
W'	Shapiro-Francia test	Section 2.3.2
W^2	Cramer-von Mises test	Section 5.1.3
z	Lin and Mudholkar's test	Section 4.4.2

usually not of interest, so a test which has high power at detecting skewed and long-tailed symmetric alternatives need only be considered. Therefore, it is important to be able to identify which tests are competitively powerful under certain specific situations, in case some information is known concerning the alternative.

It is also important to know which tests have decent power under all types of alternatives, for those instances where no *a priori* information is available. It is also useful to have tests which can be used as substitutes for each other. Therefore, we have compiled the results of many studies into a summary of the power of tests for normality. We then make recommendations for testing based on different scenarios.

7.1.1 Background of Power Comparison Simulations

The advent of computers, along with the seminal papers on tests for normality by Shapiro and Wilk (1965) and Shapiro, Wilk and Chen (1968), essentially set the standards for the development and presentation of new tests for normality (as well as tests for other distributions). In general, theoretical power calculations for specific tests are either difficult or intractable; in cases where power could be estimated, it was usually based on asymptotic approaches (e.g., Geary, 1947). Thus, simulation became the vehicle of convenience for estimating power and the comparison of tests.

At the time, there were relatively few tests for normality: besides W, there were only four moment-type tests ($\sqrt{b_1}$, b_2, u, and a) and the more general goodness of fit tests (e.g., χ^2 and EDF tests). The choice of test was actually more limited than that since the χ^2 and EDF tests were only valid for simple hypotheses, which is not a common practical situation. Also, W had only been developed for sample sizes up to 50.

Shortly after the introduction of W, Lilliefors (1967) presented some distributional results for the Kolmogorov-Smirnov test, D^*, for a composite normal null hypothesis. In hindsight it can be stated that this was not useful since this test is almost universally not recommended as a test for normality because of its poor power properties.

In 1971, D'Agostino introduced his D statistic, for use as an omnibus test in samples of over size 50. Shapiro and Francia (1972), Weisberg and Bingham (1975) and Filliben (1975) suggested correlation tests similar in construction to W which also overcame the sample size limitation of W. Between the introduction of W in 1965 and the probability plot correlation test in 1975, there were essentially no other new tests for normality introduced, making Filliben's (1975) simulation the last word in power comparisons at the time, since he included all of the well-known tests (excluding EDF tests) for normality. The only exceptions seem to be those tests developed by Uthoff (1968; 1973).

The use of EDF tests as tests for normality did not become popular until about that time, when Stephens (1974) not only developed null distributions for composite EDF tests for the normal distribution, but also identified relationships of critical values with functions of sample size, making these tests more widely available and applicable. A comparison of the power of these tests with W showed that at least some of the EDF tests were useful as tests for normality. This nearly doubled the number of tests that could be used in power comparisons.

The complexity of power comparisons, which were to become almost mandatory when presenting new tests, was also increased by the number of alternatives used to compare tests in the earlier studies. Shapiro, Wilk and Chen (1968) used 45 parameterizations of 12 different alternative distributions. Pearson, D'Agostino and Bowman (1977) presented power estimates for 58 parameterizations of 12 different distributions. While both of these studies only included a small number of useful tests (Pearson, D'Agostino and Bowman included only four omnibus tests and four directional tests) and did not include composite hypothesis EDF tests, they set a standard which would be difficult to measure up to, given space limitations in journals. Not only would a power comparison use up a lot of space when it included all tests (or at least all that had shown some useful characteristics), but very little additional information would be gained on ensuing publications, which would also have to include all tests (plus one new one) and the large range of alternatives.

These difficulties gave rise to the practice of comparing a new test with a small subset of tests for normality and/or alternatives, during the time when the development of tests for normality flourished, from 1975 to the middle of the 1980's. For example, Locke and Spurrier (1976) only compared their two tests (T_{1n} and T_{2n}) with $\sqrt{b_1}$. Although this is an extreme example of the limitations on power comparisons, there were very few large scale comparisons which could be used to directly compare a large number of tests for a broad range of alternatives.

In addition, there was no common standard for the design of the power comparisons. Different studies used different sample sizes and α levels. Reliability of the estimated power differed between studies, because different numbers of replications were used. Tests were sometimes used as two-tailed and sometimes as one-tailed tests; sometimes it was not stated how many tails were used. In some comparisons the estimated power of a new test was based on a new simulation, while the power estimates for the comparison tests were obtained from a previously published study.

7.1.2 Power Comparisons: Long-Tailed Symmetric Alternatives

Shapiro and Wilk (1965) compared W, $\sqrt{b_1}$, b_2, and u. They included the χ^2 and EDF tests in their comparison but assumed known parameters so that the tests could be used with a simple hypothesis; therefore, they will not be discussed here. Their simulation only included 200 samples of size

20. For the three long-tailed symmetric alternatives they used, W was seen to be generally competitive with b_2, although neither was decisively better. In their more extensive simulation, Shapiro, Wilk and Chen (1968) used the same tests but included ten long-tailed symmetric alternatives and five sample sizes between 10 and 50. W and b_2 were again competitive, with b_2 tending to be more powerful for the larger sample sizes.

D'Agostino (1971) compared D with the simulation results of Shapiro, Wilk and Chen (1968). Using only samples of size 50, he determined that D was competitive with W and b_2 for long-tailed symmetric distributions. D'Agostino and Rosman (1974) only compared Geary's test, W and D, but used samples of size 20, 50 and 100; they found that both D and a worked well for long-tailed symmetric alternatives. Hogg (1972) showed virtually no difference between Uthoff's (1968) U, asymptotically equivalent to a, and b_2 for logistic and double exponential alternatives. Smith (1975), using the same tests as Hogg, only considered symmetric long-tailed stable alternatives, and suggested that U be used, especially as tail heaviness increases.

Csorgo, Seshadri and Yalovsky (1973) showed that a Kolmogorov-Smirnov test based on a sample characterization of normality yields results comparable to those of W for small samples. Stephens (1974) compared composite hypothesis EDF tests to W and D and showed that the Anderson-Darling and Cramer-von Mises tests were comparable to both tests, with A^2 being slightly better than W and W^2, and not quite as good as D. The Kolmogorov-Smirnov test D^* always performed worse than the other tests. Green and Hegazy (1976) compared D, W and some modified EDF tests, with D nearly always being most powerful for the Cauchy and double exponential alternatives for samples between 5 and 80.

Filliben (1975) showed little difference between a, D, W, W', b_2 and r for samples of size 20; for samples of size 50, b_2 and W did not seem competitive, and a was marginally the best test. Gastwirth and Owens (1977) indicated that a seemed better than b_2 for long-tailed symmetric alternatives. For samples of size 20, Spiegelhalter (1977) showed that the MPLSI test for a double exponential alternative dominated W and b_2 for several long-tailed symmetric alternatives.

Of the tests introduced by LaBreque (1977), F_1 was always most powerful when compared to W, A^2, a and b_2, sometimes by an appreciable amount. His results were based only on samples of size 12 and 30.

Pearson, D'Agostino and Bowman (1977) showed that for omnibus tests, the combined skewness and kurtosis test K_s^2 was more powerful than W or D; however, when used in a directional manner, D was better than all of the omnibus tests, but there was no real difference between D and a directional (upper-tailed) b_2 test. White and MacDonald (1980) also indicated that the power of D slightly exceeded both W and b_2. Spiegel-halter (1980) showed a possible slight advantage of his omnibus test S_s over W and b_2 in some circumstances. Oja (1981) showed T_2 and Locke and Spurrier's (1977) T^* tests to be essentially equivalent in power, and better than W and b_2 for samples of size 20. A modification of T_2 (Oja, 1983) which is easier to calculate showed slight loss of power over T_2, but still exceeded that of the other tests. The MPLSI test for a Cauchy alternative (Franck, 1981) had better performance than other tests for stable alternatives, at least for samples of size 20; for samples of size 50 not much difference was demonstrated.

For scale contaminated normal alternatives, which have population kurtosis greater than 3, kurtosis and other absolute moment tests with exponent greater than 3 had higher power, on average, than other tests, including D, a, U and W (Thode, Smith and Finch, 1983). However, D, a and u had nearly equivalent power to the absolute moment tests for the heavier tailed mixtures.

Thode (1985) showed that a, D and U were the best tests for detecting double exponential alternatives.

Looney and Gulledge (1984; 1985) showed essentially no difference in power among correlation tests based on different plotting positions except, notably, W, which had slightly lower power than the others. Gan and Koehler (1990) also showed that the correlation test based on the plotting position $i/(n + 1)$ had slightly higher power than A^2 and W. Tests based on normalized spacings (Lockhart, O'Reilly and Stephens, 1986) were not as good as either A^2 or W.

7.1.3 Power Comparisons: Short-Tailed Symmetric Alternatives

For samples from size 10 to 50, Shapiro and Wilk (1965) and Shapiro, Wilk and Chen (1968) showed that u usually dominated both b_2 and W when the alternative was short-tailed and symmetric. This may not be surprising since u is the likelihood ratio test for a uniform alternative to normality.

Using samples of size 50, D'Agostino (1971) indicated that D usually had lower power than u, W and b_2. D'Agostino and Rosman (1974) only compared Geary's test, W and D, using samples of size 20, 50 and 100; they found that a single-sided a worked best for short-tailed symmetric alternatives, compared to the other two tests. Hogg (1972) showed dominance of u for a uniform alternative.

EDF tests based on characterizations of normality (Csorgo, Seshadri and Yalovsky, 1973) were never as powerful as W for the short-tailed alternatives they used; however, they did not use any other tests for normality in their comparison. Using only the uniform distribution as a short-tailed alternative, Stephens (1974) showed that W dominated D and the EDF tests. Green and Hegazy (1976) compared D, W and some modified EDF tests which were more powerful than both D and W; however, again only the uniform alternative was used, and u was not included.

Filliben (1975) showed that u was most powerful for all of the short-tailed alternatives he used in his simulation study, with b_2 being a somewhat distant second. For samples of size 20, Spiegelhalter (1977) showed that the MPLSI test for a uniform alternative dominated W and b_2 for the two short-tailed alternatives he used; however, this test is equivalent to u.

Of the tests compared by LaBreque (1977), u was always most powerful; however, his only short-tailed symmetric alternative was the uniform, and his results were based only on samples of size 12 and 30.

Pearson, D'Agostino and Bowman (1977) showed that for omnibus tests, W was somewhat better than K_s^2 and R (their notation for the rectangular bivariate joint skewness and kurtosis test, Section 3.2.3), while a two-tailed D had poor power. As a directional test, b_2 had appreciably higher power than D; however, u was not considered in this comparison. Spiegelhalter's (1980) omnibus test S_s was comparable to W and b_2 for the two alternatives (uniform and Tukey(0.7)) he included in his study for samples of size 20 and 50. Oja (1981) showed T_2 and Locke and Spurrier's (1977) T^* tests to be essentially equivalent in power, and better than W for samples of size 20. A modification of T_2 (Oja, 1983) showed a slight increase in power over T_2.

For selected sample sizes, Thode (1985) showed that for a uniform alternative the best tests were u, the lower-tailed Grubbs' T, and absolute moment tests with moment greater than 2 (including b_2, the absolute fourth moment test). No other short-tailed symmetric alternatives were examined.

Of all correlation tests based on different plotting positions, W had the highest power (Looney and Gulledge, 1984; 1985). Tests based on normalized spacings (Lockhart, O'Reilly and Stephens, 1986) were not as good as either A^2 or W. W had noticeably higher power than EDF tests, including A^2, and correlation tests based on P-P plots, such as k^2 (Gan and Koehler, 1990).

7.1.4 Power Comparisons: Asymmetric Alternatives

For samples of size 10 to 50, Shapiro and Wilk (1965) and Shapiro, Wilk and Chen (1968) showed a possible advantage of W over other tests, including the commonly used $\sqrt{b_1}$, against asymmetric alternatives. D'Agostino (1971) showed this also for samples of size 50, with D having poor power relative to W and $\sqrt{b_1}$.

Tests based on characterization of normality (Csorgo, Seshadri and Yalovsky, 1973) had poor power for asymmetric alternatives, and here also W was slightly better than $\sqrt{b_1}$ for samples up to size 35. Stephens (1974) showed that W had higher power than composite EDF tests, although A^2 was somewhat competitive; D fared poorly in his comparison. For samples of size 90, Stephens used W' rather than W, and showed that it was slightly more powerful than A^2. For all sample sizes, D and the Kolmogorov-Smirnov tests did poorly, with W^2, Kuiper's V and U^2 having intermediate power.

Filliben (1975) showed that for samples of size 20 and 50, the best tests were W, r and $\sqrt{b_1}$, in that order. Locke and Spurrier (1976) compared T_{1n}, T_{2n} and $\sqrt{b_1}$ for asymmetric alternatives, breaking them down into distributions with both tails light (e.g., beta distributions), both tails heavy (e.g., Johnson U), and one light tail and one heavy tail (e.g., gamma). For both tails heavy, $\sqrt{b_1}$ was best, while for other alternatives T_{1n} was best, particularly for those alternatives with both tails light.

For samples of size 12 and 30, LaBreque's (1977) F_2 was better than W when $\sqrt{\beta_1}$ was 2 or less, and W was slightly better otherwise. Slightly less powerful than these two tests, and essentially equivalent to each other, were A^2, $\sqrt{b_1}$ and LaBreque's F_1.

Of the omnibus tests compared by Pearson, D'Agostino and Bowman (1977), W was by far the most powerful; it was also somewhat competitive with the single-tailed $\sqrt{b_1}$ and right angle tests, which were about equal in power. Against stable alternatives, Saniga and Miles (1979) showed that $\sqrt{b_1}$ was more powerful than W for samples of sizes between 10 and 100; they also included b_2, D, u and a joint skewness/kurtosis test for comparison. Using a χ^2 and lognormal distribution as alternatives, White and MacDonald (1980) showed that W and W' were equivalent in power for samples from 20 to 50, and were more powerful than $\sqrt{b_1}$. For samples of size 100, W' was the most powerful test. For samples of size 20 and 30, Lin and Mudholkar (1980) showed the highest power for Vasicek's (1976) K_{mn} for beta distributions, while for other alternatives either W or z had the highest power. EDF tests (Kolmogorov-Smirnov and Cramer-von Mises) and $\sqrt{b_1}$ were also compared in this simulation. Spiegelhalter's (1980) omnibus test was better than W for samples of size 20, while W had higher power than S_s for samples of size 50; $\sqrt{b_1}$ was less powerful than both tests for both sample sizes. Oja (1981) showed T_1 and Locke and Spurrier's (1977) T_{1n} tests to be essentially equivalent in power, and better than W and $\sqrt{b_1}$ for samples of size 20. A modification of T_1 (Oja, 1983) which is easier to calculate showed power equivalent to T_1.

In their comparison of correlation tests and W, Looney and Gulledge (1984; 1985) showed a slight advantage in power for W, although all correlation tests were about equivalent in power. For samples of size 20 and 40, the A^2 test based on normalized spacings showed a slight but consistent advantage over W (Lockhart, O'Reilly and Stephens, 1986). W and A^2 outperformed the P-P plot correlation test k^2 (Gan and Koehler, 1990).

7.1.5 Recommendations for Tests of Univariate Normality

On the basis of power, the choice of test is directly related to the information available or assumptions made concerning the alternative. The more specific the alternative, the more specific and more powerful the test will usually be; this will also result in the most reliable recommendation. A test should also be based on ease or practicality of computation, and necessary tables (coefficients, if applicable, and critical values) should be available. All recommendations were based on the assumption that the parameters of the distribution are unknown.

Regardless of the degree of knowledge concerning the distribution, it should be common practice to graphically inspect the data. Therefore, our first recommendation is to always inspect at least a histogram and a probability plot of the data.

If the alternative is completely specified up to the parameters, then an optimal test for that distribution should be used, i.e., either a likelihood ratio test or a MPLSI test (Chapter 4), assuming such a test exists. For example, for an exponential distribution Grubbs' statistic is the MPLSI test. If no alternative-specific test can be used, then a related test may be available, e.g., Uthoff's U is the MPLSI test for a double exponential alternative, but since critical values are not readily available a single-sided a could be used in its place. The next choice would be a directional test for the class of the alternative, e.g., a one-tailed $\sqrt{b_1}$ for a gamma alternative.

If the shape and the direction of the shape (e.g., skewed to the left; symmetric with long tails) are assumed known in the event the alternative hypothesis is true, but a specific alternative is not, then usually a one-tailed test of the appropriate type will be more powerful than omnibus or bidirectional tests. Grubbs' statistic, W or one-tailed $\sqrt{b_1}$ will usually be among the most powerful of all tests for detecting a skewed distribution in a known direction. These are also the tests of choice (using the appropriate choice of critical values for Grubbs' statistic and $\sqrt{b_1}$) for skewed alternatives in which the direction of skewness is not prespecified.

Uthoff's U is the MPLSI test for a double exponential alternative, and Geary's test is asymptotically equivalent to U; therefore, these tests would be likely candidates for detecting long-tailed symmetric alternatives. D'Agostino's D is based on a kernel which indicates that it might also be appropriate for long-tailed symmetric alternatives. Many of the power comparisons described above have shown that these tests should be used, in a one-tailed manner, under these circumstances. LaBreque's F_1 needs to be investigated further for this class of alternatives. For short-tailed symmetric distributions, theoretical and simulation results indicate the best tests are u, one-tailed b_2 or Grubbs' statistic.

If there is no prior knowledge about the possible alternatives, then an omnibus test would be most appropriate. A joint skewness and kurtosis test such as K_s^2 provides high power against a wide range of alternatives, as does the Anderson-Darling A^2. The Wilk-Shapiro W showed relatively high power among skewed and short-tailed symmetric alternatives when compared to other tests, and respectable power for long-tailed symmetric alternatives.

Tests to be avoided for evaluation of normality include the χ^2 test and the Kolmogorov-Smirnov test D^*. The χ^2 test, however, has often been shown to have among the highest power of all tests for a lognormal alternative. Half-sample methods (Stephens, 1978) and spacing tests also have poor power for testing for normality.

7.2 Power of Outlier Tests

Relative to tests for normality, there have been few power comparisons of tests for outliers. One reason may be because there are relatively few outlier tests, and of the outlier tests each has a specific function. For example, Grubbs' (1950, 1969) and Dixon's (1950, 1951) outlier tests are used to detect a single outlier, whereas L_k is a test for $k > 1$ outliers; therefore, a comparison between T_n and L_3, say, would be meaningless. Similarly, L_k and sequential procedures would not be comparable since for the former the number of outliers is prespecified while for the latter the number of outliers is tested for sequentially. Some of the tests that are usually not thought of as outlier tests ($\sqrt{b_1}$, b_2, u, tests for normal mixtures) are not often compared to tests labeled as outlier tests.

7.2.1 Power Comparisons of Tests

Ferguson (1961) compared the power of b_2 and $\sqrt{b_1}$ with Grubbs' and Dixon's outlier tests. He used normal random samples and added a fixed constant to one observation in each sample. He found virtually no difference in power between Grubbs' outlier test and $\sqrt{b_1}$, with Dixon's test being only slightly less powerful. Similarly, he computed b_2, T and Dixon's r for the same samples in order to determine the power when two-tailed tests were required. Here he found that b_2 and T were virtually identical and again Dixon's test was only slightly lower in power.

In a second experiment Ferguson added a positive constant to two observations in the samples to determine the power of T and b_2, in particular because of the possibility of the masking effect on T. Kurtosis did significantly better than Grubbs' outlier test when there was more than one outlier in the sample; Dixon's test was not included in this experiment.

Thode, Smith and Finch (1983) showed T to be one of the most powerful tests at detecting scale contaminated normal distributions, a commonly used model for generating samples with outliers. Samples were generated differently than those of Ferguson: contaminating observations were generated randomly so that none, one or more than one contaminating observation may have existed in each sample. T was shown to have 92% relative power to kurtosis over all parameterizations studied, where relative power was defined as the ratio of sample sizes needed for each test (b_2 to T) in order to obtain the same power. T significantly outperformed $\sqrt{b_1}$, the Dixon test and u, which had only 65% relative power to b_2.

Whereas in the above three studies the measure of performance of a test was simply how often the null hypothesis was rejected, Tietjen and Moore (1972) and Johnson and Hunt (1979) examined other characteristics of outlier tests. Specifically, they were interested in the performance of the tests in detecting which and how many observations were identified as outliers using E_k or (sequentially) other tests for outliers.

Johnson and Hunt (1979) claimed that T was superior to the Tietjen-Moore, W and Dixon tests when there was one extreme value in the sample tested. T did show loss of performance, especially compared to more general normality tests, when there was more than one outlier (Ferguson, 1961; Johnson and Hunt, 1979). In a comparison of a number of tests for normality and goodness of fit in the context of normal mixtures, Mendell, Finch and Thode (1993) showed that $\sqrt{b_1}$ was the most powerful test when more than one outlier was present (and they were all in the same direction).

7.2.2 Recommendations for Outlier Tests

When there is the possibility of a single outlier in a sample, T has consistently been shown to be better than other procedures. For more than one outlier, $\sqrt{b_1}$ and b_2 should be used when there are outliers in one or in both directions, respectively. However, if identification of the number of outliers is of concern, then a sequential procedure should be used.

References

Csorgo, M., Seshadri, V., and Yalovsky, M. (1973). Some exact tests for normality in the presence of unknown parameters. Journal of the Royal Statistical Society B 35, 507-522.

D'Agostino, R.B. (1971). An omnibus test of normality for moderate and large size samples. Biometrika 58, 341-348.

D'Agostino, R.B., and Rosman, B. (1974). The power of Geary's test of normality. Biometrika 61, 181-184.

Dixon, W. (1950). Analysis of extreme values. Annals of Mathematical Statistics 21, 488-505.

Dixon, W. (1951). Ratios involving extreme values. Annals of Mathematical Statistics 22, 68-78.

Ferguson, T.S. (1961). On the rejection of outliers. Proceedings, Fourth Berkeley Symposium on Mathematical Statistics and Probability, University of California Press, Berkeley, 253-287.

Filliben, J.J. (1975). The probability plot coefficient test for normality. Technometrics 17, 111-117.

Franck, W.E. (1981). The most powerful invariant test of normal versus Cauchy with applications to stable alternatives. Journal of the American Statistical Association 76, 1002-1005.

Gan, F.F., and Koehler, K.J. (1990). Goodness-of-fit tests based on P-P probability plots. Technometrics 32, 289-303.

Gastwirth, J.L., and Owens, M.E.B. (1977). On classical tests of normality. Biometrika 64, 135-139.

Geary, R.C. (1947). Testing for normality. Biometrika 34, 209-242.

Green, J.R., and Hegazy, Y.A.S. (1976). Powerful modified-EDF goodness of fit tests. Journal of the American Statistical Association 71, 204-209.

Grubbs, F. (1950). Sample criteria for testing outlying observations. Annals of Mathematical Statistics 21, 27-58.

Grubbs, F. (1969). Procedures for detecting outlying observations in samples. Technometrics 11, 1-19.

Hogg, R.V. (1972). More light on the kurtosis and related statistics. Journal of the American Statistical Association 67, 422-424.

Johnson, B.A., and Hunt, H.H. (1979). Performance characteristics for certain tests to detect outliers. Proceedings of the Statistical Computing Section, Annual Meeting of the American Statistical Association, Washington, D.C.

LaBreque, J. (1977). Goodness-of-fit tests based on nonlinearity in probability plots. Technometrics 19, 293-306.

Lilliefors, H.W. (1967). On the Kolmogorov-Smirnov test for normality with mean and variance unknown. Journal of the American Statistical Association 62, 399-402.

Lin, C.-C., and Mudholkar, G.S. (1980). A simple test for normality against asymmetric alternatives. Biometrika 67, 455-461.

Locke, C., and Spurrier, J.D. (1976). The use of U-statistics for testing normality against non-symmetric alternatives. Biometrika 63, 143-147.

Locke, C., and Spurrier, J.D. (1977). The use of U-statistics for testing normality against alternatives with both tails heavy or both tails light. Biometrika 64, 638-640.

Lockhart, R.A., O'Reilly, F.J., and Stephens, M.A. (1986). Tests of fit based on normalized spacings. Journal of the Royal Statistical Society B 48, 344-352.

Looney, S.W., and Gulledge, Jr., T.R. (1984). Regression tests of fit and probability plotting positions. Journal of Statistical Computation and Simulation 20, 115-127.

Looney, S.W., and Gulledge, Jr., T.R. (1985a). Probability plotting positions and goodness of fit for the normal distribution. The Statistician 34, 297-303.

Mendell, N.R., Finch, S.J., and Thode, Jr., H.C. (1993). Where is the likelihood ratio test powerful for detecting two component normal mixtures? Biometrics 49, 907-915.

Oja, H. (1981). Two location and scale free goodness of fit tests. Biometrika 68, 637-640.

Oja, H. (1983). New tests for normality. Biometrika 70, 297-299.

Pearson, E.S., D'Agostino, R.B., and Bowman, K.O. (1977). Tests for departure from normality: comparison of powers. Biometrika 64, 231-246.

Saniga, E.M., and Miles, J.A. (1979). Power of some standard goodness of fit tests of normality against asymmetric stable alternatives. Journal of the American Statistical Association 74, 861-865.

Shapiro, S.S., and Francia, R.S. (1972). Approximate analysis of variance test for normality. Journal of the American Statistical Association 67, 215-216.

Shapiro, S.S., and Wilk, M.B. (1965). An analysis of variance test for normality (complete samples). Biometrika 52, 591-611.

Shapiro, S.S., Wilk, M.B., and Chen, H.J. (1968). A comparative study of various tests for normality. Journal of the American Statistical Association 62, 1343-1372.

Smith, V.K. (1975). A simulation analysis of the power of several tests for detecting heavy-tailed distributions. Journal of the American Statistical Association 70, 662-665.

Spiegelhalter, D.J. (1977). A test for normality against symmetric alternatives. Biometrika 64, 415-418.

Spiegelhalter, D.J. (1980). An omnibus test for normality for small samples. Biometrika 67, 493-496.

Stephens, M.A. (1974). EDF statistics for goodness of fit and some comparisons. Journal of the American Statistical Association 69, 730-737.

Stephens, M.A. (1978). On the half-sample method for goodness of fit. Journal of the Royal Statistical Society B 40, 64-70.

Thode, Jr., H.C. (1985). Power of absolute moment tests against symmetric non-normal alternatives. Ph.D. dissertation, University Microfilms, Ann Arbor, MI.

Thode, Jr., H.C., Smith, L.A., and Finch, S.J. (1983). Power of tests of normality for detecting scale contaminated normal samples. Communications in Statistics - Simulation and Computation 12, 675-695.

Tietjen, G.L., and Moore, R.H. (1972). Some Grubbs-type statistics for the detection of outliers. Technometrics 14, 583-597.

Uthoff, V.A. (1968). Some scale and origin invariant tests for distributional assumptions. Ph.D. thesis, University Microfilms, Ann Arbor, MI.

Uthoff, V.A. (1973). The most powerful scale and location invariant test of the normal versus the double exponential. Annals of Statistics 1, 170-174.

Vasicek, O. (1976). A test for normality based on sample entropy. Journal of the Royal Statistical Society B 38, 54-59.

Weisberg, S., and Bingham, C. (1975). An approximate analysis of variance test for non-normality suitable for machine calculation. Technometrics 17, 133-134.

White, H., and MacDonald, G.M. (1980). Some large-sample tests for nonnormality in the linear regression model. Journal of the American Statistical Association 75, 16-31.

CHAPTER 8

TESTING FOR NORMALITY WITH CENSORED DATA

"If a goodness-of-fit test can lead the investigator to accept a certain parametric mathematical or physical model, then this can enable him to glean some information about the tail of the response time distribution, which is often important in reliability studies. Non-parametric methods usually reveal little about the tail behaviour."

Turnbull and Weiss, 1978

Censored data refers to a set of observations where some of the values are known only up to a boundary value. Censored data occur commonly in lifetime experiments, where the lower bound is often observed, and is called "right censoring". Right censoring will occur, for example, when unit lifetimes (e.g., light bulbs, mice) are being measured, and at the end of the experiment some of the units have not yet "died"; the lower bound on the survival time for these observations is the time at which the experiment ends, while the actual survival time is unknown.

The simplest (and most common) forms of censoring, called singly Type I (time) and Type II (failure) censoring, result from different experimental protocols. In singly Type I (right) censoring, n units are observed and those which have a lifetime above a predetermined limit are censored. This would be the case in which an experiment is conducted over a fixed time period and all failures occurring prior to the time limit are observed.

This form of censoring also covers the case where there is a lower time limit on observation, and all units failing prior to that time are (left) censored. In either case, singly Type I censoring consists of a fixed time interval while the number of observations which are censored is random.

Singly Type II censoring occurs when the number of uncensored observations is fixed and the time interval is random. This type of censoring occurs when an experiment consists of observing n units until a fixed number $k < n$ fail.

Other types of censoring exist, but are less common. Doubly censored data are obtained when there is a fixed lower and upper limit, either on time (Type I) or on the number of failures (Type II). Multiply censored data occur when different censoring limits are placed on different units. Grouped data are an example of multiple right and left censoring. Multiply right Type I censored data would also occur, for example, if units are placed into service at different times during the experiment, but the experiment ends at one predetermined time for all units. Random censoring occurs when observations are randomly censored; this might occur, for example, in a competing risks situation. Arbitrary censoring occurs when none of Type I, Type II or random censoring is used, but rather an arbitrary method is used to censor the data.

Here we will focus our attention on tests using singly and doubly censored data; tests which can be used under other censoring schemes will be identified. The test statistics described below can be calculated for both Type I and Type II censoring situations; however, critical values for small samples are only available for Type II censoring for most tests, since the number of uncensored observations is fixed. For these tests only asymptotic results can be used for Type I censoring, since the number of uncensored observations, and therefore the censoring proportion k/n, is a random variable.

Most of the tests described below are adaptations of tests used to assess normality and goodness of fit for complete samples. Most can also be used (with appropriate modifications to formulas and/or critical values) for null distributions other than the normal.

Notation: There are a wide variety of notations used in the censored data literature cited here. This results in a dilemma when trying to both avoid confusion within this text, yet keeping as true as possible to original notation so that the reader may refer back to the cited literature with as little confusion as possible. However, we find it to be more beneficial in this chapter to use a uniform notation. Therefore, when referring to either Type I or Type II censoring, we will let b be the index of the greatest uncensored order statistic from a sample, and let a denote the lowest uncensored order statistic. The number of uncensored observations in a sample will be denoted by $k = b - a + 1$. So, for example, the set of uncensored observations in a right censored sample will be denoted $x_{(a)}, \ldots, x_{(b)}$, with $a = 1$ and $b = k$.

8.1 Probability Plots and Correlation Tests

Probability plots have been modified in order to accommodate censoring of data points. Plots can be used for both Type I and Type II censoring since in general the subjective nature of the plot is not greatly affected by the randomness of the censoring proportion in Type I censoring.

8.1.1 Probability Plots

For singly right censored samples, Q-Q plots are easily obtained since the $b = k < n$ uncensored observations are the first k order statistics of a normal sample of size n. Therefore, plots of $x_{(i)}, i = 1, \ldots, k$, against the first k expected order statistics from a normal sample of size n should still approximate a straight line. Censored observations are usually not included in the plot. For singly left-censored data, the uncensored observations are plotted against the $k = n - a + 1$ highest expected order statistics for a sample of size n. For data that are doubly censored, the appropriate k central order statistics are used.

Example 8.1. Suppose that, for the mouse sarcoma latency data (Data Set 7) the experiment had been halted after 650 days (Type I right censoring); then 13 of the 38 observations would have been censored. Figure 8.1 is the probability plot of these censored data using the plotting position $p = 1/(n+1)$; the censored observations are included here for informational purposes only. Three very noticeable outliers indicate that these data are probably not from a normal distribution.

For more complicated censoring schemes, probability plots can also be constructed; however, plotting positions must be modified for the possibility that a censored observation would have been observed between two uncensored observations, had the uncensored value been available. Based on the Kaplan-Meier (1958) estimate of the distribution function, the probabilities

$$p_i^{KM}(r) = 1 - \prod_{j \in S, j \leq i} \frac{n-j}{n-j+1}$$

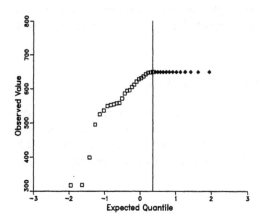

Figure 8.1 Plot of mouse sarcoma latency data (Data Set 7) censored at 650 days (n = 25 uncensored observations, 13 censored observations).

could be used for multiply right censored data, where S is the set of uncensored observations and r indicates right censoring. A generalized plotting position for multiply right censored samples is

$$p_i(r) = 1 - \frac{n - c + 1}{n - 2c + 1} \prod_{j \in S, j \leq i} \frac{n - j - c + 1}{n - j - c + 2} \qquad (8.1)$$

where $0 \leq c \leq 1$ (Michael and Schucany, 1986). Herd (1960) proposed the set of probabilities based on $c = 0$. Michael and Schucany (1986) suggested using $c = 0.3175$, while Chambers, Cleveland, Kleiner and Tukey (1983) used $c = 0.5$. For complete samples, (8.1) reduces to $(i-c)/(n-2c+1)$. For multiply left censored samples, the plotting positions are $p_i(l) = 1 - p_i(r)$.

Waller and Turnbull (1992) suggested the use of empirically rescaled probability (ERP) plots for censored data. They contended that with censored data the points on probability plots tend to clump together, resulting in a subjective evaluation which is highly influenced by few observations. For complete data this problem is observed in Q-Q plots, where the points clump in the middle for normal data. For complete data P-P plots the points are equally spaced over the abscissa, which eliminates this problem. However, for censored data even P-P plots may result in bunched observations. On the other hand, bunching is less of a problem for censored normal data than for censored data coming from a skewed null distribution (Waller and Turnbull, 1992); therefore, ERP plots will not be discussed here.

8.1.2 Correlation Tests

For complete samples, under normality the probability plot should be approximately linear over the entire range of the sample. In contrast, for censored data the probability plot should be approximately linear only over that portion of the data which is uncensored. Under right censoring, for example, the probability plot should be linear for the observations which were observed to fail up to the time the experiment ended. Therefore, the proposed correlation tests for testing normality with censored data are calculated only over the uncensored portion of the sample.

For some defined "expected" normal order statistic m_i, correlation test statistics for censored samples have the same basic form as those for complete samples (Section 2.3.2)

$$r_c = \frac{\sum_{i=a}^{b}(x_{(i)} - \bar{x}_c)(m_i - \bar{m}_c)}{\left(\sum_{i=a}^{b}(x_{(i)} - \bar{x}_c)^2 \sum_{i=a}^{b}(m_i - \bar{m}_c)^2\right)^{1/2}} \qquad (8.2)$$

where differences in the summation subscripts are based on the amount and type of censoring. In (8.2), the subscripts c indicate that the values are based on censored estimates. In this case,

$$\bar{x}_c = \sum_{i=a}^{b} x_{(i)}/(n\delta)$$

and

$$\bar{m}_c = \sum_{i=a}^{b} m_i/(n\delta).$$

For Type I censored data the summation limits and the divisor of the means are based on the random number of uncensored observations, while for Type II censoring, the summations are over the (fixed) number of uncensored observations.

For complete samples several choices of the m_i, the normal scores used for comparison to the data, have been investigated (Chapter 2). For censored samples Smith and Bain (1976) used the De Wet and Venter (1972) form of $m_i = \Phi^{-1}(i/(n+1))$ while Gerlach (1980) used the expected values of the normal order statistics. Verrill and Johnson (1987, 1988) showed the asymptotic equivalence of using these two choices for m_i in (8.2), as well as the medians of normal order statistics (Filliben, 1975), $\Phi^{-1}((i-0.375)/(n+0.25))$ (Weisberg and Bingham, 1975), and the censored version of the Wilk-Shapiro test (Shapiro and Wilk, 1965).

Example 8.2. Suppose that the patients receiving chemotherapy (Data Set 9) were followed up until 80% (16 of 20 patients) developed leukemia (here we artificially assume that all patients received chemotherapy at the same time). The probability plot of these data would be Figure 5.2 censored at observation 16. The correlation between the 16 uncensored latency time observations and the expected value of the normal order statistics (Harter, 1961) is 0.9715. Since this test value is not below the 95% critical value of 0.94407 (Verrill and Johnson, 1988), the hypothesis of normality cannot be rejected.

Verrill and Johnson (1987, 1988) also showed that the null distribution of r_c is asymptotically equivalent under both Type I and Type II censoring. Stephens (1986b) and Verrill and Johnson (1983, 1988) gave critical values for the null distribution of the Shapiro-Francia (1972) test under Type II censoring for selected sample sizes and censoring proportions. These are provided in Table B26.

8.1.3 A Test for Nonlinearity

Similar to LaBreque (1977), LaRiccia (1986) tested for quadratic and cubic forms of deviation from linearity in a probability plot regression, but for censored data. Let

$$m_i = E(Z_{(i)})$$

for those values of i such that $x_{(i)}$ is not a censored observation, where Z is the ith order statistic from a standard normal distribution (LaRiccia used ξ to denote the expected values); then

$$E(X_{(i)}) = \mu + m_i\sigma$$

for a random variable X from a normal distribution with mean and variance given by μ and σ^2. Therefore,

$$E(\vec{X}_k) = A_0 \begin{pmatrix} \mu \\ \sigma \end{pmatrix}$$

under the null hypothesis of normality, where $\vec{X}_k$ is the vector of those order statistics from a normal random sample corresponding to the uncensored observations and

$$A_0 = \begin{pmatrix} 1 & m_a \\ 1 & m_{a+1} \\ \vdots & \vdots \\ 1 & m_b \end{pmatrix}.$$

The alternative hypothesis, that there are nonlinear components to the probability plot, is tested using the Hermite polynomials $H_2(x) = x^2 - 1$ and $H_3(x) = x^3 - 3x$ to form the matrix

$$
\mathbf{A} = \begin{pmatrix} 1 & m_a & H_2(m_a) & H_3(m_a) \\ \vdots & \vdots & \vdots & \vdots \\ 1 & m_b & H_2(m_b) & H_3(m_b) \end{pmatrix}
$$

so that the alternative is given by

$$
E(\vec{\mathbf{X}}_k) = \mathbf{A} \begin{pmatrix} \mu \\ \sigma \\ \alpha_1 \\ \alpha_2 \end{pmatrix}.
$$

The proposed test is based on the null hypothesis that $\alpha_1 = \alpha_2 = 0$. Using the estimated values of α_1 and α_2 obtained from

$$
\hat{\theta} = \begin{pmatrix} \hat{\mu} \\ \hat{\sigma} \\ \hat{\alpha}_1 \\ \hat{\alpha}_2 \end{pmatrix} = \left(\mathbf{A}' \Sigma_k^{-1} \mathbf{A} \right)^{-1} \mathbf{A}' \Sigma_k^{-1} \vec{x}_k
$$

where

$$
\mathbf{C} = \hat{\sigma}^{-2} cov(\hat{\theta}) = \left(\mathbf{A}' \Sigma_k^{-1} \mathbf{A} \right)^{-1}.
$$

the general test statistic is given by

$$
T_3 = n[\hat{\alpha}_1 \hat{\alpha}_2](\mathbf{C}^{22})^{-1} \begin{pmatrix} \hat{\alpha}_1 \\ \hat{\alpha}_2 \end{pmatrix} \hat{\sigma}^{-2}
$$

where $\mathbf{C}^{22}$ is the lower 2×2 matrix of $\mathbf{C}$, i.e.,

$$
\mathbf{C} = \begin{pmatrix} \mathbf{C}^{11} & \mathbf{C}^{12} \\ \mathbf{C}^{21} & \mathbf{C}^{22} \end{pmatrix}.
$$

The $k \times k$ matrix Σ_k has entries σ_{ij}. As LaRiccia (1986) noted, the difficulty with this test statistic lies in obtaining the constants ξ_i and σ_{ij} and calculating the inverse of Σ_k. He suggested substituting the constants with large sample approximations given by David (1981) when $n \geq 50$. (These approximations are given in Chapter 13 where $r_i = 1 - \lambda_i$.) The $\lambda_i, i = a, \ldots, b$ are related to the observed data percentile by $p = n\lambda + 1$ for the pth percentile x_p. For example, the index for the lowest uncensored observation is $\lambda_a = (a - 1)/n$.

Tests for α_1 assuming that $\alpha_2 = 0$ (T_1), and for α_2 assuming that $\alpha_1 = 0$ (T_2) are obtained as above, with the appropriate column deletion from matrix $\mathbf{A}$ and entry deletion from the parameter vector $\hat{\theta}$. For symmetric censoring, T_1 and T_2 are uncorrelated and $T_3 = T_1 + T_2$.

For samples down to size 25, LaRiccia (1986) showed that the asymptotic critical values of T_1, T_2 and T_3 can be used without affecting the significance level of the test, regardless of the amount or type of censoring. These critical values are obtained from a χ_1^2 distribution for T_1 and T_2, while those for T_3 are obtained from a χ_2^2 distribution.

It should be noted that there are some errors in the LaRiccia (1986) manuscript, specifically the formulas for estimation of the expected values of normal order statistics and σ_{ij} (see Chapter 13 and David, 1981). If these formulas were used for computations performed in the preparation of this manuscript, we expect that this would affect at most the test size and power tables he provided. It should not affect the basic conclusion that this test is asymptotically χ^2.

Parameter Estimation

For censored samples estimates of μ and σ are needed for use in the calculation of the EDF and χ^2 tests. Here we only consider censored sample estimation methods which are recommended for use in the goodness of fit tests described in this chapter.

For EDF tests Pettitt (1976) recommended Gupta's (1952) method of parameter estimation using linear combinations of order statistics. For right-censored samples these are of the form

$$\hat{\mu}^* = \sum_{i=a}^{b} \beta_i x_{(i)} \tag{8.3}$$

and

$$\hat{\sigma}^* = \sum_{i=a}^{b} \gamma_i x_{(i)} \tag{8.4}$$

using the appropriate coefficients β_i and γ_i. For Type II censored samples and $n \le 10$, Gupta provided the necessary coefficients. For larger samples, the coefficients can be calculated as

$$\beta_i = \frac{1}{k} - \frac{\overline{m}(m_i - \overline{m})}{\sum_{i=a}^{b}(m_i - \overline{m})^2}$$

$$\gamma_i = \frac{m_i - \overline{m}}{\sum_{i=a}^{b}(m_i - \overline{m})^2}$$

where m_i is the expected value of the ith standard normal order statistic and $\overline{m} = \sum_{i=a}^{b} m_i/k$.

Chernoff, Gastwirth and Johns (1967) gave estimators of μ and σ which are linear combinations of order statistics. Their estimators were proposed for both uncensored and multiply Type II censored data; here we will provide the estimates for the more specific doubly Type II censored case. These estimates, which are asymptotically equivalent to the maximum likelihood estimators, are given by

$$(\hat{\mu}, \hat{\sigma}) = (E_1, E_2)\mathbf{J}^{-1} \tag{8.5}$$

where

$$E_1 = n^{-1}\sum_{i=a}^{b} x_{(i)} + x_{(a)}(z_\alpha\phi(z_\alpha) + \phi^2(z_\alpha)/\alpha)$$
$$+ x_{(b)}(\phi^2(z_\beta)/(1-\beta) - z_\beta\phi(z_\beta))$$

$$E_2 = 2n^{-1}\sum_{i=a}^{b}\Phi^{-1}\left(\frac{i}{n+1}\right) x_{(i)} + x_{(a)}\phi(z_\alpha)(z_\alpha^2 - 1 + z_\alpha\phi(z_\alpha)/(\alpha))$$
$$+ x_{(b)}\phi(z_\beta)(1 - z_\beta^2 + z_\beta\phi(z_\beta)/(1-\beta))$$

and the entries of J are

$$J_{11} = \int_{z_\alpha}^{z_\beta} [\phi'(y)]^2/\phi(y)dy + \phi^2(z_\alpha)/\alpha + \phi^2(z_\beta)/(1-\beta)$$

$$J_{12} = J_{21} = \int_{z_\alpha}^{z_\beta} \phi'(y)[1 + y\phi'(y)/\phi(y)]dy + z_\alpha\phi^2(z_\alpha)/\alpha + z_\beta\phi^2(z_\beta)/(1-\beta)$$

$$J_{22} = \int_{z_\alpha}^{z_\beta} [1 + y\phi'(y)/\phi(y)]^2\phi(y)dy + z_\alpha^2\phi^2(z_\alpha)/\alpha + z_\beta^2\phi^2(z_\beta)/(1-\beta)$$

where $\phi(\cdot)$ and $\Phi(\cdot)$ are the standard normal probability density and distribution functions, respectively, and z is a standard normal percentile. Here, α and β denote the lower and upper censoring proportions, respectively.

8.3 EDF Tests

As in the full sample case (Chapter 5), EDF tests modified for censored data were first developed for the simple hypothesis case: for example, Barr and Davidson (1973), Koziol and Byar (1975) and Dufour and Maag (1978) presented critical values for the Kolmogorov-Smirnov test; Koziol (1980b) considered Kuiper's statistics; and Pettitt and Stephens (1976) discussed the Cramer-von Mises, Anderson-Darling and Watson's U^2 tests for Type I and Type II censored data in the simple hypothesis case. Michael and Schucany (1979) and O'Reilly and Stephens (1988) proposed transforming a censored data set into a complete uniform ordered set, followed by a goodness of fit test. Pettitt (1976) provided one of the first sets of critical values for selected EDF tests for Type I and Type II censored data when parameters needed to be estimated.

Using the parameter estimates (8.3) and (8.4), EDF tests for normality are obtained by using the estimated uniform variates

$$p_{(i)} = \Phi\left(\frac{x_{(i)} - \hat{\mu}^*}{\hat{\sigma}^*}\right)$$

for the uncensored observations.

8.3.1 Anderson-Darling Test - Type II Right Censored Data

When there are $k < n$ uncensored observations, where the data are singly right censored, the Anderson-Darling test statistic is

$$_rA^2 = -\frac{1}{n}\sum_{i=1}^{k}(2i-1)[\log(p_{(i)}) - \log(1-p_{(i)})] - 2\sum_{i=1}^{k}\log(1-p_{(i)})$$

$$-\frac{1}{n}[(k-n)^2\log(1-p_{(k)}) - k^2\log(p_{(k)}) + n^2 p_{(k)}] \qquad (8.6)$$

where $r = k/n$ for fixed k under Type II censoring. Large values of $_rA^2$ would result in the rejection of the null hypothesis. Pettitt (1976) gave tables of critical values of $_rA^2$ for a limited number of censoring proportions, for small sample sizes as well as asymptotic values. Stephens (1986a) gave expanded tables of critical values for the $_rA^2$ test (Table B27).

> *Example 8.3. Using the data from Example 8.2, the parameter estimates from (8.3) and (8.4) are*
>
> $$\mu^* = 53.45$$
>
> $$\sigma^* = 26.16.$$

[In comparison, the mean and standard deviation estimates from the full sample are 60.4 and 40.0, respectively.] Using (8.6) gives

$$_rA^2 = -(1.47) - (-19.30) - (17.58) = 0.25$$

which is compared to the Anderson-Darling 5% critical value with an r = 0.8 censoring ratio. The critical value for a sample size of 20 is 0.473, so that the null hypothesis should not be rejected.

8.3.2 Cramer-von Mises Test - Type II Right Censored Data

For singly right Type II censored data the Cramer-von Mises test statistic is

$$_rW^2 = \sum_{i=1}^{k}\left(p_{(i)} - \frac{2i-1}{2n}\right)^2 + \frac{k}{12n^2} + \frac{n}{3}\left(p_{(k)} - \frac{k}{n}\right)^3 \qquad (8.7)$$

where the test subscript is as described in the previous section. Large values of $_rW^2$ would cause rejection of the null hypothesis. Pettitt (1976) gave tables of critical values of $_rW^2$ for a limited number of censoring proportions, for small sample sizes and asymptotic values. Stephens (1986a) gave expanded tables of critical values for the $_rW^2$ test (Table B27).

8.3.3 Watson's U^2 - Type II Right Censored Data

For singly right Type II censored data Watson's test statistic is

$$_rU^2 =_r W^2 - np_{(k)}\left(\frac{k}{n} - \frac{p_{(k)}}{2} - \frac{k\bar{p}}{np_{(k)}}\right) \qquad (8.8)$$

where

$$\bar{p} = \Sigma_{i=1}^{k}p_{(k)}/k.$$

The test subscript is as described in Section 8.3.1. Large values of the test statistic would result in rejection of the null hypothesis. Pettitt (1976) gave tables of the asymptotic critical values of $_rU^2$ for a limited number of censoring proportions. These are provided in Table B28.

8.3.4 EDF Tests for Type II Left Censored Data

For Type II left censored data, Equations (8.6), (8.7) and (8.8) and their associated critical values are used to test for normality. However, the transformation to uniform variates for this type of data is given by $p_{(i)} = 1 - \Phi([x_{(i)} - \hat{\mu}^*]/\hat{\sigma}^*)$.

8.3.5 EDF Tests for Doubly Type II Censored Data

Pettitt and Stephens (1976) provided a Cramer-von Mises test for doubly Type II censored normal data under a simple hypothesis. This procedure has not been extended to the composite hypothesis case.

8.3.6 EDF Tests for Type I Censored Data

Stephens (1986a) gave an approximate procedure for testing Type I censored data. Define $p_{(k+1)} = \Phi^{-1}([t* - \mu*]/\sigma*)$, where $t*$ is the experimental censoring time. Include $p_{(k+1)}$ in the formulas (8.6), (8.7) or (8.8) as if there were $k + 1$ uncensored observations. The tables of critical values, however, can only be used to provide an approximate test.

8.3.7 Anderson-Darling Test on Normalized Spacings

The Anderson-Darling test on normalized spacings (Section 5.3.3) can also be used for censored data (Lockhart, O'Reilly and Stephens, 1986). In the more general case where data are doubly censored, the $z_{(i)}$ are calculated as shown in Section 5.3.3 for $i = k, \ldots, k + j$. Critical values of the test A_S^2 for selected censoring proportions are given in Table B16 (Lockhart, O'Reilly and Stephens (1986); Stephens (1986a).

8.4 Other Tests of Censored Data

Other types of goodness of fit tests for censored data have also been developed. Mehrotra (1982) proposed a class of tests based on spacings for a simple hypothesis. Tests developed for composite hypotheses include tests using spacings (Tiku, 1980) and χ^2 tests (Mihalko and Moore, 1980).

8.4.1 Tiku's Test Based on Spacings

Tiku (1980) presented a goodness of fit test and critical values for complete and for Type II singly and doubly censored data based on the spacings of the ordered uncensored observations. Let $d_i = m_{i+1} - m_i$ where m_i

is the expected value of the ith order statistic from a standard normal distribution, and

$$G_i = (x_{(i+1)} - x_{(i)})/d_i \qquad i = a, \ldots, b-1.$$

Then the test statistic

$$Z^* = \frac{2\sum_{i=a}^{b-2}(b-i-1)G_i}{(b-a-1)\sum_{i=a}^{b-1} G_i}$$

can be used to test goodness of fit, with large values indicating non-normality. Tiku (1980) presented critical values of Z^* for a very limited number of sample sizes and censoring limits (a, b). These critical values are given in Table B29.

Example 8.4. Continuing with data from Example 8.2, the numerator is given by

$$\sum_{i=1}^{14}(b-i-1)G_i = 2812.89$$

and the denominator is

$$14\sum_{i=1}^{15} G_i = 5393.78$$

so that $Z^* = 1.04$. This is well below the approximate 95% critical value of 1.28 (Table B29), so that the null hypothesis of normality is not rejected.

8.4.2 A χ^2 Test

Chi-squared tests have been developed for both Type I and Type II censoring situations. In general, χ^2 tests should be applied by fixing the cell or interval length, calculating the expected probabilities associated with interval under the null hypothesis, and comparing the observed and expected number of observations in the usual way. For Type I censored data this is straightforward, since the rightmost interval is predefined by the cutoff time limit used for the experiment (in the case of right censored data); similarly for left or doubly censored data, the intervals in the extremes are predetermined. Therefore, the usual χ^2 test can be applied in the case of Type I censoring.

For Type II censoring, however, the intervals cannot be predetermined since the experiment length is random. Mihalko and Moore (1980) modified the Rao-Robson (1974) test for the Type II censored case to obtain a χ^2 test. For this test, we assume a singly or doubly censored sample, so that we observe the uncensored values

$$x_a, x_{a+1}, \ldots, x_b.$$

and

$$-\infty = \xi_0, \xi_1, \xi_2, \ldots, \xi_M = \infty$$

are selected order statistics used to define cell boundaries. Mihalko and Moore (1980) used the estimates (8.5) of Chernoff, Gastwirth and Johns (1967) to obtain the cell probabilities by letting

$$z_i = (\xi_i - \hat{\mu})/\hat{\sigma}$$

so that the cell i probability under the normal distribution is

$$p_i = \Phi(z_i) - \Phi(z_{i-1}).$$

If the usual full sample χ^2 test statistic is given by

$$T_2 = \mathbf{V}_n(\hat{\theta})'\mathbf{V}_n(\hat{\theta})$$

then the Rao-Robson χ^2 statistic adapted for censored data is given by

$$T_3 = T_2 + \mathbf{V}_n(\hat{\theta})'\mathbf{B}_n(\mathbf{K}_n - \mathbf{B}_n'\mathbf{B}_n)^{-1}\mathbf{B}_n'\mathbf{V}_n(\hat{\theta})$$

which is χ^2_{M-1} under the null hypothesis of normality. $\mathbf{V}_n(\hat{\theta})$ is the $M \times 1$ vector with entries given by

$$V_i = (N_i - np_i)/\sqrt{np_i}$$

where N_i is the number of observations in cell i and n is the sample size. By denoting

$$\phi_i = \phi(z_i) - \phi(z_{i-1})$$

where $\phi(.)$ is the normal density function, and

$$\nu_i = z_i\phi(z_i) - z_{i-1}\phi(z_{i-1})$$

then

$$\mathbf{V}_n(\hat{\theta})' \mathbf{B}_n = (\; \sqrt{n} \Sigma \phi_i / p_i \quad \sqrt{n} \Sigma \nu_i / p_i \;)$$

$$\mathbf{B}_n' \mathbf{B}_n = \sigma^{-2} \begin{pmatrix} \Sigma \phi_i^2 / p_i & \Sigma \phi_i \nu_i / p_i \\ \Sigma \phi_i \nu_i / p_i & \Sigma \nu_i^2 / p_i \end{pmatrix}$$

where the summations are over the M cells. $\mathbf{K}_n$ is given by $\sigma^{-2} \mathbf{J}$ where the entries of $\mathbf{J}$ are given in Section 8.2. Mihalko and Moore (1980) did not make any recommendations as to the number of cells to use.

Note that this test is also applicable to multiple Type II censoring, in which only certain percentiles of the data are actually observed. In this case, the observed order statistics, or some subset thereof, can be used to define the cells since this χ^2 test involves only functions of the cell boundaries and the number of observations within cells. Mihalko and Moore (1980) showed simplifications to the equations used to calculate the estimates and T_3 for singly right censored data; these simplifications are also provided in Schneider (1986).

8.5 Multiple and Random Censoring

Goodness of fit may also be of interest in a situation where more complicated forms of censoring occur. Multiple censoring occurs when left and/or right censoring occur at a number of different times. This would be the case if grouped or life table data were collected. Random censoring occurs when observations are censored at random, such as in a competing risks situation or when animals are sacrificed at random times to ascertain tumor development in a survival study.

O'Neill (1984) proposed a goodness of fit test for grouped data under a simple hypothesis. Koziol and Green (1976) and Koziol (1980a) considered simple hypothesis EDF tests for randomly censored data. Hollander and Proschan (1979) proposed a goodness of fit procedure for randomly censored data under a simple hypothesis. Nair (1981) suggested a procedure which compared the null survival function and the Kaplan-Meier estimate, under a simple hypothesis for randomly censored data. Hollander and Pena (1992) proposed a χ^2 test for randomly censored data. Akritas (1988) developed a Pearson type goodness of fit test for randomly censored data under a composite hypothesis.

Gail and Ware (1979) and Turnbull and Weiss (1978) proposed goodness of fit tests for grouped data with random censoring. The former test was for a simple hypothesis, while the latter was developed for a composite hypothesis.

8.6 Power Studies and Recommendations

Recommendations for using specific tests in determining goodness of fit are based on a number of factors. These include: simplicity of calculation; power (which is dependent on, among other things, the specified alternative(s)); and availability of critical values. Simplicity of calculation of the test statistics should be obvious from the descriptions in this chapter. Availability of critical values for these tests range from easily obtainable (e.g., χ^2) to those that are dependent upon both sample size and censoring proportion and which have not been well developed. There have been relatively few power studies conducted for tests with censored data. In addition, the studies which have been conducted were only based on a few tests, alternatives, sample sizes and censoring proportions, which may not be generalizable to specific situations encountered in actual practice.

Generally, the correlation test statistics are the most simple to calculate, as any statistics package can be used to compute a correlation. An expected order statistic does need to be obtained, although some of these should easily be obtainable (e.g., Φ^{-1}) and can be used with essentially no loss of power relative to the expected values. Sufficient tables of critical values, including asymptotic values for very large samples, are available to use directly or from which to extrapolate. Tiku's (1980) test statistic is also very simple to calculate, but also requires expected normal order statistics; however, tables of critical values are limited. EDF test statistics are moderately difficult to compute due to the need to obtain parameter estimates; after obtaining the estimates, calculation of the test statistic is simple and straightforward. Tables of critical values are also readily available from a single source (Stephens, 1986a) for the three EDF tests described here. LaRiccia's (1986) and the χ^2 test (Mihalko and Moore, 1980) statistics are computationally very difficult, but use the χ^2 distribution to determine significance independent of sample size or censoring proportion.

Pettitt (1976), Tiku (1980), LaRiccia (1986) and Verrill and Johnson (1988) performed limited comparative power studies for different subsets of tests, alternatives and sample size/censoring proportion combinations. Pettitt (1976) provided a short description of his power study, in which he claimed that the Anderson-Darling test was generally more powerful than the Cramer-von Mises test.

Tiku (1980) compared his test to Smith and Bain's (1976) version of the correlation test. His results showed that for a normal null distribution, power was generally slightly lower for Z^* against symmetric alternatives, and was slightly higher for skewed alternatives. This power study was based only on samples of size 20, with complete samples, 20% right single censoring and 10% (in each tail) double censoring.

In his power study, LaRiccia (1986) compared his three tests (T_1, T_2 and T_3), the Cramer-von Mises test and the Smith and Bain (1976) correlation test. This comparison included samples of size 50 with 24% right, left and double (12% each side) censoring. In general, the Cramer-von Mises test had higher power than the correlation test for singly censored data, and power was comparable for doubly censored data. It was disturbing to note that for symmetric alternatives the correlation test showed a large decrease in power for left vs right censored samples. Among the three nonlinearity tests, as would be expected, the power of T_1 was generally greater than that of T_2 for skewed alternatives and the power of T_2 was generally greater than that of T_1 for symmetric alternatives, regardless of type of censoring. The power of T_3 was generally between that of T_1 and T_2. The power of T_3 was comparable to, or slightly higher than, that of the Cramer-von Mises test for all alternatives and censoring forms. For all tests, the power under double censoring was generally much lower than for left or right single censoring, indicating the importance of tail information of a distribution for determining shape.

Verrill and Johnson (1988) provided a description of a power study they conducted. In their study they included the Wilk-Shapiro, Shapiro-Francia (1972) and Smith and Bain correlation tests, the Anderson-Darling and Cramer-von Mises EDF tests, and the χ^2 test. Sample sizes of 20, 40 and 70, and censoring proportions 0.1(0.1)1.0 were used in this study. As with complete samples, the Wilk-Shapiro test performed very well except against long-tailed symmetric alternatives. The Shapiro-Francia and EDF tests were all comparable in power, and performed well against all alternatives. The Smith and Bain version of the correlation test performed well against long-tailed symmetric alternatives, but poorly otherwise. The χ^2 test performed poorly relative to all other tests.

Based on these findings, we would suggest the appropriate test from LaRiccia (1986) if the alternative of concern is known, and otherwise T_3 as an omnibus test, if the computational difficulties are not of concern; the possibility of errors in the power study cannot be completely discounted, either (see Section 8.1.3). Otherwise, with only moderate computational difficulty, one of the EDF tests or the Shapiro-Francia correlation test should be used. The Smith and Bain (1976) test should only be used if the alternatives being considered are known to be long-tailed and symmetric. Because of the dearth of critical values, we consider Tiku's test to be of lesser value since it has power roughly equivalent to the Smith and Bain test, which should only be used in limited circumstances. The χ^2 test should be avoided due to both low power and computational difficulty. Although we do not recommend the Tiku (1980) and χ^2 tests for testing for normality, these tests have been developed, and may be valuable for testing, under other null distributions (see the appropriate reference).

8.7 Further Reading

Cohen (1991) is a good general reference on truncated and censored samples. Schneider (1986) is also a good reference, and focuses specifically on truncated and censored normal distributions. McCool (1982) provides a concise set of definitions, estimation methods and list of references.

Since much of the censored data literature is generated from life testing, texts on survival, life testing and failure testing would be good sources of information on censoring methods. Mann and Singpurwalla (1983) provide a summary of censoring in a life testing setting.

References

Akritas, M.G. (1988). Pearson type goodness of fit tests: the univariate case. Journal of the American Statistical Association 83, 222-230.

Barr, D.R., and Davidson, T. (1973). A Kolmogorov-Smirnov test for censored samples. Technometrics 15, 739-757.

Chambers, J.M., Cleveland, W.S., Kleiner, B., and Tukey, P.A. (1983). **Graphical Methods for Data Analysis**. Wadsworth Pub. Co., Monterey, CA.

Chernoff, H., Gastwirth, J.L., and Johns, M.V. (1967). Asymptotic distribution of linear combinations of functions of order statistics with applications to estimation. Annals of Mathematical Statistics 38, 52-72.

Cohen, A.C. (1991). **Truncated and Censored Samples, Theory and Applications**. Marcel Dekker Inc., New York, NY.

David, H.A. (1981). **Order Statistics**, 2nd ed. John Wiley and Sons, New York, NY.

De Wet, T., and Venter, J.H. (1972). Asymptotic distributions of certain test criteria of normality. South African Statistical Journal 6, 135-149.

Dufour, R., and Maag, U.R. (1978). Distribution results for modified Kolmogorov-Smirnov statistics for truncated or censored samples. Technometrics 20, 29-32.

Filliben, J.J. (1975). The probability plot coefficient test for normality. Technometrics 17, 111-117.

Gail, M.H., and Ware, J.H. (1979). Comparing observed life table data with a known survival curve in the presence of random censorship. Biometrics 35, 385-391.

Gerlach, B. (1980). A correlation type goodness of fit test for normality with censored sampling. Mathematische Operationsforschung und Statistik 11, 207-218.

Gupta, A.K. (1952). Estimation of the mean and standard deviation of a normal population from a censored sample. Biometrika 39, 260-273.

Harter, H.L. (1961). Expected values of normal order statistics. Biometrika 48, 151-165; Correction, Biometrika 48, 476.

Herd, G.R. (1960). Estimation of reliability from incomplete data. Proceedings, Sixth National Symposium on Reliability and Quality Control, 202-207.

Hollander, M., and Pena, E.A. (1992). A chi-squared goodness of fit test for randomly censored data. Journal of the American Statistical Association 87, 458-463.

Hollander, M., and Proschan, F. (1979). Testing to determine the underlying distribution using randomly censored data. Biometrics 35, 393-401.

Kaplan, E.M., and Meier, P. (1958). Nonparametric estimation from incomplete observations. Journal of the American Statistical Association 53, 457-481.

Koziol, J.A. (1980a). Goodness of fit tests for randomly censored data. Biometrika 67, 693-696.

Koziol, J.A. (1980b). Percentage points of the asymptotic distributions of one and two sample Kuiper statistics for truncated or censored data. Technometrics 22, 437-442.

Koziol, J.A., and Byar, D.P. (1975). Percentage points of the asymptotic distributions of one and two sample K-S statistics for truncated or censored data. Technometrics 17, 507-510.

Koziol, J.A., and Green, S.B. (1976). A Cramer-von Mises statistic for randomly censored data. Biometrika 63, 465-474.

LaBreque, J. (1977). Goodness of fit tests based on nonlinearity in probability plots. Technometrics 19, 293-306.

LaRiccia, V.N. (1986). Asymptotically chi-squared distributed tests of normality for Type II censored samples. Journal of the American Statistical Association 81, 1026-1031.

Lockhart, R.A., O'Reilly, F.J., and Stephens, M.A. (1986). Tests of fit based on normalized spacings. Journal of the Royal Statistical Society B 48, 344-352.

Mann, N.R., and Singpurwalla, N.D. (1983). Life testing. In **Encyclopedia of Statistical Sciences Vol. 4**, John Wiley and Sons, New York, NY, 632-639.

McCool, J.I. (1982). Censored data. In **Encyclopedia of Statistical Sciences Vol. 1**, John Wiley and Sons, New York, NY, 389-396.

Mehrotra, K.G. (1982). On goodness of fit tests based on spacings for Type II censored samples. Communications in Statistics - Theory and Methods 11, 869-878.

Michael, J.R., and Schucany, W.R. (1979). A new approach to testing goodness of fit for censored samples. Technometrics 21, 435-441.

Michael, J.R., and Schucany, W.R. (1986). Analysis of data from censored samples. In D'Agostino, R.B., and Stephens, M.A., eds., **Goodness of Fit Techniques**, Marcel Dekker, New York.

Mihalko, D.P., and Moore, D.S. (1980). Chi-square tests of fit for Type II censored data. Annals of Statistics 8, 625-644.

Nair, V.N. (1981). Plots and tests for goodness of fit with randomly censored data. Biometrika 68, 99-103.

O'Neill, T. (1984). A goodness of fit test for one sample life table data. Journal of the American Statistical Association 79, 194-199.

O'Reilly, F.J., and Stephens, M.A. (1988). Transforming censored samples for testing fit. Technometrics 30, 79-86.

Pettitt, A.N. (1976). Cramer-von Mises statistics for testing normality with censored samples. Biometrika 63, 475-481.

Pettitt, A.N., and Stephens, M.A. (1976). Modified Cramer-von Mises statistics for censored data. Biometrika 63, 291-298.

Rao, K.C., and Robson, D.S. (1974). A chi-square statistic for goodness of fit within the exponential family. Communications in Statistics 3, 1139-1153.

Schneider, H. (1986). **Truncated and Censored Samples from Normal Populations**. Marcel Dekker, New York.

Shapiro, S.S., and Francia, R.S. (1972). Approximate analysis of variance test for normality. Journal of the American Statistical Association 67, 215-216.

Shapiro, S.S., and Wilk, M.B. (1965). Analysis of variance test for normality (complete samples). Biometrika 52, 591-611.

Smith, R.M., and Bain, L.J. (1976). Correlation type goodness of fit statistics with censored samples. Communications in Statistics A5, 119-1132.

Stephens, M.A. (1986a). Tests based on EDF statistics. In D'Agostino, R.B., and Stephens, M.A., eds., **Goodness of Fit Techniques**, Marcel Dekker, New York.

Stephens, M.A. (1986b). Tests based on regression and correlation. In D'Agostino, R.B., and Stephens, M.A., eds., **Goodness of Fit Techniques**, Marcel Dekker, New York.

Tiku, M.L. (1980). Goodness of fit statistics based on the spacings of complete or censored samples. Australian Journal of Statistics 22, 260-275.

Turnbull, B.W., and Weiss, L. (1978). A likelihood ratio statistic for testing goodness of fit with randomly censored data. Biometrics 34, 367-375.

Verrill, S., and Johnson, R.A. (1983). The asymptotic distributions of censored data versions of the Shapiro-Wilk test of normality statistic. Technical Report No. 702, Dept. of Statistics, University of Wisconsin-Madison.

Verrill, S., and Johnson, R.A. (1987). The asymptotic equivalence of some modified Shapiro-Wilk statistics - complete and censored sample cases. Annals of Statistics 15, 413-419.

Verrill, S., and Johnson, R.A. (1988). Tables and large sample distribution theory for censored data correlation statistics for testing normality. Journal of the American Statistical Association 83, 1192-1197.

Waller, L.A., and Turnbull, B.W. (1992). Probability plotting with censored data. The American Statistician 46, 5-12.

Weisberg, S., and Bingham, C. (1975). An approximate analysis of variance test for nonnormality suitable for machine calculation. Technometrics 17, 133-134.

CHAPTER 9

ASSESSING MULTIVARIATE NORMALITY

"With multiresponse data it is clear that the possibilities for departure from joint normality are indeed many and varied. One implication of this is the need for a variety of techniques with differing sensitivities to the different types of departures; seeking a single best method would seem to be neither pragmatically sensible nor necessary."

R. Gnanadesikan, 1977

Four general strategies are defined here which can be used for assessing multivariate normality. As in the univariate case, normal probability plots are suggested both as a subjective test for multivariate normality and for assessing the direction(s) of departures from normality. The second strategy consists of objectively assessing marginal normality using univariate tests, using the necessary (but not sufficient) condition that a multivariate normal distribution is normal in each of its marginal distributions. The third strategy is the reduction of data to $m' < m$ dimensions in such a fashion that the distribution of the reduced data has some known or simple structure. Finally, the fourth strategy is the direct assessment of multivariate normality from the multivariate observations.

A variety of tests are available for each of the strategies. As with univariate tests, prior knowledge or assumptions about the structure of the alternative distribution is helpful in deciding which tests to use. Complicating factors in multivariate data include whether the component data are independent or correlated, and whether the component distributions are the same or different; the ability to detect non-normality is affected by these factors as well as the degree of departure from normality.

Since there are many types of departures from multivariate normality, a single best test does not exist. A multivariate test may dilute the effects of a single non-normal component or subset of components; conversely, marginal tests may miss departures in multivariate combinations of variables. Looney (1986) recommended an extensive protocol of tests for assessing multivariate normality.

Multivariate analyses are sometimes distinguished by whether they are invariant under linear transformations or they are dependent upon the original data coordinate system. Therefore, while most of the tests described are affine-invariant, in certain cases those tests for multivariate normality which are coordinate-dependent (e.g., the Q statistics of Cox and Small, 1978) may be more pertinent in a given situation.

Many of the tests used to assess multivariate normality are based on univariate extensions, many of which (e.g., EDF tests, the Shapiro and Wilk W, probability plots) require ordering of the data. This introduces another complicating factor of multivariate data. Barnett (1976) indicated that for multivariate distributions there is no inherent ordering of points, and suggested three possible orderings:

 (i) marginal ordering, wherein all variates are ordered independently;
 (ii) reduced ordering, where the observations are reduced to single-valued observations and are ordered based on the reduced values; and
 (iii) conditional ordering where the multivariate observations are ordered based on one of the variates.

Tests described below which require ordering are most often those of type (ii), and the most frequent reduction used is the squared radii (Section 9.1.2). Due to the large number of tests for multivariate normality, we restricted most of our attention to tests having a combination of computational ease, decent power properties, and/or methodological, theoretical or historical interest. We do, however, cite the literature regarding a number of other tests for multivariate normality which the reader may peruse as their interests may dictate.

9.1 The Multivariate Normal Distribution

The multivariate normal density function is

$$f(\mathbf{x}) = \frac{1}{(2\pi)^{m/2}|\Sigma|^{1/2}} \exp[-\frac{1}{2}(\mathbf{x} - \mu)'\mathbf{\Sigma}^{-1}(\mathbf{x} - \mu)]$$

where $\mathbf{x}$ is an m-variate observation, μ is the mean vector and Σ is the covariance matrix, $\Sigma = \{\rho_{ij}\sigma_i\sigma_j\}$. In this chapter, the notation will be as follows: define $\mathbf{x_i}$ to be an $m \times 1$ observation vector; $\overline{\mathbf{x}}$ is the sample mean vector and $\mathbf{S}$ is the $m \times m$ sample covariance matrix; x_{ij} will denote the jth entry $(j = 1, \ldots, m)$ in the ith observation vector.

9.1.1 Properties of the Multivariate Normal Distribution

Most tests for multivariate normality take advantage of one or more of the properties that are unique to the multivariate normal distribution. One property which is often used when assessing multivariate normality is the necessary, though not sufficient, condition that each of the marginal distributions of $\mathbf{x}$ are univariate normal. This permits the use of any of the univariate tests of normality previously defined to determine if there is non-normality in any of the marginals, thereby indicating multivariate non-normality. The advantage of using this approach is that there are many well-developed tests for univariate normality; however, correlation between the variates results in correlation between the m univariate tests, causing difficulty in determining their joint distribution.

Two "linearity" properties of multivariate normals have also been used to derive tests. One is that if there exists correlation between any two of the variates, then the relation between the two is strictly linear. The second is that, regardless of the correlation structure, any linear combination of the variates is normally distributed. Therefore, projections of the observations onto a reduced space $\mathbf{R}^k$, $k \leq m$, result in observations that have a normal distribution in k-space.

Another property of the multivariate normal distribution is that the quadratic form $(\mathbf{x_i} - \mu)'\Sigma^{-1}(\mathbf{x_i} - \mu) = c$ forms an ellipse of constant probability in m-space whose shape is determined by Σ. Further, the squared observation vector lengths (squared radii, see Section 9.1.2) in m space from the true mean are distributed as χ_m^2 random variables. These attributes are often assumed to hold approximately when using estimates of the mean and covariance matrix. A related property is that the angles between the marginal projection of the observation vectors onto any of the variate planes and an (arbitrary) fixed vector through the mean are uniform on $(0, 2\pi)$; further, these angles are independent of the vector lengths.

9.1.2 Scaled Residuals and Squared Radii

Since linear combinations of normal random variables are themselves normal, then for multivariate normal observations the quantities

$$\mathbf{Z}_i = \Sigma^{-1/2}(\mathbf{x_i} - \mu) \tag{9.1}$$

are also m-variate normal. These quantities are often called the scaled residuals, and are multivariate extensions of standardized variables, i.e., $Z_i \sim N(0_m, I_m)$.

The Euclidean distances of the observations from the mean μ, *which are dependent on the coordinate scales*, are given by $(x - \mu)'(x - \mu)$. However, as noted by Healy (1968), correlation between variates and unequal variances can result in having two observations which are equidistant from the mean in the Euclidean distance, while one is within an ellipse of constant probability and the other is outside of that same ellipse; also, a change of scale on one variate can change the ordering of the observations based on the Euclidean distance. Therefore, Healy suggested the use of the squared Mahalanobis distance (squared radii)

$$R_i^2 = Z_i'Z_i = (x_i - \mu)'\Sigma^{-1}(x_i - \mu) \qquad (9.2)$$

which is the squared length of the vector from the origin μ to the observation in m-space relative to the probability ellipses, i.e., if $R_i^2 = R_j^2$, then x_i and x_j are both located on the same ellipse. Further, it is known that under multivariate normality the R_i^2 follow a χ_m^2 distribution.

When the parameters are unknown, the scaled residuals and squared radii can be estimated by

$$z_i = S^{-1/2}(x_i - \overline{x}) \qquad (9.3)$$

and

$$r_i^2 = z_i'z_i = (x_i - \overline{x})'S^{-1}(x_i - \overline{x}) \qquad (9.4)$$

respectively, for some efficient estimators $\overline{x}$ of μ and S of Σ.

Example 9.1. R.A. Fisher's (1936) measurements of sepal length, sepal width, petal length and petal width of three species of iris (setosa, versicolor and virginica) are possibly the most commonly used sample data set in multivariate analysis. Data Set 12 is the 4-variate setosa data. Table 9.1 contains the squared radii obtained from this data set.

The $N(0_m, I_m)$ and χ^2 distributions of the z_i and r_i^2 will only be approximate when substituting parameter estimates for μ and Σ. Small (1978) and Mason and Young (1985) showed that there is a closer linear relation of the r_i^2 with an appropriately parameterized beta distribution than with χ_m^2 (see Section 9.2). These quantities and their distributional

Table 9.1 Squared radii for the Iris Setosa Data (Data Set 12).

0.449	1.891	5.349	1.995	1.685
2.081	2.015	2.722	4.889	12.328
1.284	2.947	11.044	7.699	4.201
1.706	7.040	7.230	5.248	12.310
0.762	10.22	9.748	1.267	8.601
3.713	7.654	3.771	3.302	2.195
3.424	5.742	2.526	5.721	2.756
0.343	0.636	0.829	3.086	1.489
2.996	5.186	1.323	3.270	1.253
3.200	1.612	2.174	0.589	0.495

properties have often been used as a basis for extending univariate tests to the multivariate case. Especially with the use of the r_i^2, reduction of the data to a single dimension makes the multivariate testing problem less problematic. Many of the plotting and testing procedures described below are based on the squared radii.

9.2 Multivariate Plots

Several plotting procedures have been suggested for exploratory examination of the multivariate normal hypothesis. These procedures take advantage of some of the properties of the multivariate normal distribution described above. They include scatterplots of the component data, probability plots of the marginal data, and probability plots of reduced data.

9.2.1 Marginal Plots

A first approach to assessing multivariate normality is the use of univariate probability plots to independently assess each of the marginal variables. Healy (1968) also proposed using scatterplots of all variables taken two at a time; although he suggested that this was a more effective way of identifying outliers, it also allows identification of other nonlinear relations between variables. A third approach includes ordering the marginal observations independently and plotting the ordered observations against each other taking the variates two at a time. Under the hypothesis of normality, these plots are equivalent to normal probability plots and should follow a linear pattern.

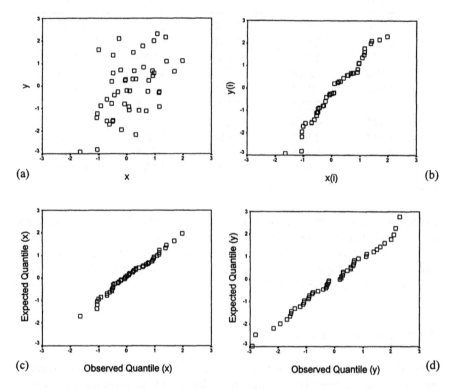

Figure 9.1 Component plots of simulated bivariate normal data, $n = 50$. (a) x vs y; (b) $x_{(i)}$ vs $y_{(i)}$; (c), (d) marginal probability plots.

Example 9.2. Figure 9.1 shows the three types of marginal probability plots for a simulated bivariate normal sample (x, y) of size 50 where the correlation between the components is $\rho = 0.45$. Panel (a) is the scatterplot of the bivariate pairs (x_i, y_i). Panel (b) is a plot of the ordered $x_{(i)}$ with the ordered $y_{(i)}$, ignoring the pairing. Panels (c) and (d) are the normal probability plots of the individual components.

9.2.2 Probability Plots Using Squared Radii and Angles

Healy (1968) suggested using probability plots of the r_i^2 compared to a χ_m^2 distribution. In the bivariate case Healy noted that the expected values of the order statistics for a sample of size n are

$$\frac{2}{n}, \frac{2}{n} + \frac{2}{n-1}, \ldots, \frac{2}{n} + \frac{2}{n-1} + \ldots + \frac{2}{1}$$

since the χ_2^2 is an exponential distribution with mean 2.

Healy also suggested transformation of the r_i^2 so that the probability plot could be made on ordinary normal probability paper rather than having to obtain quantiles of χ_m^2. Such transformations would tend to reduce the variability of extreme values in the null case, which would make them more effective at identifying outliers. Healy proposed using the square root or cube root transformations, although he did not indicate a preference for either.

For x from a multivariate normal distribution, the r_i^2 are proportional to a beta variable of the first kind (Gnanadesikan and Kettenring, 1972); in particular

$$nr^2/(n-1)^2 \sim B(m/2, (n-m-1)/2) \qquad (9.5).$$

Asymptotically, this can be approximated by a χ_m^2. For $m = 2$ and $n = 25$, the chi-squared approximation may be appropriate, but for $m = 4$ samples of size 100 may not be adequate to use this approximation. Small (1978) suggested using probability plots of the r_i^2 with beta order statistics, using Blom's (1958) general plotting position

$$\frac{i - \alpha}{n - \alpha - \beta + 1}$$

with

$$\alpha = (m - 2)/2m$$

and

$$\beta = 0.5 - (n - m - 1)^{-1}.$$

Using simulation, Small demonstrated a slightly better fit of the beta distribution for $m = 6$ and $n = 40$.

Example 9.3. Figure 9.2 is the probability plot of the squared radii for the iris setosa data (Table 9.1). The probability plot is constructed using the modified squared radii as given in (9.5), and is plotted against $B(2, 22.5)$ order statistics using Blom's position with $\alpha = 0.5$ and $\beta = 0.48$.

In the bivariate case the angles θ_i made by the observation vectors with the x_1 axis are uniform over the interval $(0, 2\pi)$ (Andrews, Gnanadesikan

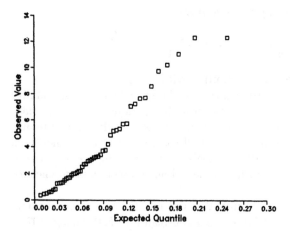

Figure 9.2 Beta probability plot of squared radii for four measurements on iris setosa (Data Set 12), n = 50.

and Warner, 1973). Therefore, a uniform probability plot of $\theta_i^* = \theta_i/2\pi$ can provide another indication of non-normality in the data. By defining $u_i = F(r_i^2)$ where F is the χ_2^2 distribution function, a bivariate plot of (u_i, θ_i^*) should be uniform over the unit square. However, in light of Small (1980) and Gnanadesikan and Kettenring (1972), F may be better defined as the appropriate beta distribution.

For $m > 2$, one of the $m - 1$ angles made between the projections of the data onto each of the variate planes and, say, the x_1 axis is uniform $(0, 2\pi)$. The remaining $m - 2$ angles have a distribution proportional to $\sin^{m-1-j}(\theta_j)$, $0 \le \theta_j \le \pi$, $j = 1, \ldots, m - 2$.

9.2.3 Kowalski's Line Test

For the bivariate case, Kowalski (1970) suggested plotting the ordered R_i^2 against the $\log_{10}$ of the quantiles of the χ_2^2. Under bivariate normality the squared radii are given by

$$R_i^2 = ((x_{1i}-\mu_1)^2/\sigma_1^2 - 2\rho(x_{1i}-\mu_1)(x_{2i}-\mu_2)/(\sigma_1\sigma_2) + (x_{2i}-\mu_2)^2/\sigma_2^2)/(1-\rho^2)$$

when the parameters are known. The distribution of the R^2 under normality is given by

$$F(R^2) = 1 - \exp(R^2/2)$$

where F is the χ_2^2 distribution function. Then

$$Y_{(i)} = \log_{10}(1 - F(R_{(i)}^2)) = -0.271 R_{(i)}^2 \tag{9.6}$$

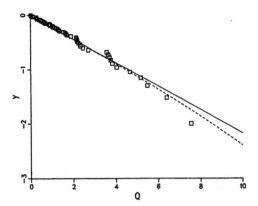

Figure 9.3 Line plot of simulated bivariate normal data, n = 50. Solid line is based on χ² distribution (9.6), dashed line on beta distribution (9.7).

where the estimate of $Y_{(i)}$ is $\log_{10}[(n-i+0.5)/n]$, so that a plot of $R^2_{(i)}$ against $Y_{(i)}$ should give a straight line which passes through the origin and has a slope of -0.217.

Mason and Young (1985) showed that if the estimates r_i^2 are used with the beta distribution approximation, then

$$y_{(i)} = 0.217(n-3)\log_e(1 - nr^2_{(i)}/(n-1)^2) \tag{9.7}$$

which reduces to (9.6) as n gets large. The line described by (9.7) also passes through the origin, but gradually curves away from the straight line with slope -0.217, with more and earlier curvature for small n. Mason and Young recommended that the χ^2_2 approximation not be used for $n < 50$, and noted that even for sample sizes up to 100 there is noticeable curvature away from linearity.

Example 9.4. Figure 9.3 is the line test plot of the simulated bivariate data shown in Figure 9.1. Both the χ² reference line (9.6) and the beta reference line (9.7) are shown. The improvement in fit from using the beta distribution is obvious here.

9.2.4 A Multivariate Q-Q Plot

Probability plots are plots of an ordered data set against an ordered reference sample, such as the expected value of the order statistics. Since there

is no inherent ordering in the multivariate case, it is difficult to define a
multivariate extension of probability plots that is multivariate in nature,
rather than taking a component-wise approach or making a reduction of
dimension (e.g., plotting the squared radii).

Easton and McCulloch (1990) presented a multivariate Q-Q plot which
addressed the issue of no inherent ordering in multivariate space. This
method was based on matching the observed data with a multivariate reference sample. Since unidimensional ordering is simply a matching of two
sets of numbers and plotting the pairs, they proposed a multidimensional
matching procedure. By assigning the Euclidean distance between two
points as the cost function of matching the observed x_i with y_j from the
reference sample, then the "best" matching is found by identifying that
permutation of the data σ^* in the set P of all permutations which solves

$$\min_{\sigma \in P} \sum_{i=1}^{n} ||y_i - x_{\sigma(i)}||^2.$$

In the univariate case, since the optimal permutation involves only the order of the data sets, any linear transformation of the data will result in
the same pairing. In essence, if the data and the reference sample have the
same shape, then there is a linear transformation which will make the two
samples nearly identical. In the multivariate case, x and y will have the
same shape if there is an $m \times m$ matrix A, an m-vector b and a permutation σ such that $Ax_{\sigma(i)} + b \sim y$. In this instance the optimal permutation
σ^* is not independent of A and b. Therefore, given a reference sample,
the matching problem can be solved by alternating between optimizing the
permutation for fixed A and b, and then reoptimizing A and b for the
current permutation σ using multivariate regression. Easton and McCulloch (1990) used the assignment algorithm by Carpaneto and Toth (1980)
to obtain the optimal permutation at each step and a multivariate random
sample from the hypothesized distribution as the reference sample. More
details on reference sample generation and optimization methods can be
found in their paper.

If $x_i^* = A^* x_{\sigma^*(i)} + b^*$ is the best (transformed) matching of the data to
the reference sample, then the first set of displays to consider are probability
plots of each of the components of x^* vs y. Since the transformation is
the one which most nearly equates the two data sets, the resulting plots
should approximate a 45° line through the origin. These plots will have
a fuzzy appearance, but the usual deviations from linearity will appear in
the presence of outliers, skewness or heavy/light tailedness of the data. In
contrast to marginal probability plots, isolated points may stand out in the
middle of a "fuzzy" Q-Q plot, and large or heterogeneous variability around
the 45° line indicates deviation from normality (or other null distribution).

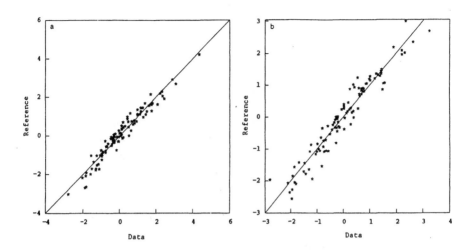

Figure 9.4 Multivariate Q-Q plots of 100 simulated bivariate normal obser-
vations: (a) fuzzy probability plot, coordinate 1; (b) fuzzy probability plot,
coordinate 2; (c) distance Q-Q plot; (d) χ^2 plot of squared radii (from Eas-
ton and McCulloch, 1990; reproduced with permission from Journal of the
American Statistical Association. Copyright 1990 by the American Statis-
tical Association. All rights reserved.).

A second display which can be obtained from this matching procedure
is a distance Q-Q plot. For this, a second reference sample **u** is drawn
from the reference distribution. Then the n Euclidean distances between
x* and **y** are plotted against the Euclidean distances obtained from the
matching of **u** and **y**, under the assumption that the distances should be
similar. The pairs of ordered distances should approximate a 45° line. To
minimize the additional variability incurred by generating a second random
sample, Easton and McCulloch recommended that repeated samples of **u**

be generated and compared to $\mathbf{y}$, and the average distance be used as a comparison of the distance between $\mathbf{x}^*$ and $\mathbf{y}$.

> *Example 9.5. Figures 9.4(a)-(d) are multivariate Q-Q plots for a simulated bivariate normal sample (x, y) of size 100 from Easton and McCulloch (1990). Panels (a) and (b) are the fuzzy Q-Q plots for each of the two coordinates; these demonstrate the fuzzy nature of this type of plot while still having the expected linearity of a probability plot. Panel (c) is the distance Q-Q plot of the simulated data. For these data the plot is not as well behaved in the tails as the χ^2 probability plot of the squared radii (Panel (d)).*

9.3 Assessing Multivariate Normality Using Marginals

A necessary but not sufficient condition for multivariate normality is that each of the marginal distributions is univariate normal. An obvious first-order method of detecting non-normality would be to test each marginal for normality; any univariate test given in prior chapters could be used. However, because the components are usually not independent, the joint distribution of the m univariate tests cannot easily be determined. D'Agostino (1986) avoided this complication by suggesting a Bonferroni approach, i.e., applying each test for marginal normality at the α/m level. Other methods of testing based on marginal properties are described in this section.

9.3.1 Marginal Skewness and Kurtosis

Small (1980) combined the marginal skewness and the marginal kurtosis values to obtain a combined skewness and a combined kurtosis test; he also proposed an omnibus test based on combining the two tests.

Let $\mathbf{B}_1$ and $\mathbf{B}_2$ be the m-vectors of marginal skewness and kurtosis values, respectively, with covariance matrices $\mathbf{V}_1$ and $\mathbf{V}_2$. If the marginal tests were normal, then $\mathbf{B}_i'\mathbf{V}_i^{-1}\mathbf{B}_i$ $(i = 1, 2)$ would be distributed as a chi-squared random variable. However, as noted in Chapter 3, the asymptotic normal distributions of skewness and kurtosis are approached very slowly.

Small applied the Johnson S_u transformations used by Bowman and Shenton (1975) for a univariate transformation to normality (3.7, 3.8) component-wise to $\mathbf{B}_1$ and $\mathbf{B}_2$ to obtain the transformed vectors $\mathbf{y}(\mathbf{B}_1)$ and $\mathbf{y}(\mathbf{B}_2)$. He then used the quadratic forms

$$Q_1 = \mathbf{y}(\mathbf{B}_1)'\mathbf{U}_1^{-1}\mathbf{y}(\mathbf{B}_1)$$

and

$$Q_2 = \mathbf{y(B_2)'U_2^{-1}y(B_2)}$$

as the test statistics. Since the $\mathbf{y(B_i)}$ are approximately standard normal, $\mathbf{U_1} = \{r_{ij}^3\}$ and $\mathbf{U_2} = \{r_{ij}^4\}$, where the r_{ij} are the sample correlations; the diagonal elements of the $\mathbf{U_i}$ are unity. Small (1980) stated that Q_1 and Q_2 are each well approximated by a χ_m^2 distribution for $29 \le n \le 100$ and $2 \le p \le 8$. An omnibus test

$$Q = Q_1 + Q_2$$

can also be used which is approximately χ_{2m}^2.

9.3.2 Combining Marginal Wilk-Shapiro Tests

Royston (1983) proposed combinations of marginal Wilk-Shapiro W test statistics for assessing multivariate normality. Using Royston's (1982) transformation of W to normality, the test statistic for each marginal is calculated using

$$z_i = ((1 - W_i)^\lambda - \mu)/\sigma$$

(2.6), where λ, μ and σ are functions of n which are given in Section 2.3.1. Letting

$$k_i = [\Phi^{-1}(\tfrac{1}{2}\Phi(-z_i))]^2$$

he noted that, since large positive values of z reject normality, then k_i will be large when z_i is large and positive. Each k_i is approximately distributed as a χ_1^2 random variable.

If the m variables were uncorrelated, then $G = \sum k_i/m$ can be used in a test for normality, where $G \sim \chi_m^2/m$; at the other extreme, if the variates were perfectly correlated, then $G \sim \chi_1^2$. For intermediate correlations, Royston suggested using $G \sim \chi_e^2/e$ where e is given by

$$e = \frac{m}{1 + (m-1)\bar{c}}$$

and

$$\bar{c} = \sum_{i=1}^{m}\sum_{j \ne i} \hat{c}_{ij}/(m^2 - m).$$

The $\hat{c}_{ij}$ are estimates of the correlation between the k_i, which for $10 \le n \le 2000$ are calculated as

$$\hat{c}_{ij} = r_{ij}^5[1 - 0.715(1 - r_{ij})^{0.715}/v]$$

where r_{ij} are the sample between-component correlations and

$$v = 0.21364 + 0.015124 \log^2(n) - 0.0018034 \log^3(n) \qquad (9.8).$$

Equation (9.8) is similar for n over the stated range, and a good compromise is obtained by using $v = 0.35$ for all n. Royston investigated the χ^2 approximation for $m = 2$ and 3 and claimed that it was adequate; larger dimensions were not investigated.

9.3.3 Component Linearity and Maximum Curvature

Since multivariate normality implies a linear relation between components, Cox and Small (1978) proposed pairwise testing for linearity between components of a multivariate distribution as a test for multivariate normality. They also developed a test which compares that pair of variables, consisting of linear combinations of the original components, which has maximum curvature when the variables are regressed on each other.

One simple pairwise test of linearity between components i and j is $Q_{i,j}$, the Student t statistic of significance for the coefficient of component x_j^2 when x_i is regressed on x_j and x_j^2. For the purpose of symmetry, the joint statistic $(Q_{i,j}, Q_{j,i})$ is used. Then (at least for large samples)

$$\max(|Q_{i,j}|, |Q_{j,i}|)$$

can be used by referring to tables of the bivariate normal distribution, or

$$\mathbf{Q'R^{-1}Q}$$

can be used as a χ_2^2 test, where $\mathbf{Q} = [Q_{i,j} \; Q_{j,i}]$ and

$$\mathbf{R} = \begin{pmatrix} 1 & r_{ij}(2 - 3r_{ij}^2) \\ r_{ij}(2 - 3r_{ij}^2) & 1 \end{pmatrix}$$

where r_{ij} is the observed correlation between components i and j.

An alternative method is a regression of each component x_i on all pairs of components x_j and x_k^2 ($j, k = 1, \ldots, m; j, k \neq i$). From these regressions the $m(m-1)$ Q values may be obtained, ordered and plotted on a normal probability plot, provided the sample size is sufficiently large (Small, 1985). Also, nonlinear functions other than a quadratic term (e.g., inverse or logarithm) can be used. Each of these linearity tests result in $m(m-1)$ statistics, which increases rapidly as m gets large.

They also proposed an invariant procedure which measures the amount of curvature present between two linear combinations of variables, and used this as a test statistic. By letting $Y = \mathbf{a}'\mathbf{x}$ and $W = \mathbf{b}'\mathbf{x}$, $\mathbf{a}'\Sigma\mathbf{a} = \mathbf{b}'\Sigma\mathbf{b} = 1$ (so that Y and W have mean 0 and variance 1), let

$$\eta^2(\mathbf{a}, \mathbf{b}) = \gamma^2/[E(W^4) - 1 - [E(W^3)]^2]$$

where γ is the least squares coefficient of Y on W^2 given W (Small, 1985). For fixed $\mathbf{b}$ this can be maximized with respect to $\mathbf{a}$ to obtain $\eta^2(\mathbf{b})$. The maximum over the sample of $\eta^2(\mathbf{b})$ must be found numerically, resulting in η^2_{max} which, for $n \geq 50$ and $m \leq 6$, is $N(\mu, \sigma^2)$, with $\mu = \log(5m^2/8n)$ and $\sigma = \log(0.53 + 3.87/m)$.

9.3.4 Marginal Transformations

Andrews, Gnanadesikan and Warner (1971, 1973) suggested performing a multivariate Box-Cox transformation (Section 4.5) in order to obtain a better behaved distribution, and also as a test for multivariate normality. They proposed simultaneous estimation of individual transformations of the marginal data, and a test that the transformations $\Lambda = \{\lambda_j\} = 1$ for the $j = 1, \ldots, m$ transformations. Given that the transformed vector $\mathbf{x}_i^{(\lambda)} = \{x_{ij}^{(\lambda_j)}\}$, where $\{x_{ij}^{(\lambda_j)}\}$ is the Box-Cox transformation of the jth component of $\mathbf{x}_i$, then the MLE of the mean and covariance matrix under the transformation is

$$\hat{\mu}_\lambda = n^{-1} \sum \mathbf{x}_i$$

$$\mathbf{S}_\lambda = n^{-1} \sum (\mathbf{x}_i - \hat{\mu}_\lambda)'(\mathbf{x}_i - \hat{\mu}_\lambda)$$

and

$$L_{max}(\lambda) = -\frac{n}{2} \log |S_\lambda| + \sum_{j=1}^{m} (\lambda_j - 1) \sum_{i=1}^{n} \log(x_{ij}) \qquad (9.9).$$

The value of $\hat{\lambda}$ which maximizes (9.9) should then be found numerically. The significance level of the test for the transformation that all λ_j are unity is obtained by referring

$$2(L_{max}(\hat{\lambda}) - L_{max}(\mathbf{1}))$$

to a χ^2_m distribution.

Kowalski (1970) described a transformation of the marginal variables (x, y) in the bivariate case

$$x' = \Phi^{-1}[F(x)]$$

$$y' = \Phi^{-1}[G(y)]$$

where F and G are the distributions of the raw marginals. He used Fourier estimators of the distributions F and G, based on a method described by Kronmal and Tarter (1968).

Hasofer and Stein (1990) suggested coordinate transformation to normality, followed by a Gram-Schmidt orthogonalization. Multivariate normality would be accepted if the transformed data were both normal and independent. Hensler, Mehrotra and Michalek (1977) proposed a test based on reducing the null composite hypothesis to an equivalent problem of testing a simple hypothesis or m univariate normal hypotheses. Details for their proposed methods can be found in the respective papers.

9.4 Direct-Data Assessment of Multivariate Normality

Most methods for direct assessment of multivariate normality are extensions of univariate techniques. Some involve reductions of the data, such as tests for whether the r_i^2 are χ^2 or beta distributed. There are also some unique methods used to extend the univariate methods. For example, Ward (1988a, 1988b) and Hawkins (1981) both used Anderson-Darling tests; however, their transformations of multivariate normal data to uniformity were quite different. Mardia (1970, 1974) and Malkovich and Afifi (1973) each presented multivariate extensions of skewness and kurtosis tests. Although further research is needed to make a determination, these different versions of the same test may provide measures of different aspects of non-normality.

9.4.1 Tests for Skewness and Kurtosis

Because of the popularity and good power properties of univariate moment tests as tests for normality, it is only natural that some of the first tests for assessing multivariate normality would be based on an attempt to extend the notion of skewness and kurtosis to a multivariate setting. Mardia (1970, 1974) and Malkovich (1971; Malkovich and Afifi, 1973) proposed the first tests of multivariate skewness and kurtosis. Similar to the development of moment tests, omnibus tests based on combinations of the skewness and kurtosis tests were also proposed.

Mardia's Tests for Skewness and Kurtosis

Perhaps the most often referenced tests for multivariate normality are sample estimates of multivariate skewness and kurtosis first presented by

Mardia in 1970. These were developed by extending aspects of robustness studies for the t statistic which involve β_1 and β_2. The test statistics are calculated from generalized versions of the squared radii,

$$r_{ij} = \mathbf{z}_i'\mathbf{z}_j = (\mathbf{x_i} - \overline{\mathbf{x}})'\mathbf{S}^{-1}(\mathbf{x_j} - \overline{\mathbf{x}})$$

Note that, using this notation, $r_{ii} = r_i^2$.

Kendall and Stuart (1977) investigated the effects of univariate non-normality on the t-test, and assuming that the fourth cumulant, κ_4, from a non-normal population is negligible, then to order n^{-1},

$$corr(\overline{x}, s^2) \sim \sqrt{\beta_1/2}.$$

By extension, assuming cumulants higher than order 3 are negligible and taking moments of $\overline{\mathbf{x}}$ and $\mathbf{S}$ to order n^{-1}, Mardia's skewness measure is

$$b_{1,m} = n^{-2} \sum_{i,j=1}^{n} r_{ij}^3.$$

Box and Andersen (1955) showed that in the univariate case for the one sample Pitman permutation test, the square of the t statistic follows a $F_{\delta,\delta(n-1)}$ distribution where

$$\delta = 1 + \frac{\beta_2 - 3}{n} + o(n^{-1})$$

so that the coefficient of n^{-1} is a univariate measure of kurtosis. Similarly in the multivariate case, using Box and Andersen (1955) and Arnold (1964),

$$(n - m)T^2/(m(n - 1)) \sim F_{\delta m, \delta(n-m)}$$

where T^2 is Hotelling's multivariate t-test and

$$\delta = 1 + \frac{E(b_{2,m}^*) - (m + 2)}{n[1 - (E(b_{2,m}^*/(n + 2)))]} \tag{9.10}$$

with

$$b_{2,m}^* = \frac{n + 2}{n^2 m} \sum_{i=1}^{n} [(x_i - \mu)S^{-1}(x_i - \mu)]^2.$$

Equation (9.10) can be rewritten as

$$\delta = 1 + n^{-1} \left(\frac{\beta_{2,m} - m(m + 2)}{m} \right) + o(n^{-1})$$

where $\beta_{2,m} = E((\mathbf{x} - \mu)\boldsymbol{\Sigma}^{-1}(\mathbf{x} - \mu))$ with the obvious estimate

$$b_{2,m} = n^{-1} \sum_{i=1}^{n} r_{ii}^2 = n^{-1} \sum_{i=1}^{n} (r_i^2)^2.$$

Under multivariate normality the exact moments of the two test statistics are

$$E(b_{1,m}) = \frac{m(m+2)((n+1)(m+1) - 6)}{(n+1)(n+3)}$$

with unknown variance, while

$$E(b_{2,m}) = m(m+2)(n-1)/(n+1)$$

$$V(b_{2,m}) = \frac{8m(m+2)(n-3)(n-m-1)(n-m+1)}{(n+1)^2(n+3)(n+5)}$$

(Mardia, 1974). Asymptotically, under multivariate normality $nb_{1,m}/6$ is χ^2 with

$$f = m(m+1)(m+2)/6$$

degrees of freedom, and

$$b_{2,m} \sim N(m(m+2), 8m(m+2)/n).$$

Mardia and Zemroch (1975) gave a FORTRAN subroutine for calculating $b_{1,m}$ and $b_{2,m}$.

Critical values for $b_{1,2}$ and $b_{2,2}$ were given in Mardia (1974) for selected sample sizes from $n = 10$ to $n = 5000$. For $b_{1,m}$ and an 0.05 level (one-tailed) test, Mardia proposed the χ^2 approximation, using $A = nKb_{1,m}/6$ where

$$K = (m+1)(n+1)(n+3)/[n((n+1)(m+1) - 6)].$$

For $b_{2,m}$ and $n > 400$, lower 2.5% critical points can be obtained using the normal approximation; this approximation is also adequate for upper 2.5% critical values for all $n \geq 50$. For $50 \leq n \leq 400$, a normal approximation with mean $m(m+2)(n+m+1)/n$ and variance $8m(p+2)/(n-1)$ can be used.

Mardia and Foster (1983) gave several omnibus tests based on combinations of $b_{1,m}$ and $b_{2,m}$. One set of omnibus tests is based on Wilson-Hilferty approximations to χ^2 variables. $C_W^2 = \mathbf{w}'\mathbf{W}^{-1}\mathbf{w}$ accounts for correlation between $b_{1,m}$ and $b_{2,m}$, where the vector $\mathbf{w}$ has elements $b_{1,m}$

and $b_{2,m}$ under transformations to normality and $\mathbf{W}$ is the correlation matrix associated with $\mathbf{w}$; entries of $\mathbf{w}$ are Wilson-Hilferty transformations of a χ^2 variable

$$W(b_{1,m}) = \frac{1}{6\sqrt{2f}}\{6(\frac{4nf^2}{3}b_{1,m})^{1/3} - 18f + 4\}$$

and

$$W(b_{2,m}) = 3\left(\frac{f_1}{2}\right)^{1/2}\left(1 - (1 - 2/f_1)/[1 + v(2/(f_1 - 4))^{1/2}]^{1/3} - 2/9f_1\right)$$

where

$$v = \frac{b_{2,m} - E(b_{2,m})}{\sqrt{V(b_{2,m})}}$$

and

$$f_1 = 6 + (8m(m+2)(m+8)^{-2})^{1/2}\sqrt{n}[(m(m+2)/)/2)^{1/2}/(m+8)\sqrt{n}$$

$$+(1 + 0.5nm(m+2)/(m+8)^2)^{1/2}].$$

$\mathbf{W}$ is a matrix with 1's on the diagonal and $c = \mathrm{Cov}(W(b_{1,m}), W(b_{2,m}))$ as the off diagonal elements, where

$$c = \left(\frac{f_1}{16f}\right)^{1/2}[-\frac{40}{9}(1 - 2/f_1)\frac{1}{f_1 - 4} + \frac{n}{3\sigma}(1 - 2/f_1)^{1/3}(2/(f_1 - 4)^{1/2}$$

$$\times Cov(b_{1,m}, b_{2,m}) + \ldots]$$

by Taylor expansion and $\sigma^2 = var(b_{2,m})$. Alternatively, the omnibus test $S_W^2 = W(b_{1,m}) + W(b_{2,m})$ can be used; although this test does not take the correlation of $b_{1,m}$ and $b_{2,m}$ into account, it is more easily calculated and does not seem to affect the size of the test.

Another set of omnibus tests is based on a normal approximation to a χ^2 variable for $b_{1,m}$ and a straightforward transformation to normality of $b_{2,m}$,

$$U(b_{1,m}) = n(b_{1,m} - 6f/n)/(72f)^{1/2}$$

and

$$U(b_{2,m}) = \sqrt{n}[b_{2,m} - m(m+2)(n-1)/(n+1)]/[8m(m+2)]^{1/2}.$$

The omnibus test statistics are

$$S_N^2 = U^2(b_{1,m}) + U^2(b_{2,m})$$

and

$$C_N^2 = \mathbf{b}'\mathbf{V}^{-1}\mathbf{b}$$

where

$$\mathbf{b} = \{b_{1,m} - 6f/n \quad b_{2,m} - m(m+2)(n-1)/(n+1)\}$$

and

$$\mathbf{V} = \begin{pmatrix} 72f/n^2 & 12m(8m^2 - 13m + 23)/n^2 \\ 12m(8m^2 - 13m + 23)/n^2 & 8m(m+2)/n \end{pmatrix}$$

Due to the asymptotic normality of all of the transformations, all four of the omnibus tests presented here are approximately χ_2^2 under the null hypothesis. Comparison of the four omnibus tests under the null hypothesis showed that for sample sizes of 50, 75 and 100 the 95% critical value of the χ_2^2 distribution gives very nearly the correct α level of each test when $m = 2$.

Bera and John (1983) suggested the test based on the statistic

$$M_3 = nb_{1,m}/6 + n(b_{2,m} - m(m+2))^2/[8m(m+2)]$$

as an omnibus test, but only considered $m = 2$, and did not provide critical values for the test.

Maximum Skewness and Kurtosis

Since $\mathbf{C}'\mathbf{X}$ is normally distributed for any constant vector $\mathbf{C}$ ($\mathbf{C} \neq 0$) when $\mathbf{x}$ is normal, Malkovich and Afifi (1973) defined the distribution of a random vector $\mathbf{X}$ to have multivariate skewness if

$$\beta_1(\mathbf{C}) = \frac{E\{(\mathbf{C}'\mathbf{X} - \mathbf{C}'E(\mathbf{X}))^3\}^2}{\{Var(\mathbf{C}'\mathbf{X}\}^3} > 0$$

for some vector $\mathbf{C}$; without loss of generality, we can assume $\mathbf{C}'\mathbf{C} = 1$. Similarly, multivariate kurtosis was defined as

$$\beta_2(\mathbf{C}) = \frac{E\{(\mathbf{C}'\mathbf{X} - \mathbf{C}'E(\mathbf{X}))^4\}}{\{Var(\mathbf{C}'\mathbf{X}\}^2} \neq 3$$

for some vector $\mathbf{C}$.

As tests for skewness and kurtosis, they suggested using b_1^*, where

$$b_1 = \frac{n[\sum(y_j - \bar{y})^3]^2}{[\sum(y_j - \bar{y})^2]^3}$$

with $y_i = \mathbf{C}'\mathbf{x}_i$. Using Roy's union-intersection principle, the hypothesis of no multivariate skewness is accepted if

$$b_1^* = \max_{\mathbf{C}} b_1(\mathbf{C}) \le K_{b_1}.$$

Similarly, a hypothesis of no multivariate kurtosis is accepted if

$$(b_2^*)^2 = \max_{\mathbf{C}} [b_2(\mathbf{C}) - K]^2 \le K_{b_2}$$

where K and K_{b_2} are appropriate constants and

$$b_2 = \frac{n \sum (y_j - \bar{y})^4}{[\sum (y_j - \bar{y})^2]^2}.$$

Since kurtosis is not symmetrically distributed, K and K_{b_2} should be chosen to weight the minimum and maximum values (over all $\mathbf{C}$) of b_2 evenly so that the probabilities of finding a significant low value or high value of kurtosis when the null hypothesis is true are each $\alpha/2$; as the sample size gets large, K converges to 3. For computational purposes, let $\mathbf{A}$ be the $m \times m$ matrix defined by

$$\mathbf{A} = \sum_{i=1}^{n} (\mathbf{x}_i - \bar{\mathbf{x}})(\mathbf{x}_i - \bar{\mathbf{x}})'$$

and define $\mathbf{A}^*$ by $\mathbf{A}^{*'}\mathbf{A}\mathbf{A}^* = \mathbf{I}$. Then, letting

$$\mathbf{y}_j = \mathbf{A}^{*'}(\mathbf{x}_j - \bar{\mathbf{x}}) \qquad \text{for } j = 1, \ldots, n$$

we can compute

$$b_1^* = \max_{\mathbf{C}'\mathbf{C}=1} n \left[\sum (\mathbf{C}'\mathbf{y}_i)^3 \right]^2$$

and

$$(b_2^*)^2 = \max_{\mathbf{C}'\mathbf{C}=1} n \left[\sum (\mathbf{C}'\mathbf{y}_i)^4 - K \right]^2.$$

Using the Lagrange multiplier method, the set of equations

$$\sum (\mathbf{C}'\mathbf{y}_i)^2 \mathbf{y}_i - \lambda \mathbf{C} = 0$$

can be solved iteratively with the restriction that $\mathbf{C}'\mathbf{C} = 1$, with the result that $b_1^* = \lambda^2$.

For kurtosis, the equations

$$\sum (\mathbf{C}'\mathbf{y}_i)^3 \mathbf{y}_i - \gamma \mathbf{C} = 0$$

can be solved iteratively so that $[\gamma - K]^2 = (b_2^*)^2$. Computational details for the iterative method of calculating b_1^* and $(b_2^*)^2$ are given in Malkovich (1971). Machado (1983) examined the distributional properties of b_1^* and b_2^* using appropriate transformations of the asymptotic distributions.

Tests for Pearson Alternatives

Under certain parametric restrictions, the multivariate form of the Pearson distribution reduces to the multivariate normal. Bera and John (1983) used Rao's (1948) score principle to test those restrictions, in principle developing tests for Pearson alternatives. These tests turn out to be, at least to some degree, related to univariate skewness and kurtosis tests for the scaled residuals, z_i. Using summary measures, they defined

$$T_i = \sum_{k=1}^{n} \frac{z_{ik}^3}{n}$$

$$T_{ii} = \sum_{k=1}^{n} \frac{z_{ik}^4}{n}$$

$$T_{ij} = \sum_{k=1}^{n} \frac{z_{ik}^2 z_{jk}^2}{n}$$

which are asymptotically independent and normal with means 0, 3 and 1, respectively, and variances $6/n$, $24/n$ and $4/n$, respectively. Since the z_i are already scaled to mean 0 and variance 1, T_i is the univariate skewness test and T_{ii} is the univariate kurtosis test (Chapter 3) for the ith component of z. Due to consistency conditions, they recommended first testing the m T_i values, using

$$C_1 = n \sum_{i=1}^{m} \frac{T_i^2}{6}$$

as a test for skewness, which is asymptotically χ_m^2. If C_1 is not significant, then use the test

$$C_2 = n \left(\frac{1}{24} \sum_{i=1}^{m} (T_{ii} - 3)^2 + \frac{1}{4} \sum_{i=1}^{m} \sum_{j=1}^{i-1} (T_{ij} - 1)^2 \right)$$

which is approximately $\chi_{m(m+1)/2}^2$. Omnibus tests for skewness and kurtosis are based on the statistics

$$C_3 = n \left(\frac{1}{6} \sum_{i=1}^{m} T_i^2 + \frac{1}{24} \sum_{i=1}^{m} (T_{ii} - 3)^2 \right)$$

or
$$C_4 = C_1 + C_2$$

which can be tested using χ^2_{2m} and $\chi^2_{m(m+3)/2}$ as the reference distributions, respectively.

Smooth Tests for Skewness and Kurtosis

Koziol (1986, 1987) considered multivariate tests based on the theory of Neyman's smooth tests. He derived tests for skewness and kurtosis based on a Gram-Charlier expansion of a multivariate kernel, with components which are products of univariate Hermite orthogonal polynomials. His smooth test statistic for skewness is algebraically equivalent to Mardia's $b_{1,m}$, i.e.,

$$\hat{U}_3^2 = \frac{n}{6} b_{1,m}$$

while $\hat{U}_4^2$, the smooth test statistic for kurtosis, contains $b_{2,m}$ as one of its components. $\hat{U}_3^2$ and $\hat{U}_4^2$ are asymptotically independent and are distributed as χ^2 variables with $\binom{m+2}{3}$ and $\binom{m+3}{4}$ degrees of freedom, respectively, under the null hypothesis.

The components of $\hat{U}_3^2$ and $\hat{U}_4^2$ can be decomposed into orthogonal components which are each asymptotically χ_1^2, and these components can be examined for detailed information on coordinate-wise departures from normality (Koziol, 1986).

Koziol (1989) also suggested another variant of $b_{2,m}$,

$$\tilde{b}_{2,m} = n^{-2} \sum_{i=1}^{n} \sum_{j=1}^{n} r_{ij}^4$$

which he termed the "next higher degree analog of $b_{1,m}$". He showed that under the assumption of multivariate normality this statistic was asymptotically distributed as a noncentral χ^2 random variable with $\binom{m+3}{4}$ degrees of freedom with noncentrality parameter $nm(m+2)/8$. He also showed that

$$24n\hat{U}_4^2 = n^2 \tilde{b}_{2,m} - 6n^2 b_{2,m} + 3n^2 m(m+2)$$

so that the calculation of $\hat{U}_4^2$ could be accomplished more easily by initially calculating the components $\tilde{b}_{2,m}$ and $b_{2,m}$. Since the distribution of $\hat{U}_4^2$ is easier to deal with than that of $\tilde{b}_{2,m}$, he recommended using $\tilde{b}_{2,m}$ only as a shorter method of calculating the former statistic.

9.4.2 Extensions of the Shapiro-Wilk W

Similar to moment tests, the popularity and good power properties of regression and correlation tests suggested the extension of those methods to the multivariate case.

Maximum W^*

Malkovich (1971; Malkovich and Afifi, 1973) presented a multivariate Wilk-Shapiro criteria based on Roy's union-intersection principle. Since $\mathbf{C}'\mathbf{X}$ is normal if and only if $\mathbf{X}$ is normal for some constant vector $\mathbf{C}$, then multivariate normality is accepted if

$$\min_{\mathbf{C}} W_C \geq K_W$$

where W_C is the univariate Wilk-Shapiro test statistic for the reduced observations given by $z_i = \mathbf{C}'\mathbf{x}$ and K_W is the appropriate constant. The vector $\mathbf{C}$ which gives a lower bound is given by

$$\mathbf{C}'(\mathbf{x_1} - \overline{\mathbf{x}}) = \frac{n-1}{na_1}$$

$$\mathbf{C}'(\mathbf{x_j} - \overline{\mathbf{x}}) = \frac{-1}{na_1}, \qquad j > 1.$$

Since there is no solution to these equations in general, an approximate solution can be found by using

$$\mathbf{C} = a_1^{-1}\mathbf{A}^{-1}(\mathbf{x_1} - \overline{\mathbf{x}})$$

where a_1 is the first coefficient for the Wilk-Shapiro test (Table B1) and, as before, $\mathbf{A}$ is the $m \times m$ matrix given by

$$\mathbf{A} = \sum_{i=1}^{n}(\mathbf{x_i} - \overline{\mathbf{x}})(\mathbf{x_i} - \overline{\mathbf{x}})'.$$

However, since any observation can be designated as $\mathbf{x_1}$, the generalized statistic W^* can be obtained by identifying $\mathbf{x_k}$ as that observation for which

$$(\mathbf{x_k} - \overline{\mathbf{x}})'\mathbf{A}^{-1}(\mathbf{x_k} - \overline{\mathbf{x}}) = \max_{1 \leq i \leq n}(\mathbf{x_i} - \overline{\mathbf{x}})'\mathbf{A}^{-1}(\mathbf{x_i} - \overline{\mathbf{x}}) \qquad (9.11).$$

Then the order statistics $U_{(i)}$ are found where

$$U_i = (\mathbf{x_k} - \overline{\mathbf{x}})' \mathbf{A}^{-1}(\mathbf{x_i} - \overline{\mathbf{x}}) \qquad (9.12)$$

and the univariate Wilk-Shapiro test is applied to the $U_{(i)}$,

$$W^* = \frac{\left(\sum_{i=1}^n a_{n,j} U_{(j)}\right)^2}{(\mathbf{x_k} - \overline{\mathbf{x}})' \mathbf{A}^{-1}(\mathbf{x_k} - \overline{\mathbf{x}})}.$$

The coefficients used to calculate W^* do not depend on m, and W^* reduces to W when $m = 1$; however, critical points of the test do depend on m.

Of the set of n vectors used in (9.11), Fattorini (1986) proposed using that vector $\mathbf{C}_l$ which minimized the Wilk-Shapiro statistic, i.e.,

$$W_F^* = \min_{1 \leq l \leq n} \frac{\left(\sum_{i=1}^n a_{n,j} U_{(j)}^l\right)^2}{(\mathbf{x_l} - \overline{\mathbf{x}})' \mathbf{A}^{-1}(\mathbf{x_l} - \overline{\mathbf{x}})}$$

so that $W_F^* \leq W^*$. For W_F^*, U_i^l is (9.12) after substituting $\mathbf{x_l}$ for $\mathbf{x_k}$.

The Ω Test

For a variable y with a known distribution function F, a transformation to standard normality is given by

$$z = \Phi^{-1}[F(y)].$$

Royston (1983) suggested using either a χ^2 or beta distribution to transform the squared radii to approximate normality by

$$r_i' = \Phi^{-1}[F(r_i^2)].$$

Using the Wilk-Shapiro W and the normality transformation (2.6), Royston proposed the Ω test for m-normality. He further proposed examining all subsets of the m variates of size k, $k = 1, \ldots, m$. Each value of k gives $K = \binom{m}{k}$ non-independent test statistics $\Omega_1, \ldots, \Omega_K$. These tests may be inspected individually or further combined into a single test for each value of k

$$\theta_k = \sum_{i=1}^K [\Phi^{-1}[\tfrac{1}{2}\Phi(-\Omega_i)]]^2$$

where $\theta_k \sim \chi_K^2$.

A Multivariate Probability Plot Correlation Test

Tsai and Koziol (1988) presented a test similar to the Shapiro-Francia (1972) W'. This test is the correlation of the squared radii with the expected order statistics of the χ^2_m distribution. Using the ordered values of the squared radii $r^2_{(1)} \le r^2_{(2)} \le \ldots \le r^2_{(n)}$, they defined the test statistic

$$r_{m;n} = \frac{\sum (r^2_{(i)} - \overline{r^2})(Q_i - \overline{Q})}{\sqrt{\sum (r^2_{(i)} - \overline{r^2})^2 \sum (Q_i - \overline{Q})^2}}$$

where the Q_i are the expected values of the χ^2_m order statistics. For this test, small values of $r_{m;n}$ would indicate deviation from normality of the original observations. Percentage points for $2 \le m \le 9$ and $n = 10(10)60$ were obtained by simulation (Table B30).

9.4.3 χ^2 Tests

The χ^2 test is easily adaptable to any null distribution, including those that are multivariate in nature. As in the univariate case, cells must be defined and the expected and observed number of observations found in each of the cells must be ascertained. The problems associated with the univariate χ^2 test, however, must still be addressed, i.e., cell size and number of cells.

Kowalski (1970) used the χ^2_2 distribution to determine cell sizes in order to perform a χ^2 goodness of fit test. If we define

$$2c^2 = (\mathbf{x} - \mu)\Sigma^{-1}(\mathbf{x} - \mu)$$

as an ellipse of constant probability in a multivariate normal distribution, then the volume of the ellipse is

$$V_c = P(R_i^2 < 2c^2) = 1 - \exp(-c^2)$$

The volume of a ring between the ellipses defined by $2c^2$ and $2(c + dc)^2$ is simply

$$V = \exp(-c^2) - \exp(-[c + dc]^2)$$

and nV observations would be expected to occur within the ring. Comparison of expected with observed number of observations within the cells is then made using the standard χ^2 formula.

Mason and Young (1985) used the beta approximation for the r_i^2

$$W_c = P(r_i^2 < 2c^2) = 1 - \left(1 - \frac{2c^2 n}{(n-1)^2}\right)^{(n-3)/2} .$$

W_c reduces to V_c as n gets large. To obtain rings of equal size, the approximate relationship

$$2c^2 = n(1 - (1 - i/k)^{2/(n-3)})$$

can be used for a specified number, k, of cells. Under multivariate normality, $n(W_{c+dc} - W_c)$ observations will be within the ring defined by c and $c + dc$. Neither Kowalski nor Mason and Young made any suggestions concerning the number or size of cells to use when performing the χ^2 test.

Moore and Stubblebine (1981) generalized the χ^2 test of Kowalski to a general number of m dimensions. They indicated that the use of a χ^2 test based on the quadratic form equal to a constant is the same as determining cells where the mean and variance are considered fixed and have the distribution $N(\bar{\mathbf{x}}_m, \mathbf{S})$. Recommending equal cell (m-dimensional ring) probabilities, they showed that these are identified as the i/m percentile points of the χ^2_m distribution. However, they also showed that the limiting distribution of $\mathbf{X}^2$ when the parameters need to be estimated is

$$\mathbf{X}^2 \sim \chi^2_{m-2} + \lambda \chi^2_1$$

where $0 < \lambda < 1$, so that the asymptotic critical points fall between the χ^2_{m-2} and χ^2_{m-1} distributions. Moore and Stubblebine stated that the $\mathbf{X}^2$ test in the multivariate case is insensitive to skewness but inspection of cell frequencies can easily indicate "peakedness, broad shoulders and heavy tails".

9.4.4 EDF Type Tests

Empirical distribution function (EDF) tests (Chapter 5) were originally based on a transformation to uniformity of a set of observations under the null hypothesis. To this point, there is no EDF procedure which has been extended to test for uniformity in m-space; all of the multivariate EDF tests described below are based on a reduction of the multivariate data to univariate uniformity.

EDF Tests of Squared Residuals

Univariate EDF tests have been proposed which are based on the ordered squared residuals $r^2_{(i)}$ and their χ^2_m approximation. The Cramer-von Mises test (Section 5.1.3) has been suggested by Malkovich (1971;

Malkovich and Afifi, 1973), Koziol (1982) and Paulson, Roohan and Sullo
(1987),

$$J_n = 1/12n + \sum (u_{(i)} - (i - 0.5)/n)^2$$

where $u_{(i)} = F_m(r_{(i)}^2)$, F_m being the χ_m^2 distribution function. Malkovich
(1971; Malkovich and Afifi, 1973) also suggested the Kolmogorov-Smirnov
test (Section 5.1.1). Paulson, Roohan and Sullo (1987) suggested using the
Anderson-Darling test (Section 5.1.4) and gave finite sample 15%, 10%,
5% and 1% percentage points for normal data up to 5 dimensions, and
asymptotic percentage points for the Anderson-Darling and Cramer-von
Mises test for up to 25 dimensions (Table B31).

In the bivariate case, O'Reilly and Medrano (1988) investigated the
power of the Anderson-Darling test to test the fit of the squared residuals
to a χ^2 distribution and the uniformity of the angles, combining the two
tests using the Bonferroni inequality.

Rayleigh's Test for Uniformity of Angles

Koziol (1983) also presented a test of the uniformity of the angles θ_i
using Rayleigh's test. Defining the vector

$$\hat{\mathbf{l}}_i = r_i^{-1} \mathbf{z}_i$$

then Rayleigh's statistic is

$$\hat{\mathbf{R}} = n^{-\frac{1}{2}} \sum_{i=1}^{n} \hat{\mathbf{l}}_i.$$

Under the null hypothesis, $\hat{\mathbf{R}}$ is normal with mean vector $\mathbf{0}$ and covariance
given by $\mathbf{V} = v\mathbf{I}$ with

$$v = m^{-1}[1 - \frac{2}{m}(\frac{\Gamma((m+1)/2)}{\Gamma(m/2)})^2].$$

Koziol then used

$$R_y = \hat{\mathbf{R}}' \mathbf{V}^{-1} \hat{\mathbf{R}}$$

as a test statistic, and compared it to a χ_m^2 distribution to obtain probability levels. Since they are independent, then if $p_1 = P(x > J_n)$ and
$p_2 = P(x > R_y)$ where the probabilities are obtained from the appropriate
distribution for the observed test statistics J_n and R_y, then $-2(\log(p_1) +
\log(p_2)) \sim \chi_4^2$.

An EDF Test Using Jackknifed Estimates

Hawkins (1981) provided a test statistic W_T which would simultaneously determine multivariate normality and heteroscedasticity for $g > 1$ samples. However, since it is not our intention to present tests for comparison of several samples, we will only consider the test based on W_1, i.e., the special case of $g = 1$.

Hawkins used the univariate Anderson-Darling test on data reduced to uniformity based on the distribution of the n observations using independent estimates of the mean and covariance matrix using the remaining $n - 1$ observations. The quadratic form of (9.4), using $\bar{x}_i$ and S_i, the estimated mean and covariance matrix of x excluding the observation x_i, follows a Hotelling's T^2 distribution. Rather than calculate n means and inverting n covariance matrices, some algebra shows that the T^2 values can be calculated as

$$F_i = \frac{n(n - m - 1)r_i^2}{m((n - 1)^2 - nr_i^2)}$$

which follows an F distribution with m and $n - m - 1$ degrees of freedom. Defining

$$A_i = P(F > F_i),$$

the A_i are uniformly distributed under the null hypothesis, although they are not independent. Using the Anderson-Darling formula,

$$W_1 = n - n^{-1} \sum_{i=1}^{n} (2i - 1)[\log(A_{(i)}) + \log(1 - A_{(n-i+1)})]$$

Hawkins suggested using the asymptotic theory of the Anderson-Darling test to determine critical values even though the A_i are not independent, based on consistency of simulation results. However, since Hawkins' results were based on several samples, and he also suggested caution when using W_T (the multi-sample version of W_1), care should be used in the interpretation of this statistic.

Foutz' F_n

Booker, Johnson and Beckman (1984) investigated a goodness of fit procedure for multivariate normality based upon an empirical probability measure when parameters are unknown. This test, denoted F_n, had been proposed by Foutz (1980) for both univariate and multivariate goodness of fit for samples of size $n - 1$, although he only considered the simple hypothesis case. This method involves defining $n-1$ "cutting functions" to construct n statistically equivalent blocks, i.e., blocks $B_i, i = 1, \ldots, n$, such that the probabilities of observations occurring in blocks are distributed as the n spacings observed from a random sample of size $n-1$ from $U(0,1)$. In the univariate case, the order statistics of the observations can be used to determine the blocks. In the multivariate case, Booker, Johnson and Beckman (1984) suggested using the ordered r_i^2 to identify the blocks, with the probability measure coming from the beta distribution given by (9.5).

The test statistic is calculated as

$$F_n = \sum_{i=1}^{n} \max[0, 1/n - D_i]$$

where $D_i = P(X \in B_i | F)$ and F is the hypothesized null distribution (Franke and Jayachandran, 1984). Booker, Johnson and Beckman (1984) found that even for small sample sizes F_n was reasonably approximated by its asymptotic distribution, which is normal with mean e^{-1} and variance $n^{-1}(2e^{-1} - 5e^{-2})$; further, the distribution is independent of m.

EDF Tests of the Cumulative Distribution Function

Ward (1988a) proposed two tests using an estimate of the cumulative distribution function of $\mathbf{x}$,

$$y_i = \hat{F}(\mathbf{x_i}) = \prod_{j=1}^{m} \Phi(z_{ij})$$

He then used the (univariate) Kolmogorov-Smirnov D (Section 5.1.1) and Anderson-Darling A^2 tests (Section 5.1.4) to test the goodness of fit of the y_i to the distribution

$$g(y) = \frac{-\log^{m-1}(y)}{\Gamma(m)} \qquad 0 < y < 1.$$

Note that by using the y_i, the observations have an inherently different ordering than that obtained from using the r_i^2. This is seen in that the r_i^2 are minimized at the point $\mathbf{x} = \bar{\mathbf{x}}$, while y_i tends towards its minimum value of 0 as each of the m variates goes away from $\bar{\mathbf{x}}$, e.g., as each variate goes towards $-\infty$ if the variates are independent or positively correlated. It would be interesting to compare the strengths and weaknesses of EDF tests obtained using these two (as well as any other) orderings.

9.4.5 Projections and Transformations

In addition to the reduction of multivariate data to squared radii, additional tests based on data transformation and projection have been proposed. These transformations are generally more computationally involved than those needed to obtain the r_i^2, and examination of the power properties of the tests described here have not been performed.

Directional Tests of Normality

Andrews, Gnanadesikan and Warner (1971, 1973) projected the data along directions that are determined to some degree by the data itself and which are chosen to be sensitive to particular types of non-normality. They proposed that, since non-normality in the data may result in non-normal clustering of points, the vector

$$\mathbf{d}_\alpha = \frac{\sum_{i=1}^n w_i \mathbf{z}_i}{\|\sum_{i=1}^n w_i \mathbf{z}_i\|}$$

may be used to point to these clusters, where $w_i = \|\mathbf{z}_i\|^\alpha$ and α is a constant to be chosen which will determine the region of sensitivity. In particular, if $\alpha < 0$ then $\mathbf{d}_\alpha$ points in the direction of non-normal clusters near the mean, while for $\alpha > 0$ then $\mathbf{d}_\alpha$ points to clusters far from the mean.

As a test for normality, they suggested that the observations be projected onto the direction identified by $\mathbf{d}_\alpha^* = \mathbf{S}^{1/2} \mathbf{d}_\alpha$. The lengths of the projections of the original observations onto $\mathbf{d}_\alpha^*$ will, under the null hypothesis, form a univariate normal sample which can be tested using any univariate test for normality. They did not recommend using the values of $\mathbf{d}_\alpha$ as formal tests for multivariate normality.

Rosenblatt's Multivariate Transformation

Rosenblatt (1952) developed a transformation of an absolutely continuous m-variate distribution with known parameters into the uniform distribution on the m-dimensional hypercube. If $F(\mathbf{x} = \{x_1, \ldots, x_m\})$ is continuous, then the transformation is given by

$$z_1 = P(X_1 \leq x_1) = F_1(x_1)$$

$$z_2 = P(X_2 \leq x_2 | X_1 = x_1) = F_2(x_2 | x_1)$$

up to

$$z_m = F_m(x_m | x_1, \ldots, x_{m-1}).$$

If the distribution is normal, the transformation assumes a particularly simple form (see Rosenblatt, 1952). In the bivariate normal case the transformation is

$$F_1(x) = \Phi\left(\frac{x_1 - m_1}{\sigma_1}\right)$$

$$F_2(x_2 | x_1) = \Phi\left(\frac{x_2 - m_2 + \rho\sigma_1/\sigma_2(x_1 - m_1)}{\sigma_2\sqrt{1 - \rho^2}}\right).$$

Similar to conditional probability integral transformations, which are dependent upon the order of the variables (Chapter 5), this transformation is dependent upon the order of the components within the observation vectors.

The uniform hypercube can be divided into k^m cells, each with equal probability in order to perform a χ^2 test (Rosenblatt, 1952; Kowalski, 1970). The effects of estimated parameters on the distribution of the z_i and on the proposed χ^2 test are unknown.

O'Reilly and Medrano (1988) suggested using a generalization of the Rao-Blackwell estimate of the normal distribution function (Moore, 1973) which is applied to the multivariate case using Rosenblatt's transformation. The Rao-Blackwell estimator of the multivariate normal distribution is an inverted t distribution (Ghurye and Olkin, 1969)

$$INVt^{(m)}(\mathbf{S}, \frac{n-1}{n}, n - k - 1, \bar{\mathbf{x}})$$

which is used to transform each observation vector $\mathbf{x_i}$ into a uniform vector $\mathbf{u_i}$ using Rosenblatt's transformation. The components of the $\mathbf{u_i}$ are m independent $U(0, 1)$ variables. Computer programs for Rosenblatt's transformation for $m = 2, 3$ are available from O'Reilly and Medrano (language unspecified). They used an Anderson-Darling test, $\tilde{A}^2$, on the nm components from the transformation.

Durbin's Multivariate Sample Transformation

Wagle (1968) extended Durbin's (1961) method for transforming a univariate normal sample with unknown parameters to a normal sample with mean 0 and variance unity. If $\mathbf{X}$ denotes the $m \times n$ matrix of observations, then calculate

$$\mathbf{Z} = \mathbf{T}^{-1}\mathbf{X}$$

where $\mathbf{V} = \mathbf{TT}'$ is the sample dispersion matrix, i.e., $\mathbf{V} = (n-1)\mathbf{S}$.

If $\hat{\mathbf{X}}$ and $\mathbf{A}$ are observations of random matrices which are independent of $\mathbf{X}$ and distributed as the sample mean and covariance matrix from a sample of n observations from an m-variate normal distribution with mean vector $\mathbf{0}$ and covariance matrix $\mathbf{I}$, then

$$\tilde{\mathbf{X}} = \hat{\mathbf{X}} + \frac{n-1}{n}\mathbf{A}^{1/2}\mathbf{Z}$$

has a multivariate normal distribution, $\tilde{\mathbf{X}} \sim N(\mathbf{0}_m, \mathbf{I}_m)$. Wagle recommended that, since each of the m variates is identically distributed normally and has known parameters, the nm observations can be considered as being from a univariate $N(0,1)$ population, and any univariate test of normality or goodness of fit based on a simple hypothesis can be used.

As with many of the transformation procedures, if the same test were performed twice on the same set of observations then different answers may result. For the other transformations the sequence of observations determines the value of the test statistic, while for this transformation it is due to the random process of selecting a mean vector and covariance matrix.

Nearest Distance Test

Andrews, Bickel, Hampel, Huber, Rogers and Tukey (1972) proposed the nearest distance test for ascertaining joint normality. The initial step consists of transforming the data to the unit hypercube. This can be accomplished by using the standardized residuals $\mathbf{z}_i$ and calculating the vector $\mathbf{y}_i$ where the entries are defined by $y_{ij} = \Phi(z_{ij})$, $j = 1, \ldots, m$. After calculating the distances

$$d(i, i') = \max_k[\min\{|y_{ki} - y_{ki'}|, ||y_{ki} - y_{ki'}| - 1|\}]$$

the nearest distance is found,

$$d_{min} = \min_{i' \neq i} d(i, i').$$

These distances are further transformed to standard normal deviates: for each $\mathbf{y}_i$ let

$$w_i = \Phi\left(\frac{1 - \exp[-n(2d_{min}(i))^m]}{1 - \exp(-1)}\right)$$

iff $d_{min}(i) < 1/2n^{1/m}$ and $d(i, i') > 1/2n^{1/m}$, $i' < i$. Under the null hypothesis, the transformed distances are independent of the coordinates from which they are measured; this independence can be measured using multiple regression. For all of the $n' \leq n$ points that follow these two conditions, fit the regression model

$$w_i = \beta_0 + \sum_{j=1}^{m} \beta_j x_{i,j} + \sum_{j=1}^{m} \sum_{k=1}^{m} \beta_{jk} x_{ij} x_{ik}.$$

The regression sum of squares should be compared with a χ^2 distribution with $(m+1)(m+2)/2$ degrees of freedom, rejecting normality for large values of the sum of squares.

9.4.6 Other Methods of Assessing Multivariate Normality

The three tests described below evaluate the multivariate nature of data directly, but do not fall into any of the previous categories.

Comparison of a Nonparametric Density Estimate

Loh (1986) proposed using the test statistic

$$T_n = \sum_{i=1}^{n} \log[\hat{g}(\mathbf{x}_i)/f(\mathbf{x}_i, \hat{\theta})]$$

where $\hat{\theta}$ is the vector of maximum likelihood estimates of the normal parameters, f is the multivariate normal distribution, and $\hat{g}$ is a nonparametric estimate of the density of $\mathbf{x}$. T_n should be small for a normal null hypothesis, and large otherwise.

For the bivariate case, Loh used a product normal kernel to estimate $\hat{g}(\mathbf{x})$ with common individual bandwidth $h_j = s_j n^{-1/6}$, where s_j is the sample standard deviation of the jth vector component. The proposed test statistic is

$$T_n = \sum_{i=1}^{n} \log[(nh_1 h_2)^{-1} \sum_{j=1}^{n} \prod_{k=1}^{2} \phi[(x_{ik} - x_{jk})/h_k]]$$

$$+ n \log(2\pi s_1 s_2 \sqrt{1 - R^2}) + n$$

where R is the sample correlation coefficient. The distribution of T_n under the null hypothesis depends only on the true value of the correlation coefficient, ρ.

Empirical Characteristic Function Tests

Henze and Zirkler (1990) presented a class of invariant consistent tests for composite multivariate normality, based on the weighted integral of the difference between the empirical characteristic function (ECF) and its pointwise limit. The test statistics are calculated as

$$T_\beta = n(4I_{(\mathbf{S}\ singular)} + D_{n,\beta}I_{(\mathbf{S}\ nonsingular)})$$

where $\mathbf{S}$ is the sample covariance matrix, I is the indicator function and

$$D_{n,\beta} = \int_{R^d} \left| \psi_n(\mathbf{t}) - \exp(-\frac{1}{2}||\mathbf{t}||^2) \right|^2 \phi_\beta(\mathbf{t})dt$$

with ψ_n being the ECF and ϕ_β being a weighting function with parameter β,

$$\phi_\beta(\mathbf{t}) = (2\pi\beta^2)^{-m/2} \exp\left(-\frac{||\mathbf{t}||^2}{2\beta^2}\right), \quad \mathbf{t} \sim \mathbf{R}^m \qquad (9.11).$$

Since $D_{n,\beta}$ is undefined when $\mathbf{S}$ is singular, T_β is set to its maximum value of 4, causing rejection of the null hypothesis. Small values of β for this weighting function would be sensitive to heavy-tailed alternatives.

Using the weighting function (9.11) allows a relatively straightforward equation for calculating $D_{n,\beta}$,

$$D_{n,\beta} = n^{-2} \sum_{j,k=1}^{n} \exp\left(-\frac{\beta^2}{2}||\mathbf{z}_j - \mathbf{z}_k||^2\right) + (1 + 2\beta^2)^{-m/2}$$

$$-2(1+\beta^2)^{-m/2}n^{-1} \sum_{j=1}^{n} \exp\left(-\frac{\beta^2}{2(1+\beta^2)}r_j^2\right)$$

where

$$||\mathbf{z}_j - \mathbf{z}_k||^2 = (\mathbf{x}_j - \mathbf{x}_k)'\mathbf{S}^{-1}(\mathbf{x}_j - \mathbf{x}_k).$$

T_β is used as a one-tailed test which rejects the null hypothesis when the test value is too large. They indicated that, for the choice of weighting function, a good choice of β would depend on n and m by

$$\beta = \frac{1}{\sqrt{2}} \left(\frac{n(2m+1)}{4}\right)^{1/(m+4)} \qquad (9.12)$$

although T_1 and $T_{0.5}$ performed well relative to this value of β for their limited power study (see Section 9.5) regardless of the values of m and n.

9.5 Power of Tests for Multivariate Normality

For tests for univariate normality, the introduction of a new test almost invariably includes a power comparison with other tests. Compared to the univariate case, however, in the multivariate case realistic alternative distributions are not so well defined and tests are much more computationally complex. Therefore, power comparisons have been limited in both scope and in size. Because of the diversity of alternatives, which are affected by the number of components, whether the components are correlated and the magnitude of the correlation, and whether the components have the same or different distributions among themselves, it is very difficult to compare power of tests obtained from different studies.

Most of the simulations for power of multivariate tests for normality include bivariate alternatives, probably because test calculations are faster, variate generation is easier, and bivariate data are more common than larger dimensions. Indeed, some of the power studies were restricted to the bivariate case.

Malkovich (1971) compared b_1^*, b_2^*, W^* and the Kolmogorov-Smirnov and Cramer-von Mises tests as well as univariate tests of normality using the squared residuals for sample sizes of 10, 25 and 50 and dimensions of 2, 3 and 5. Alternatives investigated included independent and correlated lognormal, uniform and t distributions and scale and location normal mixtures. His general findings were that b_1^* and b_2^* generally had the highest power. As with the univariate case, W^* tended to more powerful relative to other tests when b_1^* had high power. The EDF tests were better relative to other tests at detecting symmetric non-normal distributions. The univariate tests using squared residuals had poor power compared to other tests.

Mardia and Foster (1983) presented results for Mardia's skewness and kurtosis statistics as well as the omnibus tests they proposed; while they referred to a larger scale study, they only presented power for two alternatives and $m = 2$. For a bivariate distribution with independent symmetric margins (i.e., a Tukey distribution with $\lambda = 0.7$), $b_{2,2}$ clearly was the most powerful test. For independent Johnson S_b components, C_N^2 and C_W^2 were slightly better than the other tests.

For a variety of sample sizes (20, 35, 50, 100 and 200), Bera and John (1983) compared C_1, C_2, C_3, C_4, $b_{1,m}$, $b_{2,m}$ and M_3. A variety of bivariate distributions with independent and correlated components were used in the simulation. In general C_1 was more powerful than $b_{1,m}$, C_2 was more powerful than $b_{2,m}$ and C_4 was more powerful than M_3; a more general statement is difficult to make since some tests were better than others over the set of alternatives studied and the omnibus test was not always more powerful than the individual moment tests.

Booker, Johnson and Beckman (1984) compared W_T, F_n and the multivariate Kolmogorov-Smirnov test (Malkovich and Afifi, 1973) for samples of size 50 and a variety of alternatives (Pearson Type II and VII, distributions based on Khintchine's theorem, Johnson translation system, Plackett, and Morgenstern distributions). Their results indicated that F_n had very poor power compared to the other tests, and the Kolmogorov-Smirnov test was usually noticeably better than W_T.

Loh (1986) compared T_n with $b_{1,m}$, $b_{2,m}$ and Q for some correlated and independent bivariate distributions using only a sample size of 30. For the distributions with uncorrelated components (χ^2, double exponential and normal-lognormal), T_n was better than the other tests; for these same distributions with moderate correlations between components ($\rho = 0.7$), $b_{1,m}$ was best, even for the symmetric double exponential. For quadratic normal and double exponential distributions, which are correlated by definition, Q was generally the best test, followed by $b_{1,m}$.

For samples of size 20 and 50 and dimensions 2 and 5, Paulson, Roohan and Sullo (1987) showed essentially no differences between the Anderson-Darling and Cramer-von Mises tests based on the estimated χ^2 distribution of the squared residuals. Comparisons were made for the lognormal and various χ^2, t and Dirichlet distributions. Results not published in this comparison were those for the χ^2 test (Moore and Stubblebine, 1981), which they excluded because of relatively poor power performance of the χ^2 test.

Ward (1988b) included 3 sample sizes (25, 50 and 100) and dimensions of 2(1)10 in his power simulation study. Alternatives considered were χ^2 with independent and correlated components, a scale contaminated normal mixture (i.e., equal means but different covariance matrices), and a symmetric location contaminated normal mixture. For his published results, α was set at 0.20. Seven tests were compared: Mardia's skewness and kurtosis tests as well as S_W^2, W_T, W^* and Ward's version of D and A^2. For the χ^2 alternatives, $b_{1,m}$ and A^2 had the highest power while for the bivariate location mixture A^2 had the highest power. W_T had appreciably greater power for the scale mixture compared to the other tests. S_W^2 did relatively well against all alternatives. It is notable that as the dimension of the distribution increased, the power of W^* decreased relative to the other tests. In fact, the power of W^* decreased as the dimension increased for every alternative while for all of the other tests the power increased or remained constant with increasing dimension for the χ^2 distributions and the scale normal mixture. For the location mixture, none of the tests' power increased with the dimension.

For samples of size 20 and 50, Henze and Zirkler (1990) compared

W^*, W_F^*, $b_{1,m}$, and $b_{2,m}$ with four parameterizations of their proposed test T_β: $\beta = 0.5$, 1, 3 and the "optimal" for each sample size (9.12). They used a wide variety of alternative distributions, including exponential, t, lognormal, gamma, χ^2, beta and logistic with independent components; a normal and independent non-normal component; normal mixtures; spherically symmetric Pearson Type II and Type VII distributions; and other spherically symmetric distributions of various types. Generally, the power of $T_{0.5}$ mimicked, and was often a couple of percentage points better than, $b_{1,m}$. For samples of size 20, W^* and W_F^* were similar, while for $n = 50$ there were some instances where W_F^* had a very large advantage over W^*.

For the skewed independent component distributions, the best tests were $T_{0.5}$ and $b_{1,m}$, while for the symmetric case the best tests were $T_{0.5}$ and $b_{2,m}$. For those distributions with one non-normal component, $T_{0.5}$ and W_F^* were best. For normal mixtures with independent components, W_F^* and T_3 usually had the highest power, while for correlated ($\rho = 0.9$) components T_1 was often the best test. For the spherically symmetric Pearson distributions, $b_{2,m}$ usually had the highest power while for other spherically symmetric distributions T_3 often had appreciably more power than all other tests.

Henze and Zirkler (1990) also compared the power of the tests for distributions with $m = 5$ (excluding normal mixtures). Most of the conclusions concerning power in the bivariate case continued to hold in the 5 dimensional case. As in the bivariate case, the power of $T_{0.5}$ was usually similar to that of $b_{1,m}$. For samples of size 20, W^* and W_F^* were similar, while for $n = 50$ there were some instances where W_F^* had a very large advantage over W^*.

For the skewed independent component distributions, the best tests were $T_{0.5}$ and $b_{1,m}$. For symmetric components with heavy tails, $b_{1,m}$ and $T_{0.5}$ had slightly higher power than $b_{2,m}$, W_F^* and W^*, while for components with light tails $b_{2,m}$ was clearly superior. For those distributions with one non-normal component, $T_{0.5}$ and W_F^* were best. For the spherically symmetric Pearson distributions, $b_{2,m}$ usually had the highest power while for other spherically symmetric distributions $b_{2,m}$ always had at or near the highest power over all tests for all alternatives.

For uncorrelated bivariate distributions where the components have the same distribution, the results of O'Reilly and Medrano (1988) showed that C_N^2 was generally the most powerful test. C_N^2 was also usually most powerful for distributions with one normal and one non-normal component. When correlation was introduced into the distributions, their proposed test based on the combined EDF test of the squared residuals and the vector angles was noticeably more powerful than all other tests. Other tests which were included in their study were S_N^2, S_W^2, C_W^2, Q, W^*, and an omnibus test using Mardia's skewness and kurtosis tests, with critical values obtained using Bonferroni's inequality. All of their power simulations were for samples of size 40.

Hasofer and Stein (1990) and Tsai and Koziol (1988) did some limited power simulations for bivariate distributions, but did not compare their proposed tests with any other tests.

9.6 Recommendations

Although we do not recommend a protocol for detecting multivariate normality as extensive as that suggested by Looney (1986), we do recommend a comprehensive inspection of the data when multivariate normality is an issue. The following steps and suggestions should be considered a minimal procedure for assessing multivariate normality:

(1) Univariate probability plots and bivariate plots of the marginals taken two at a time should be inspected.

(2) Univariate tests of normality should be performed on each of the marginal distributions; critical values can be determined using the Bonferroni rule.

(3) One or more of the tests based on dimension reduction (e.g., a test of the squared radii) should be used to assess multivariate normality.

(4) One or more of the multivariate tests should be used to directly assess multivariate normality; it is suggested that more than one test be used, especially if the characteristics of non-normality are not known *a priori*.

(5) For all plots and tests using the squared radii, where feasible the beta distribution should be used as the reference distribution rather than the χ^2.

Regardless of the increase in computing power over the last decade, many of the testing procedures are difficult to program, and are not included in standard statistical packages; also, percentage points are not always available to use for critical values. Therefore, only a limited number of the tests described in this chapter are practical for use. In addition, the lack of definitive power studies make it difficult to evaluate the relative usefulness of the tests presented herein. However, some suggestions for the better multivariate tests, based on the compilation of the power studies and the ease of use of the tests, can be made:

- The Bera and John (1983) tests and Mardia's skewness and kurtosis tests seems to have relatively high power against a variety of alternatives; for unspecified alternative types, one of the omnibus forms of these tests should be used.

- The T_β tests (Henze and Zirkler, 1990), and in particular $T_{0.5}$, seem to have good power properties over a wide variety of alternatives.
- Of the multivariate Wilk-Shapiro type tests, W^* (Malkovich and Afifi, 1973) is not recommended because of its poor power performance; W_F^* (Fattorini, 1986) may be useful but needs to be investigated further.
- F_n (Foutz, 1980) does not have good power properties, and should not be used.
- If specific characteristics of alternative distributions are of concern, an appropriate test(s) should be chosen which seems to have high power for those characteristics.

References

Andrews, D.F., Bickel, P.J., Hampel, F.R., Huber, P.J., Rogers, W.H., and Tukey, J.W. (1972). **Robust Estimates of Location: Survey and Advances**. Princeton University Press, New Jersey.

Andrews, D.F., Gnanadesikan, R., and Warner, J.L. (1971). Transformations of multivariate data. Biometrics 27, 825-840.

Andrews, D.F., Gnanadesikan, R., and Warner, J.L. (1973). Methods for assessing multivariate normality. In P.R. Krishnaiah, ed., **Multivariate Analysis**, 3rd ed., Academic Press, New York.

Arnold, H.J. (1964). Permutation support for multivariate techniques. Biometrika 51, 65-70.

Barnett, V. (1976). The ordering of multivariate data (with discussion). Journal of the Royal Statistical Society A 139, 318-354.

Bera, A., and John, S. (1983). Tests for multivariate normality with Pearson alternatives. Communications in Statistics - Theory and Methods 12, 103-117.

Blom, G. (1958). **Statistical Estimates and Transformed Beta Variables**. John Wiley and Sons, New York.

Booker, J.M., Johnson, M.E., and Beckman, R.J. (1984). Investigation of an empirical probability measure based test for multivariate normality. LA-UR 84-1499, Los Alamos National Laboratory, New Mexico.

Bowman, K.O., and Shenton, L.R. (1975). Omnibus test contours for departures from normality based on $\sqrt{b_1}$ and b_2. Biometrika 62, 243-250.

Box, G.E.P., and Andersen, S.L. (1955). Permutation theory in the derivation of robust criteria and the study of departures from assumptions. Journal of the Royal Statistics Society B 17, 1-34.

Carpaneto, G., and Toth, P. (1980). Solution to the assignment problem (Algorithm 548). Transactions on Mathematical Software 6, 104-111.

Cox, D.R., and Small, N.J.H. (1978). Testing multivariate normality. Biometrika 65, 263-272.

D'Agostino, R.B. (1986). Tests for the normal distribution. In D'Agostino, R.B., and Stephens, M.A., eds., **Goodness of Fit Techniques**, Marcel Dekker, New York.

Durbin, J. (1961). Some methods in constructing exact tests. Biometrika 48, 41-55.

Easton, G.S., and McCulloch, R.E. (1990). A multivariate generalization of quantile-quantile plots. Journal of the American Statistical Association 85, 376-386.

Fattorini, L. (1986). Remarks on the use of the Shapiro-Wilk statistic for testing multivariate normality. Statistica 46, 209-217.

Fisher, R.A. (1936). The use of multiple measurements in taxonomic problems. Annals of Eugenics 7, 179-188.

Foutz, R.V. (1980). A test for goodness of fit based on an empirical probability measure. Annals of Statistics 8, 989-1001.

Franke, R., and Jayachandran, T. (1984). Tables for a new multivariate goodness of fit test. Journal of Statistical Computation and Simulation 20, 101-114.

Ghurye, S.G., and Olkin, I. (1969). Unbiased estimation of some multivariate probability densities and related functions. Annals of Mathematical Statistics 40, 1261-1271.

Gnanadesikan, R. (1977). **Methods for Statistical Data Analysis of Multivariate Observations**. John Wiley and Sons, New York.

Gnanadesikan, R., and Kettenring, J.R. (1972). Robust estimates, residuals, and outlier detection with multiresponse data. Biometrics 28, 81-124.

Hasofer, A.M., and Stein, G.Z. (1990). Testing for multivariate normality after coordinate transformation. Communications in Statistics - Theory and Methods 19, 1403-1418.

Hawkins, D.M. (1981). A new test for multivariate normality and homoscedasticity. Technometrics 23, 105-110.

Healy, M.J.R. (1968). Multivariate normal plotting. Applied Statistics 17, 157-161.

Hensler, G.L., Mehrotra, K.G., and Michalek, J.E. (1977). A goodness of fit test for multivariate normality. Communications in Statistics - Theory and Methods 6, 33-41.

Henze, N., and Zirkler, B. (1990). A class of invariant consistent tests for multivariate normality. Communications in Statistics - Theory and Methods 19, 3595-3617.

Kendall, M., and Stuart, A. (1977). **The Advanced Theory of Statistics**, Vol. 1, 4th ed., MacMillan, New York.

Kowalski, C.J. (1970). The performance of some rough tests for bivariate normality before and after coordinate transformations to normality. Technometrics 12, 517-544.

Koziol, J.A. (1982). A class of invariant procedures for assessing multivariate normality. Biometrika 69, 423-427.

Koziol, J.A. (1983). On assessing multivariate normality. Journal of the Royal Statistical Society B 45, 358-361.

Koziol, J.A. (1986). Assessing multivariate normality: a compendium. Communications in Statistics - Theory and Methods 15, 2763-2783.

Koziol, J.A. (1987). An alternative formulation of Neyman's smooth goodness of fit tests under composite alternatives. Metrika 34, 17-24.

Koziol, J.A. (1989). A note on measures of multivariate kurtosis. Biometrical Journal 31, 619-624.

Kronmal, R., and Tarter, M.E. (1968). The estimation of probability densities and cumulatives by Fourier series methods. Journal of the American Statistical Association 63, 925-952.

Loh, W.-Y. (1986). Testing multivariate normality by simulation. Journal of Statistical Computation and Simulation 26, 243-252.

Looney, S.W. (1986). A review of techniques for assessing multivariate normality. Working paper 860105, College of Business Administration, Louisiana State University.

Machado, S.G. (1983). Two statistics for testing multivariate normality. Biometrika 70, 713-718.

Malkovich, J.F. (1971). Tests for multivariate normality. Ph.D. dissertation, University Microfilms, Ann Arbor, Michigan.

Malkovich, J.F., and Afifi, A.A. (1973). On tests for multivariate normality. Journal of the American Statistical Association 68, 176-179.

Mardia, K.V. (1970). Measures of multivariate skewness and kurtosis with applications. Biometrika 57, 519-530.

Mardia, K.V. (1974). Applications of some measures of multivariate skewness and kurtosis in testing normality and robustness studies. Sankhya 36, 115-128.

Mardia, K.V., and Foster, K. (1983). Omnibus tests of multinormality based on skewness and kurtosis. Communications in Statistics - Theory and Methods 12, 207-221.

Mardia, K.V., and Zemroch, P.J. (1975). Algorithm AS 84. Measures of multivariate skewness and kurtosis. Applied Statistics 24, 262-265.

Mason, R.L., and Young, J.C. (1985). Re-examining two tests for bivariate normality. Communications in Statistics - Theory and Methods 14, 1531-1546.

Moore, D.S. (1973). A note on Srinivasan's goodness-of-fit test. Biometrika 60, 209-211.

Moore, D.S., and Stubblebine, J.B. (1981). Chi-square tests for multivariate normality with application to common stock prices. Communications in Statistics - Theory and Methods 10, 713-738.

O'Reilly, F.J., and Medrano, L.G. (1988). On a test for multivariate normality. Unpublished manuscript, presented at the 1988 Joint Statistical Meetings, New Orleans, LA.

Paulson, A.S., Roohan, P., and Sullo, P. (1987). Some empirical distribution function tests for multivariate normality. Journal of Statistical Computation and Simulation 28, 15-30.

Rao, C.R. (1948). Large sample tests of statistical hypotheses concerning several parameters with applications to problems of estimation. Proceedings of the Cambridge Philosophical Society 44, 50-55.

Rosenblatt, M. (1952). Remarks on a multivariate transformation. Annals of Mathematical Statistics 23, 470-472.

Royston, J.P. (1982). An extension of Shapiro and Wilk's W test for normality to large samples. Applied Statistics 31, 115-124.

Royston, J.P. (1983). Some techniques for assessing multivariate normality based on the Shapiro-Wilk W. Applied Statistics 32, 121-133.

Shapiro, S.S., and Francia, R.S. (1972). Approximate analysis of variance test for normality. Journal of the American Statistical Association 67, 215- 216.

Small, N.J.H. (1978). Plotting squared radii. Biometrika 65, 657-658.

Small, N.J.H. (1980). Marginal skewness and kurtosis in testing multivariate normality. Applied Statistics 29, 85-87.

Small, N.J.H. (1985). Multivariate normality, testing for. In S. Kotz and N.L. Johnson, eds., *Encyclopedia of Statistical Sciences*, Vol. 6, John Wiley and Sons, New York.

Tsai, K.-T., and Koziol, J.A. (1988). A correlation procedure for assessing multivariate normality. Communications in Statistics - Simulation and Computation 17, 637-651.

Wagle, B. (1968). Multivariate beta distribution and a test for multivariate normality. Journal of the Royal Statistical Society B 30, 511-516.

Ward, P.J. (1988a). Goodness of fit tests for multivariate normality. Ph.D. dissertation, University of Alabama, Tuscaloosa AL.

Ward, P.J. (1988b). A comparison of powers of some tests for multivariate normality. Proceedings of the Statistical Computing Section, American Statistical Association, Joint National Meetings, August 1988, New Orleans, LA.

CHAPTER 10

TESTING FOR MULTIVARIATE OUTLIERS

"Despite the apparent complexity of the [multivariate outlier] problem, one can still characterize outliers by the fact that they are somewhat isolated from the main cloud of points. They may not "stick out on the end" of the distribution as univariate outliers must, but they must "stick out" somewhere."

Rohlf, 1975

The need for detecting outliers in a multivariate sample is similar to that for detecting outliers in univariate samples (Chapter 6), but the ability to detect outliers becomes more difficult and complex with increasing dimension. In a univariate sample the definition of an outlier is obvious: an outlier is an observation which is separated from the remainder of the data. A plot of the data, whether a box plot, stem-and-leaf plot, histogram or probability plot, will show potential outliers in one tail and/or the other. For multivariate data, outliers can be more difficult to identify, because of the number of ways in which they can manifest. Multivariate outliers can increase correlations among variables, or decrease correlations; they can inflate variances, similar to univariate outliers; they can be due to a large error in one component, or small errors in a number of the components. For these reasons, as Gnanadesikan and Kettenring (1972) stated, "it would be fruitless to search for a truly omnibus outlier protection procedure".

In this chapter we first suggest preliminary methods for detecting outliers. We then describe more formal methods of outlier detection based
on distances. Methods related to robust procedures are then presented;
robust estimation will be discussed in more detail in Chapter 12. Other
formal and informal methods for detecting outliers will then be described,
followed by recommendations for detecting multivariate outliers based on
the methods described herein.

As in the previous chapter, vectors and matrices will be designated in
bold type; vectors will be indicated by lower case letters, and matrices will
be indicated by uppercase. Sample size and dimensionality will be denoted
by n and m, respectively. Whenever possible, notation will be consistent
with the references cited, however, some changes were necessary to avoid
confusion.

10.1 Preliminary Assessments

As with any other type of analysis, examination of the data should begin
with simple graphical, exploratory and inferential methods before complex
analyses are undertaken.

10.1.1 Marginal Assessment

As an initial step in outlier detection, components should be evaluated
singly and in pairs. Plots of the marginal components of the sample should
precede any more complex analysis. These should include univariate plots
(e.g., histograms, probability plots or stem-and-leaf plots) of the individual
components as well as bivariate scatterplots of all pairs of components, and
three dimensional scatterplots of all triplets of components. Univariate
tests for outliers by component can also be conducted using any of the
tests described in Chapter 6.

10.1.2 Multivariate Normality Goodness of Fit

The methods for assessing multivariate normality described in Chapter 9
can also be used to identify outliers in multivariate samples. However, those
tests will generally signify that a sample is not normally distributed, and
the type of non-normality will not be indicated. The methods described
in this chapter have the ability to identify the outlier(s), if that were indeed the cause of the non-normality. Probability plots of the squared radii
(Section 9.2) can be used as an informal method for detecting outliers.

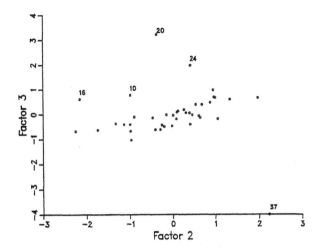

Figure 10.1 Scatterplot of principal component variables 2 and 3 for cerebrospinal fluid gas measurements on acidosis patients (Data Set 14), n = 40. Labeled observations are identified as potential outliers.

10.1.3 Principal Components

Principal component analysis is a commonly used multivariate procedure that is available in virtually all statistical packages, and so the details will not be provided here. In summary, if Z is the matrix of the projection of the data matrix X onto the principal components, then the principal component variables of Z can be used to detect outliers in place of the original variables. In particular, for the case of multivariate normality, since Z is a linear transformation of X, if X is multivariate normal then so is Z; therefore any tests for multivariate normality can also be used on Z. Univariate tests, bivariate scatterplots, multivariate normality tests and multivariate outlier tests described in this chapter can be used on both the original and principal components variables.

Of particular importance in the assessment of outliers are the first few and last few principal components. The former are sensitive to outliers which inflate variances and covariances, while the last few are sensitive to outliers which add spurious dimensions to or hide collinearities in the data (Gnanadesikan and Kettenring, 1972).

Example 10.1. Hartigan (1975) presented data on blood and cerebrospinal fluid gas measurements in 40 acidosis patients; only the

Table 10.1 Principal components factor loadings of cerebrospinal fluid gases data, n = 40 (Example 10.1).

Variable	Factor 1	Factor 2	Factor 3
CO_2	0.905	0.406	0.130
HCO_3	0.980	0.190	-0.060
pH	0.264	0.964	0.015

data for cerebrospinal fluid are used here (Data Set 14). Measurements include pH, HCO_3 and CO_2. Skewness and kurtosis values for each of the marginal components do not indicate any non-normality (Appendix A).

A principal components analysis based on the covariance matrix was performed on these data; the rotated factor loadings are presented in Table 10.1. A plot of the second and third component variables (Figure 10.1) shows that observation 37 has a relatively high score for factor 3 and an extremely low score for factor 2. Observation 20 shows a high score for factor 2. Observations 16, 10 and 24 also seem to be somewhat separated from the remainder of the observations on factor 2. In the plot of pH and CO_2 (Figure 10.2), observation 37 is on the border of the point cloud, but is not an extreme outlier, while observation 20 is in the center of the data. Plots of pH with HCO_3 (not shown) and HCO_3 with CO_2 (see Figure 10.5) similarly show no obvious individual outliers.

Additional flexibility can be obtained by basing the principal components analysis on the sample correlation matrix of **X** rather than on the sample covariance matrix, or by replacing the covariance (or correlation) matrix with a robust estimate (Barnett and Lewis, 1994). The correlation matrix is most useful when the variables are on widely different scales (Gnanadesikan and Kettenring, 1972).

10.2 Distance Measures

In univariate samples, the concept of identifying an outlier is easily understood. Since an outlier is an observation that is separated from the other observations, a distance can be determined between the potential outlier and the location of the other observations. A distance measure can be defined in a number of ways, such as the distance from the mean, median

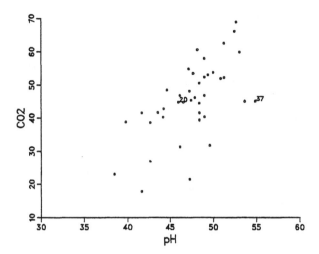

Figure 10.2 Scatterplot of cerebrospinal fluid gas measurements of pH and CO_2 on acidosis patients (Data Set 14), n = 40. Observations 37 and 20 are identified as in Figure 10.1.

or nearest observation, or from some other robust estimate of location. In order to remain scale invariant, the distance is usually also standardized (e.g., using the sample standard deviation).

10.2.1 Generalized Distances

The use of distance of an observation from a location is intuitively appealing in multivariate samples also. As in the univariate case, distance must be defined based on some measure of location and, usually, scale. An obvious first consideration of location and scale are the mean and covariance matrix, respectively. Gnanadesikan and Kettenring (1972) proposed two general classes of observation distance measures based on the estimated mean and the covariance matrix which are useful for identifying multivariate outliers in normal samples:

$$d_{1i}(b) = (\mathbf{x}_i - \bar{\mathbf{x}})'\mathbf{S}^b(\mathbf{x}_i - \bar{\mathbf{x}})$$

$$d_{2i}(b) = d_{1i}(b)/(\mathbf{x}_i - \bar{\mathbf{x}})'(\mathbf{x}_i - \bar{\mathbf{x}})$$

The different classes, d_{1i} and d_{2i}, together with varying values of b, are related to principal components and can be used to identify different types of outliers. In particular, for $b \geq 1$ the class of d_{1i} measures is related to the first few principal components, and therefore are sensitive to outliers which inflate variances and covariances, while for $b \leq -1$, d_{1i} measures are related to the last few principal components and are sensitive to outliers which add spurious dimensions or hide collinearities. For d_{2i} measures, the same relationship hold as $b > 0$ or $b < 0$.

For particular cases, consider

$$d_{1i}(0) = (\mathbf{x}_i - \overline{\mathbf{x}})'(\mathbf{x}_i - \overline{\mathbf{x}})$$

which is the squared Euclidean distance. Generally, $d_{1i}(0)$ is less useful for statistical purposes because it is not scale invariant, but can be used to identify observations that inflate the overall scale. The measure

$$d_{1i}(-1) = (\mathbf{x}_i - \overline{\mathbf{x}})'\mathbf{S}^{-1}(\mathbf{x}_i - \overline{\mathbf{x}}) = r_i^2$$

is the squared radii. Ferguson (1961) showed that, when the mean and covariance matrix are unknown, a best test for identifying a single outlier when the underlying distribution is a scale contaminated normal (i.e., for some value of i and some constant c, $\mathbf{V}(\mathbf{x}_i) = c\mathbf{\Sigma}, c > 1$) is also based on the maximum of the squared radii. Furthermore, this test is uniformly best over all values of c. Jennings and Young (1988) provided critical values for identifying a single multivariate outlier for samples with dimensions $m = 2(1)10, 12, 15, 20$ and selected sample sizes from 5 to 500 (Table B32).

Alternate measures of location can be used to provide extra flexibility in identifying outliers. Since an outlier directly affects the estimates of the mean and covariance matrix, a standardized distance may be defined based on location and scale which is not influenced by the potential outlier. For example, the location and scale may be calculated using the mean and covariance matrix estimated from all observations except the potential outlying observation,

$$T_i^2 = \frac{n-1}{n}(\mathbf{x}_i - \overline{\mathbf{x}}_{(i)})'\mathbf{S}_{(i)}^{-1}(\mathbf{x}_i - \overline{\mathbf{x}}_{(i)})$$

where $\overline{\mathbf{x}}_{(i)}$ and $\mathbf{S}_{(i)}$ are the sample mean and covariance matrix estimated using the $n - 1$ observations excluding $\mathbf{x}_i$. The maximum of the T_i^2 also provides a test for a single outlier, as it compares each observation to a robust (against that observation) estimate of the parameters. However, this test statistic is equivalent to $d_{1i}(-1) = r_i^2$, since

$$r_i^2 = \frac{(n-1)^2 T_i^2}{n(n-2) + nT_i^2}$$

(Jennings and Young, 1988). Similar distance measures suggested for detecting outliers but based on other robust estimates of location and scale are presented in Section 10.3.

10.2.2 Wilks' Extreme Deviate Test for a Single Outlier

Probably the best known test for multivariate outliers was developed by Wilks (1963). When Wilks' method was initially developed, tables of lower bounds of critical values were made available for detecting one or two outliers in up to 5 dimensions. Since that time others have extended this procedure to include more than two outliers and larger number of dimensions, as well as produced a sequential and a graphical procedure based on the Wilks' test.

For an m-variate sample of n observations from a multivariate normal distribution, the sample mean and any m of the observations form a simplex in m-dimensional Euclidean space. If the sums of squares and cross products matrix is

$$\mathbf{A} = \sum_{i=1}^{n}(\mathbf{x}_i - \overline{\mathbf{x}})'(\mathbf{x}_i - \overline{\mathbf{x}}) \tag{10.1}$$

then the sum of squares of the volumes of all possible simplexes is

$$V = (m!)^{-2}|\mathbf{A}|$$

where $|\mathbf{A}|$ is the determinant of $\mathbf{A}$ (Wilks, 1963). If an observation $\mathbf{x}_k$ was eliminated from the sample, then the sum of squares of all possible volumes from the sample of $n-1$ observations is

$$V_{(k)} = (m!)^{-2}|\mathbf{A}_{(k)}|$$

where $\mathbf{A}_{(k)}$ is the sums of squares and cross products matrix (10.1) obtained from the sample after eliminating observation $\mathbf{x}_k$. The amount of reduced volume attained upon eliminating x_k can then be determined from the ratio

$$\Lambda_{(k)} = \frac{|\mathbf{A}_{(k)}|}{|\mathbf{A}|}$$

for each observation $k = 1, \ldots, n$.

Similarly, the amount of volume reduction attained upon eliminating two observations i_1 and i_2 can be determined by

$$\Lambda_{(i_1, i_2)} = \frac{|\mathbf{A}_{(i_1)}|}{|\mathbf{A}|} \frac{|\mathbf{A}_{(i_1, i_2)}|}{|\mathbf{A}_{(i_1)}|}$$

$$= \frac{|\mathbf{A}_{(i_1, i_2)}|}{|\mathbf{A}|}$$

More generally, for any set of k observations indexed by $t = (i_1, i_2, \ldots, i_k)$ the decrease in volume reduction can be calculated using

$$\Lambda_{(t)} = \frac{|\mathbf{A}_{(t)}|}{|\mathbf{A}|}$$

The greatest outlier is that observation which leads to the largest reduction from $|\mathbf{A}|$ to $|\mathbf{A}_i|$, so that Wilks' test statistic for a single outlier is given by

$$\Lambda = \min_i(\Lambda_{(i)}).$$

Wilks showed that the $\Lambda_{(i)}$ are identically distributed as $Be(\frac{n-m-1}{2}, \frac{m}{2})$. Since the $\Lambda_{(i)}$ are not independent, however, their joint distribution is intractable. Wilks determined the upper bound of the distribution function, from which lower bounds of critical values for testing Λ could be obtained.

Wilks' test is equivalent to identifying the maximum of the squared radii (Section 9.1.2), since

$$\Lambda_{(i)} = 1 - \frac{n}{n-1}(\mathbf{x}_i - \bar{\mathbf{x}})'\mathbf{A}^{-1}(\mathbf{x}_i - \bar{\mathbf{x}})$$
$$= 1 - \frac{n}{(n-1)^2}r_i^2$$

noting that $\mathbf{A} = (n-1)\mathbf{S}$. Therefore, $\max(r_i^2)$ or Λ can be used as tests for a single multivariate outlier using the same table of critical values.

> *Example 10.2. After applying a logarithmic transformation to the iris setosa data (Data Set 12), the squared radii were calculated. The largest observed value was 16.88, for the observation 43. This is larger than the 5% critical value for $n = 50$ and $m = 4$, which is 15.89, but smaller than the 1% value of 18.27 (Table B32). Equivalently, the value of Wilks' test is*
>
> $$\Lambda = 1 - \frac{50}{49^2}16.88$$
> $$= 0.648$$
>
> *which gives the minimum value of Λ_i over all observations in this data set.*

Wilks' test for a single outlier can also be derived as the likelihood ratio test that all observations come from the same multivariate normal distribution against the alternate hypothesis that one observation comes from a normal distribution with the same covariance matrix but different mean, where the mean vector, covariance matrix, and index of the potential outlier are unknown. In particular, the maximum (negative) log-likelihood under this null hypothesis is given by

$$L_0(\mathbf{x}) = -\frac{n}{2} \log |\mathbf{A}|$$

and the likelihood under the alternative hypothesis is

$$L_1(\mathbf{x}) = -\frac{n}{2} \log |\mathbf{A}_{(i)}|$$

where the omitted observation $\mathbf{x}_i$ is chosen to maximize $L_1(\mathbf{x})$. This is, of course, simply $-\log(\Lambda)$.

Wilks (1963) provided lower bounds for critical values of Λ for multivariate samples up to dimension 5, and for selected sample sizes from 5 through 500. These bounds were obtained by noting that Hotelling's T^2 distribution can be written in the form

$$(n - m - 1)T^2/[m(n - 2)] \sim F_{m,n-m-1}$$

and the $\Lambda_{(i)}$ can be written in terms of a T^2 variable

$$\Lambda_{(i)} = \left(1 + \frac{T_i^2}{n - 2}\right)^{-1}$$

which follows a beta distribution. Percentage points for Λ were approximated by using Bonferroni bounds obtained from the lower α/n percentage points of the appropriate beta distribution. Better availability of percentage points can be obtained by using the upper $1 - \alpha/n$ percentage points of the F distribution, given by

$$\left(1 + \frac{m}{n - m - 1} F_{m,n-m-1}\right)^{-1}$$

(Caroni and Prescott, 1992).

Barnett and Lewis (1994) provided tables of critical values for $\max(r_i^2)$. Jennings and Young (1988) extended Wilks' table of critical values in the number of dimensions, also using the maximum squared radii, by providing a table for $m = 2(1)10, 12, 15, 20$ and selected sample sizes $5 \leq n \leq 500$ (Table B32).

10.2.3 Wilks' Test for More than One Outlier

When the number of outliers being tested for is specified and is greater than one, a generalization of Wilks' Λ can be used. Specifically, when $k \geq 2$ outliers are of concern, then the values

$$\Lambda_{(t)} = \frac{|\mathbf{A}_{(t)}|}{|\mathbf{A}|}$$

can be calculated for all subsets of k observations indexed by t. As with the single outlier situation, the test statistic is given by

$$\Lambda^k = \min \Lambda_{(t)}$$

where the minimum is taken over all possible subsets t of k observations. Λ^k is also equivalent to the generalized likelihood ratio test for k outliers under Ferguson's (1961) "Model A", i.e., k of the observations come from a normal distribution with a different mean but same covariance matrix as the other $n - k$ observations. $\Lambda_{(t)}$ follows the distribution of

$$\prod_{i=1}^{k} Y_i$$

where

$$Y_i \sim Be(\frac{n - m - i}{2}, \frac{m}{2})$$

(Wilks, 1962). This distribution is known as Wilks' Λ distribution. For the specific case of two outliers, Wilks (1962) also showed that the $\Lambda_{(i_1, i_2)}$ are distributed identically to Z^2 where

$$Z \sim Be(n - m - 2, 2)$$

For the cases of two or three outliers, Fung (1988) provided formulas for simpler computing of the $\Lambda_{(i_1, i_2)}$ and $\Lambda_{(i_1, i_2, i_3)}$ based on the generalized radii. For the two outlier case,

$$\Lambda_{(i_1, i_2)} = 1 - [(n - 1)(r_i^2 + r_j^2) - nr_i^2 r_j^2 + nr_{ij}^2 + 2r_{ij}]/(n - 2)$$

where r_{ij} are the generalized radii defined in Section 9.4.1. For the three outlier case,

$$\begin{aligned}
\Lambda_{(i_1, i_2, i_3)} = 1 - [&(n - 2)(r_i^2 + r_j^2 + r_k^2) + 2(r_{ij} + r_{jk} + r_{ki}) \\
& - (n - 1)(r_i^2 r_j^2 + r_j^2 r_k^2 + r_k^2 r_i^2) + (n - 1)(r_{ij}^2 + r_{jk}^2 + r_{ki}^2) \\
& + 2(r_{ij} r_{jk} + r_{jk} r_{ki} + r_{ki} r_{ij} - r_i^2 r_{jk} - r_j^2 r_{ki} - r_k^2 r_{ij}) \\
& + 2nr_{ij} r_{jk} r_{ki} - n(r_i^2 r_{jk}^2 + r_j^2 r_{ki}^2 + r_k^2 r_{ij}^2)]/(n - 3) + nr_i^2 r_j^2 r_k^2
\end{aligned}$$

Obviously, the computational effort required to find the maximum of the $\Lambda_{(i_1,i_2,...,i_k)}$ over all possible combinations of k observations increases significantly with k, m and n. Fung (1988) stated that in practice it may suffice to take the largest (in the case of trying to detect two outliers) or two largest (in the case of three outliers) squared radii and searching for that observation which gives Λ^k. However, this is stated without proof. Note that this recommendation implies taking a sequential approach to identifying the prespecified number of outliers.

Example 10.3. The squared radii probability plot (Section 9.2.2) for the cerebrospinal fluid data (Data Set 14) seem to indicate two possible outliers (Figure 10.3). Observation 37 gives the largest value of r_i^2 (and, therefore, gives the value of Wilks' test for a single outlier). Using this observation and pairing it with all other observations shows that the two observations giving the minimum value $\Lambda^2 = 0.2116$ are observations 37 and 20. This result is well below the 0.01 (lower bound) significance level of 0.4045. These two observations were also identified as possible outliers using the principal components plot in Section 10.1.3.

Fung (1988) also provided critical values for Λ^2 and Λ^3 for selected sample sizes $10 \leq n \leq 50$ and for dimensions up to $m = 5$ (Table B33).

10.2.4 The Sequential Wilks' Test

The most common situation encountered in outlier detection is when it is suspected that a sample of data contains outliers, but the number of outliers is unknown. In the previous sections, tests for outliers and their associated probability levels are described for those situations where the number of outliers in a sample is assumed known.

Caroni and Prescott (1992) presented a sequential application of Wilks Λ test for multivariate outliers wherein the significance level of the test is adjusted at each step to account for outliers detected up to the previous step. For this procedure, the maximum number of possible outliers k is presumed known, and the test is used to identify any number of outliers up to k.

In this procedure, Λ is used to test for a single outlier. If Λ proves to be significant, the outlier is set aside, resulting in a sample of size $n - 1$. A

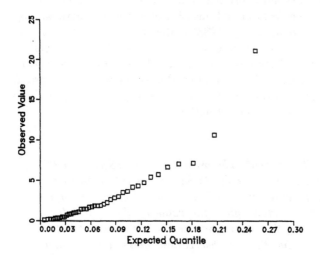

Figure 10.3 Beta probability plot of squared radii for three cerebrospinal fluid gas measurements on acidosis patients (Data Set 14), n = 40.

search for a second outlier is then conducted, using

$$\Lambda_{(ij)} = \frac{|\mathbf{A}_{(ij)}|}{|\mathbf{A}_{(i)}|}$$

where x_i is fixed as the first outlier; the second outlier is identified as that observations which produces

$$\Lambda_2 = \min(\Lambda_{(ij)})$$

where the minimum is taken over all observations after excluding the ith. The sequential procedure continues by identifying a single outlier at a time from the remaining observations after removal of those outliers determined to be significant, until a non-significant test statistic is obtained.

To return an overall α test level, the critical values for this sequential test are obtained by using the selected α level critical value for a sample of n to identify the first outlier, the selected α level critical value for a sample of size $n-1$ to identify the second outlier, etc., up to the kth outlier being identified using the critical value for a sample of $n-k$.

Example 10.4. In the cerebrospinal fluid data (Data Set 14), two outliers are apparent in the probability plot of r_i^2 (Figure 10.3),

*Table 10.2 Sequential test statistics for outliers in cerebrospinal fluid gases,
n = 40 (Example 10.4).*

# outliers (k)	Observations	test statistic	$\alpha = 0.05$ cut-off
1	37	21.14	13.36
2	37 20	19.39	13.26
3	37 20 24	13.13	13.16
4	37 20 24 16	14.15	13.06
5	37 20 24 16 10	18.40	12.96
6	37 20 24 16 10 14	8.15	12.85

*and were identified using Wilks' Λ^2 (Example 10.3). Table 10.2
shows that the sequential test identifies observations 37 and 20 as
the most discordant values, producing highly significant results in
both cases; test statistics are the maximum r_i^2 at each step. A third
observation, observation 24, does not quite attain a significance
level of 0.05.*

*However, if we accepted observation 24 as being sufficiently
near the critical value to be considered an outlier and continued the
sequential test, two more highly significant outliers (observations
16 and 10) would be identified. A sixth step in the sequential
procedure does not show additional outliers.*

10.2.5 Sequential Probability Plot Using Wilks' Statistic

Bacon-Shone and Fung (1987) developed a sequential graphical method for
detecting univariate and multivariate outliers. This method is similar to
a probability plotting procedure for the $\Lambda_{(i_1,i_2,\ldots,i_k)}$. For $k > 1$ outliers,
the marginal distribution of $\Lambda_{(i_1,i_2,\ldots,i_k)}$ can be written as the product of
beta distributions. Although the quantiles of this distribution are difficult
to obtain, Box (1949) provided the approximation

$$W = -[n - \frac{1}{2}(m + k + 3)] \log(\Lambda_{(i_1,i_2,\ldots,i_k)}) \qquad (10.2)$$

which is approximately distributed as a χ^2_{mk} random variable. The approximation is good for moderate to large sample sizes; for small n, a correction
factor is needed. However, for fixed n, m and k, this correction factor is
nearly constant so that the linearity of the probability plot is not affected.

In this procedure, at the kth step the $(n - k + 1)$ largest values of W from (10.2) are plotted against values which are proportional to the approximate expected quantiles. Bacon-Shone and Fung (1987) used the χ_m^2 distribution with cumulative probabilities

$$P_{(j)} = 1 - j/(n - k + 2), \qquad j = 1, \ldots, n - k + 1$$

as the comparison quantiles in the plot. Under the null hypothesis of no outliers, this plot should be approximately linear with no extreme outliers.

This procedure is used sequentially to identify sets of outliers by assigning $k = 1, 2, \ldots, m$ until no outlier sets are observed. Since at each step the plotted $n - k + 1$ values of W do not necessarily have the same set of common indices, some curvature of the plot may occur, especially in the lower left of the plot. This can be ignored, since the primary objective of this procedure is to detect outliers. The larger values of W will generally have $k - 1$ indices in common.

Example 10.5. For the cerebrospinal fluid data (Data Set 14), the probability plot of the transformed Wilks' statistic (10.2) shows that observation 37 is an outlier (Figure 10.4). The second step in this procedure indicates the pair of observations (37, 20) are distinct outliers. Additional steps do not indicate any obvious extreme sets of observations.

At each step in this procedure, $\binom{n}{k}$ values of $\Lambda_{(t)}$ must be calculated. The calculations are relatively simple, but for larger values of n and k the number of calculations becomes large. Based on experience only, it is suggested that one might assume that any set of outliers identified at step k will include those $k - 1$ values identified in the previous steps. This modified procedure would be similar to the sequential procedure discussed in the previous section.

10.3 Robust Methods for Detecting Multivariate Outliers

The presence of extreme observations will generally affect estimates of location, scale and/or correlation. A cornerstone of robust estimation is the determination of influence, which is the effect an observation has on a parameter estimate (see Chapter 12). Measures of influence have been suggested as one means of identifying outliers.

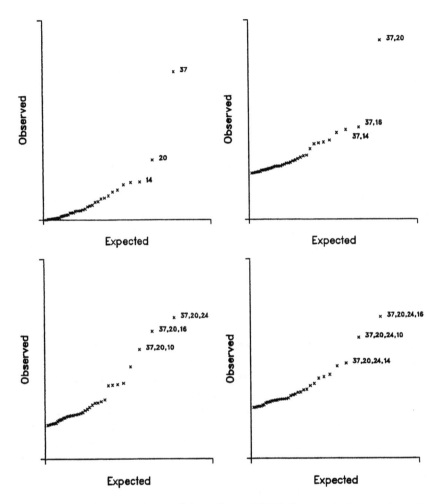

Figure 10.4 Sequential probability plots of Wilks' statistic for cerebrospinal fluid data (Data Set 14). The three observation sets with largest values are labeled at each step.

Wilks Λ and other methods described in the previous section are based on the distance of each observation from some "center" of the data cloud. However, since outliers tend to bias the estimates towards themselves, they decrease the observed distance when using standard estimates such as the mean and covariance matrix as the basis for the distance.

One purpose for the development of robust estimation was to dampen the effects of discordant observations on estimation. These estimates were then suggested for use in defining distance. Robust methods also have the advantage of being resistant to masking and swamping effects of outliers (Rousseeuw and van Zomeren, 1990), and have some ability to detect clusters.

10.3.1 Influence Functions

The influence function (Hampel 1968, 1973, 1974) is a useful measure for determining the effects of individual observations on a parameter estimate. In particular, the sample influence function (Devlin, Gnanadesikan and Kettenring, 1975) is a measure of the effect an individual observation has on the estimate of a parameter of interest. Observations which have a large effect on the estimate(s) of interest rate particular attention with respect to their outlier status.

The sample influence function for a parameter θ is defined as

$$I_-(\mathbf{x}; \hat{\theta}) = (n-1)(\hat{\theta} - \hat{\theta}_i)$$

where $\hat{\theta}$ is the full sample estimate of θ and $\hat{\theta}_i$ is the estimate using the sample of $n-1$ observations which exclude observation $\mathbf{x}_i$. The sample influence function can be calculated for each of the n observations, and the largest values of $I_-(\mathbf{x}; \hat{\theta})$ can be used to identify potential outliers.

Hampel (1968) described the influence function in the context of estimating mean and variance, while Devlin, Gnanadesikan and Kettenring (1975) described the influence function with respect to bivariate correlation. The influence function of the Pearson bivariate correlation is given by

$$I_-(\mathbf{x}; r) = (n-1)(r - r_{-i})$$

while that of Fisher's correlation transform is

$$I_-(\mathbf{x}; z(r)) = (n-1)(z(r) - z(r_{-i}))$$

where r and r_{-i} are the estimated correlations using the full sample and the full sample except for observation $\mathbf{x}_i$, respectively. Devlin, Gnanadesikan and Kettenring (1975) suggested plotting the bivariate data along with contours of constant influence, although the choice of which constants to use are subjective, based on "sample size and other considerations relevant to the particular application" (Gnanadesikan, 1977). Probability plots of $I_-(\mathbf{x}; z(r))$ have also been suggested, where the reference distribution is that of the product of two independent standard normal variables (Gnanadesikan, 1977).

Chernick (1983) suggested using large values of the influence function to identify outliers; however, a definition of "large" needs to be constructed before the influence function can be used as a formal test for multivariate outliers. Influence functions for the mean and variance of multivariate observations also need to be devised; although influence functions for these parameters for each marginal can be examined, an effect on the overall estimates cannot be determined. Similarly, when $m > 2$, influence functions for all pairs of variables can be examined, but an overall effect cannot be determined. Of course, the relevant influence function (i.e., mean, variance or correlation) would have to be chosen depending upon the application.

In a similar vein, Gnanadesikan and Kettenring (1972) suggested examining the correlations r_{-i} for each pair of the m variables to determine the effects of each observation on the correlations. Note that this is identical to the function $I_-(\mathbf{x}; r)$ for a given sample.

10.3.2 Influence Weights

Campbell (1980) presented a method of mean and covariance robust estimation (M-estimation, see Chapter 12) using case weights derived from the squared radii. The robust estimates are defined by

$$\overline{\mathbf{x}}_M = \sum_{i=1}^{n} w_i \mathbf{x}_i / \sum_{i=1}^{n} w_i$$

and

$$\mathbf{V}_M = \sum_{i=1}^{n} w_i^2 (\mathbf{x}_i - \overline{\mathbf{x}}_M)'(\mathbf{x}_i - \overline{\mathbf{x}}_M) / (\sum_{i=1}^{n} w_i^2 - 1).$$

The M-estimates and case weights are calculated iteratively. The influence (weight) of an observation on these estimates increases linearly with its standardized distance from the center when this distance is not large, then levels off and decreases as the observation becomes more distant. In this procedure, influence is defined as the function

$$\omega(r_i) = \begin{cases} r_i & r_i \leq r_0 \\ r_0 \exp(-\frac{1}{2}(r_i - r_0)^2/b_2^2) & r_i > r_0 \end{cases}$$

where

$$r_0 = m^{\frac{1}{2}} + b_1/2^{\frac{1}{2}}$$

Table 10.3 Observations with non-unity weights using Campbell's (1980) M-estimation method for cerebrospinal fluid gases data, n = 40 (Example 10.6).

Observation #	Weight	p
$b_1 = 2, b_2 = 1.25$		
37	0	< 0.001
20	0.28×10^{-26}	< 0.001
16	0.000019	< 0.001
24	0.000078	< 0.001
10	0.00053	< 0.001
$b_1 = 2, b_2 = \infty$		
37	0.57	< 0.01
24	0.84	< 0.01

for some constants b_1 and b_2, and the r_i at each of the iteration steps are based on the current M-estimates $\bar{\mathbf{x}}_M$ and $\mathbf{V}_M$. The case weights are defined as

$$w_i = \omega(r_i)/r_i$$

so that cases within r_0 of the mean estimate have weight 1, and extreme observations have a weight less than 1. For this influence function, the weight of an extreme observation can go to 0, depending on the value of b_2, and so will have little effect on the estimates. Campbell (1980) suggested using either $(2, \infty)$ or $(2, 1.25)$ for (b_1, b_2). The iterative process can be terminated when estimate differences between consecutive iterations are small. Matthews (1984) suggested terminating the procedure when none of the component variable means or entries in the dispersion matrix change by more than 1%.

Matthews (1984) then suggested using the final weights, w_i, to identify outliers. Significant outliers can be identified since

$$P(w_i < 1.0) = 0.02$$
$$P(w_i < 0.9) = 0.01$$
$$P(w_i < 0.5) < 0.001$$

for a multivariate normal population.

Example 10.6. Estimates of the mean and covariance matrix and final weights were obtained using the method described by Campbell (1980) for the cerebrospinal fluid data (Data Set 14), using

$b_1 = 2$ and $b_2 = 1.25$. *Nine iterations were required to obtain the final solution. Five observations (10, 16, 20, 24, 37) were identified as being significant outliers (Table 10.3), with the largest non-unity weight for all observations being 0.00053, so that all outliers were significant at the 0.001 level. These observations were also identified as outliers using the sequential Wilks' test (Example 10.4).*

For comparison, only the two outliers identified most often in the other examples given in this chapter were identified using influence weights when using $b_1 = 2$ and $b_2 = \infty$, neither weight reaching the 0.001 level of significance. Estimates of the means were not greatly different from the usual means, being 47.45, 21.92 and 45.58 for pH, HCO_3 and CO_2, respectively.

10.3.3 Robust Principal Components

Campbell (1980) suggested a similar method of estimating robust principal components. Rather than using a robust estimate of the covariance matrix as the basis for the analysis, he suggested M-estimation of the mean and variance applied to each principal component, which results in obtaining a weight for each observation for each of the principal components.

Components are computed sequentially, beginning with the largest eigenvalue of the unweighted dispersion matrix. The principal component scores for this component variable are obtained as

$$y = \mathbf{u}_1' \mathbf{x}$$

where $\mathbf{u}_1$ is the first eigenvector corresponding with the largest eigenvalue. Using the estimation procedure described in Section 10.3.2, M-estimates and weights for this univariate sample are obtained. Campbell (1980) suggested using the median and $(0.74 \times IQR)^2$ as initial estimates of location and scale. After each iteration, the weights are defined as the minimum of the current weights and those from the previous iteration to prevent oscillation of the solution.

The mean and variance of the original sample are calculated using the final weights obtained from the principal component scores. The first eigenvalue and eigenvector of the new covariance matrix are then obtained, and the above analysis is repeated until successive eigenvalues are sufficiently close (Matthews (1984) suggested differences less than 0.01%).

To determine weights in other directions, form

$$\mathbf{x}^j = (\mathbf{I} - \mathbf{U}_{j-1}\mathbf{U}_{j-1}')\mathbf{x}^{j-1}$$

where $\mathbf{U}_{j-1} = (\mathbf{u}_1, \ldots, \mathbf{u}_{j-1})$ is the set of first $j - 1$ eigenvectors. This projects the data onto the space orthogonal to that spanned by the previous $j - 1$ eigenvectors. Replacing $\mathbf{x}$ with $\mathbf{x}^j$ in the univariate analysis, and determine the first eigenvector $\mathbf{u}$. The principal component scores are given by

$$\mathbf{u}'\mathbf{x}^j = \mathbf{u}'(\mathbf{I} - \mathbf{U}_{j-1}\mathbf{U}'_{j-1})\mathbf{x}.$$

This method produces m sets of weights for the observations, which can be examined to determine the directions in which the observations are extreme.

10.3.4 Minimum Volume Ellipsoid

While M-estimation provides robustness against outliers and the masking effect, this method can only tolerate at most a proportion of outliers equal to $1/(m + 1)$. For small data dimensions this is acceptable (e.g., for 4 dimensional data the breakdown point is 20%). For higher dimensions, Rousseeuw and van Zomeren (1990) proposed using the minimum volume ellipsoid (MVE) estimator. The robust estimate of the mean is obtained by the center of the minimum volume ellipsoid covering 50% of the observations. The dispersion matrix is obtained using the same subset of observations, but is multiplied by a correction factor to obtain consistency. Robust distances are then computed for each observation using these estimates. Rousseeuw and van Zomeren (1990) used the relation

$$\sqrt{RD_i^2} > \sqrt{\chi^2_{m,0.975}}$$

to identify significant outliers, where RD_i^2 are the robust distances, i.e., the squared radii based on the robust estimates.

Rousseeuw and van Zomeren (1990) provided two approximation algorithms for computing the MVE. The most straightforward is the resampling method (Rousseeuw and Leroy, 1987). For this method, subsamples of different observations of size $m + 1$ are drawn. For each subsample J, the sample mean $\mathbf{t}_J$ and covariance matrix $\mathbf{C}_J$ are calculated. The robust distances for all observations are calculated relative to the subsample estimates, and the ellipsoid is inflated or deflated in order to contain exactly h points, where $h = [(n + m + 1)/2]$ observations are within that radius ($[\cdot]$ is the greatest integer function). The squared volume of the resulting ellipsoid is proportional to $m_J^{2m}\det(\mathbf{C}_J)$, with the "best" subset providing the minimum, where m_J^2 is the $100h/n$th percentile of the current set of RD_i^2. From this "best" subset, the MVE estimators are obtained as

$$\mathbf{t}(\mathbf{X}) = \mathbf{t}_J$$

and

$$\mathbf{C}(\mathbf{X}) = c_{n,m}^2 m_J^2 \mathbf{C}_J / \chi_{m,0.50}^2$$

where the correction factor

$$c_{n,m}^2 = (1 + (15/(n-m)))^2$$

is a small sample correction factor.

Since the random search method for the MVE may be computationally slow (Rousseeuw and van Zomeren, 1990; Woodruff and Rocke, 1993), Atkinson (1994) developed a simple forward search algorithm which has higher speed but may result in less precise estimates. However, as Atkinson notes, "if the purpose of estimation is to detect outliers, very precise estimates of the parameters may not be necessary."

10.3.5 An Approximate Minimum Volume Ellipsoid

Hadi (1992) proposed a simple iterative method for detecting multivariate outliers, using a stopping rule which approximates the MVE estimate. Similar to the methods of Campbell (1980) and Rousseeuw and van Zomeren (1990), outliers are identified by examining the distance r_i of the individual observations from the center of the data as determined from robust estimates of location and scale.

The first step in this procedure is to arrange the observations in ascending order based on the distance measure

$$D_i = [(\mathbf{x}_i - \mathbf{c}_R)' \mathbf{S}_R^{-1} (\mathbf{x}_i - \mathbf{c}_R)]^{1/2}$$

for an initial set of robust estimates $\mathbf{c}_R$ and $\mathbf{S}_R$. The sample is then divided into a "basic" subset of the first $m+1$ observations and a "non-basic" subset of the remaining $n-m-1$ observations. The first iteration of the procedure is performed by computing the set of distances

$$D_i = [(\mathbf{x}_i - \mathbf{c}_b)' \mathbf{S}_b^{-1} (\mathbf{x}_i - \mathbf{c}_b)]^{1/2} \tag{10.3}$$

where $\mathbf{c}_b$ and $\mathbf{S}_b$ are the mean and covariance matrix of the basic subset. The observations are reordered according to the current values of the distances in (10.3). New basic and non-basic subsets are defined as the first $m+2$ and the remaining $n-m-2$ observations in the new ordering. The second step is then repeated using the new subsets; this iteration is repeated using a new basic subset of $m+i$ observations at iteration i. After the iteration procedure is terminated, final distances are computed as

$$D_i = [(\mathbf{x}_i - \mathbf{c}_b)'(k\mathbf{S}_b)^{-1}(\mathbf{x}_i - \mathbf{c}_b)]^{1/2} \tag{10.4}$$

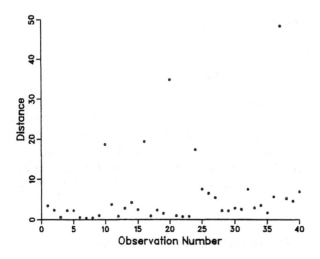

Figure 10.5 Plot of Hadi's (1992) distances (without correction factor) by case number for cerebrospinal fluid data (Data Set 14), $n = 40$.

where

$$k = c_{nmh} m_j / \chi^2_{m,0.50}$$

and

$$c_{nmh} = (1 + h/(n - m))^2$$

and m_j is defined as in the previous section. Note that Rousseeuw and van Zomeren (1990) used a constant value $c^2_{n,m}$ as an approximate small sample correction factor for all sample sizes rather than the c_{nmh} used here.

Hadi suggested that the process be terminated after $h - m$ iterations, where $h = [(n + m + 1)/2]$, which would result in a subset of h observations in the basic subset and so be consistent with the number of observations in the MVE. An iteration count is used to define a stopping rule rather than a cutoff criteria since the distribution of the distances is difficult to derive (Hadi, 1992).

As starting values for parameter estimates in the first step, Hadi (1992) suggested starting with the vector of marginal medians, c_m, and defining a covariance matrix

$$S_m = (n - 1)^{-1} \sum_{i=1}^{n} (x_i - c_m)'(x_i - c_m).$$

After computing observation distances based on these initial estimates, the observations are put in ascending order and a weight function v_i is defined which is equal to 1 if $i \leq [(n + m + 1)/2]$ and equal to 0 otherwise. The estimates used in the first step of the iterative process are then obtained as

$$\mathbf{c}_R = \sum_{i=1}^{n} v_i \mathbf{x}_i / \sum_{i=1}^{n} v_i$$

and

$$\mathbf{V}_R = \sum_{i=1}^{n} v_i (\mathbf{x}_i - \mathbf{c}_R)'(\mathbf{x}_i - \mathbf{c}_R)/(\sum_{i=1}^{n} v_i - 1).$$

Example 10.7. Hadi's distances (10.4) for the cerebrospinal fluid data (Data Set 14) are plotted by case number in Figure 10.5. Since this method is graphic rather than a formal testing procedure, distances are plotted without the correction factor c_b, which is constant for a given sample.

Observations 37 and 20 stand out as extreme values, with observations 24, 16 and 10 clustering at a second level but standing out from the remaining observations. These results agree with other methods used on this data set, where the former two observations are always identified as outliers, and the latter three observations show some indications of being outliers.

Hadi (1992) also gave a method for performing this procedure when the basic subset is not of full rank.

10.3.6 Stalactite Plots

Atkinson and Mulira (1993) proposed the use of stalactite plots as a graphic tool for identifying outliers. The stalactite plot shows the number of outliers detected at each step of an iterative search procedure which uses robust estimates of the mean and covariance matrix at each step.

Similar to the resampling method of Rousseeuw and van Zomeren (1990) for obtaining the MVE, a random sample of $m+1$ of the observations is chosen as the basic subset. Beginning with this basic subset, the forward search procedure described in the previous section (Hadi, 1992) is used to build a robust subset of the observations; however, rather than stopping when h observations comprise the basic subset, here $n - m - 1$ steps are completed, i.e., until all observations are included in the basic subset at the last step.

At each step, the distance of each observation based on the current set of estimates is compared to a cutoff value to determine which observations are outliers. Atkinson (1993) used χ^2_{m,p_0} as a cutoff, where $p_0 = (n-0.5)/n$. The stalactite plot is presented as an $n - m - 1$ (step) by n (observation) display of the observations which are identified at each step (as indicated by the basic subset size, k), where an outlier is represented by an "x" (or other symbol), and is left blank otherwise. Observations which are consistently outliers across all values of k can reasonably be suspected of being outliers. Identification of the steps (i.e., which subsets of observations) which could best be used to obtain robust estimates may also be inferred by examining the patterns in the plot.

> *Example 10.8. A stalactite plot of the cerebrospinal fluid data (Data Set 14) is shown in Figure 10.6; every other step (k) is displayed in the interests of space. Since the starting subset of observations used to obtain initial estimates for the plot seems to be irrelevant with respect to the final result (Atkinson, 1993), the first iteration was based on Hadi's (1992) estimates for this example.*
>
> *Observations 37 and 20 are identified as outliers at all stages of the procedure, even when they are included in the estimation process. Observations 16 and 24 are identified up until k = 39 (not shown), and observation 10 is identified up until k = 38, indicating that the inclusion of observations 20 and 37 changes the estimates enough to mask their outlying status.*

As might be expected, the stability of the identified outliers is poor in the early stages of the plot in that many observations are too far from the current estimate. Atkinson (1993) therefore suggested the use of normalized distances to stabilize the pattern of outliers; use of normalized distances is equivalent to defining the cutoff as $D_k \chi^2_{m,p_0}/(m(n-1))$, where D_k is the sum of the squared radii calculated at the step defined by k.

10.4 Adaptations from Regression Problems

In this text, the primary focus is on the distribution of unstructured samples. Structured data, such as that described by regression, presents a variety of different schemes for identifying outliers based on the assumed structure. However, some of the regression diagnostics used to evaluate

```
                    1111111111222222222233333333334
     k              1234567890123456789012345678901234567890

     6        xx xx    xxxx xxxx xxxxxxxxxxxxxxxxxxxxxxxx
     8        xx xx     xx xxxx xxxxxxxxxxxxxxxxxxxxxxxx
    10        xx xx     xx xxxx xxx  xxxxxxxxxxxxxxxxxxx
    12        xx x      xx xxxx xxx  xxxxxxxxxxxxxxxxxx
    14        x  x      xx xxxx x x  xxxxxxxxxxxxxxxxxx
    16        x  x      xx  xxx  x   xxxxxxxxxxxxxxxxxx
    18        x         xx  xxx  x   xxxxxxxxxxx xxxxx
    20        x         xx  xxx  x   xxxx  x xxx xxxxx
    22        x         xx  x x  x   xxxx    x x xxxxx
    24                  x   x x  x   xxxx      x  xxxxx
    26                  x  xx x  x   xxxx      x  xxxxx
    28                  x   x   x  x   xxxx      x  xxxxx
    30                  x   x   x   xxx      x    x  x
    32                  x   x   x   x          x
    34                  x   x   x   x          x
    36                  x   x   x   x          x
    38                  x   x   x   x          x
    40                          x                 x
                    1111111111222222222233333333334
                    1234567890123456789012345678901234567890

                             Observation Number
```

Figure 10.6 Stalactite plot for cerebrospinal fluid data (Data Set 14), n = 40.

the effects of individual observations on regression equations are based on measures obtained solely from the independent variables, which essentially ignores the structure imposed by their relationship to the dependent variable. Therefore, these types of measures can be used in an unstructured sample context to detect influential observations or outliers.

10.4.1 The Hat Matrix

The "hat" (or prediction) matrix has long been used to identify influential points in regression analysis (e.g., Belsley, Kuh and Welsch, 1980). The hat matrix is defined by

$$\mathbf{H} = \tilde{\mathbf{X}}(\tilde{\mathbf{X}}'\tilde{\mathbf{X}})^{-1}\tilde{\mathbf{X}}'$$

which does not involve the dependent variable; here $\tilde{\mathbf{X}}$ indicates the data matrix $\mathbf{X}$ augmented with a column of 1's, which would be the case in usual least squares regression. The diagonal values of $\mathbf{H}$, h_i are denoted the leverage values, and large values of h_i are indicative of values distant from the mean vector. Gillespie (1993) suggested using the h_i for identifying multivariate outliers in a non-regression context, using a cutoff of $2.5m/n$ to determine significance. However, since

$$r_i^2 = (n-1)(h_i - 1/n)$$

this test is also identical to Wilks' test (and therefore could also be classified as a distance test). Critical values for h_i can be determined using the F distribution since

$$\frac{n - m}{m - 1} \frac{[h_i - 1/n]}{1 - h_i} \sim F_{m-1, n-m}$$

(Belsley, Kuh and Welsch, 1980). The availability of leverage values in most statistical software with linear regression procedures makes this a practical test procedure.

10.4.2 A Collinearity Measure of X

The problem of regression collinearity is said to exist if there is an approximate linear relationship among the predictor variables. This results in well documented problems for regression analyses (e.g., Belsley, Kuh and Welsch, 1980). In the unstructured multivariate problem described in this text, collinearity *per se* is not of concern. However, an individual observation which substantially changes the nature of the observed collinearity structure of the data is an indication that such an observation could be an outlier. Single observations can hide or create collinearities (Hadi, 1987). One measure of collinearity among predictor variables in regression problems is the condition index (condition number), where a high value indicates collinearity.

Hadi (1987) proposed the use of the scaled change in the condition index

$$\delta_i = \frac{\kappa_i - \kappa}{\kappa}$$

where κ is the condition index using all of the observations and κ_i is the condition index of $\mathbf{X}$ after deleting observation i. Two complications of this measure arise, however: (a) the condition index is not invariant to column scaling of $\mathbf{X}$, and (b) there is a great deal of computational effort in calculating the $n + 1$ sets of eigenvalues.

To avoid these difficulties, Hadi (1987) developed a scaled measure of collinearity to evaluate the influence of each observation on the collinearity structure of $\mathbf{X}$. Noting an approximation of κ_i can be obtained using

$$\hat{\kappa}_i^2 = \frac{1 + \sqrt{1 - 4/C_i}}{1 - \sqrt{1 - 4/C_i}}$$

where C_i is the squared trace of $\mathbf{X}'_{(i)} \mathbf{X}_{(i)}$ divided by the determinant of $\mathbf{X}'_{(i)} \mathbf{X}_{(i)}$. If C_i is large, then deletion of $\mathbf{x}_i$ increases the condition index, i.e., $\mathbf{x}_i$ hides a collinearity.

However, C_i is also not scale invariant to the columns of $\mathbf{X}$, so Hadi provided a scaled measure of C_i given by

$$D_i = \frac{(m - \mathbf{x}_i \mathbf{G}^{-1} \mathbf{x}_i') det(\mathbf{G})}{(1 - h_i) det(\mathbf{X'X})}$$

where h_i is the leverage of $\mathbf{x}_i$ and $\mathbf{G} = diag(\mathbf{X'X})$.

Although this measure is related to leverage (and, therefore, to Wilks' test), collinearity-influential values do not always have high leverage, and vice versa (Gillespie, 1993). Therefore, this condition index measure may be useful as an adjunct to Wilks' test (or the leverage values) as a test specific to correlation influence. No specific criteria for formally identifying a significant value of D_i has been suggested.

10.5 Other Tests for Multivariate Outliers

Although not used as often as Wilks' test and its derivatives as tests for detecting multivariate outliers, a number of other tests have been presented. Some of these (Rohlf, 1975; Schwager and Margolin, 1982) have been developed as formal tests for outliers. Others have been suggested as informal procedures (Andrews, 1972; Berkane and Bentler, 1988).

10.5.1 The Generalized Gap Test

Rohlf (1975) considered a generalized gap test for detecting individual and clusters of multivariate outliers. Rohlf used a minimum spanning tree (shortest simply connected graph) to define adjacent ("nearest neighbor") points, i.e., points that are connected in the minimum spanning tree. The gap between two adjacent points is their interpoint distance. A single outlier is identified by having only one other adjacent point, with a relatively large interpoint distance. A cluster of outlying points is a set of points with relatively small interpoint distances among them, but with a relatively large distance between one of the cluster points and one point outside of the cluster. Details for calculating a minimum spanning tree can be found in Section 13.1.3 and the references cited therein.

The proposed gap test is performed by initially conducting univariate outlier tests on each of the component variables in order to obtain "good" estimates of component standard deviations. Each variable is then standardized by

$$x_{ij}' = x_{ij}/s_j$$

where x_{ij} is the jth component of observation i, and s_j is some robust univariate estimate of the standard deviation of the jth component. The distance between each pair of observations i and j is defined as

$$d_{ij} = \left(\sum_{k=1}^{m} (x'_{ik} - x'_{jk})^2 / m \right)^{1/2}$$

which is then used to define the minimum spanning tree (Section 13.1.3); let $d_i, i = 1, \ldots, n-1$, be used to denote the $n-1$ edges of the minimum spanning tree. It is then left to determine whether any of the d_i is too large.

Rohlf (1975) suggested two methods of identifying large values of d_i using the squared distances d_i^2: either constructing a gamma quantile plot of the squared distances or testing the ratio

$$G = \max(d_i^2)/\overline{d}^2$$

to see if it is too large.

The d_i^2 appear to fit a gamma distribution quite well when $\mathbf{X}$ is normal, indicating that they may be, at least almost, independent; however, since they have been chosen to minimize the total distance, they are not random. Therefore, the expected gamma distribution parameters of $\eta = m/2$ and $\lambda = 1/4$ cannot be used, and estimates of the parameters must be obtained. Some estimation methods are described in Wilk, Gnandesikan and Huyett (1962) and Greenwood and Durand (1960) (see Section 13.1.4).

The test for the maximum edge only requires an estimate of the shape parameter η. Upper bounds used to test for this maximum are given in Rohlf (1975) for selected sample sizes and values of η (Table B34); other tables are provided in Pearson and Hartley (1966) and Cochran (1941). For large n, $\max(d_i^2)/\hat{\lambda}$ can be compared to the distribution of the largest order statistic of a standardized gamma distribution.

Example 10.9. The cerebrospinal fluid data (Data Set 14) is used to demonstrate the gap test, using only the HCO_3 and CO_2 variables to keep the technique visually simple for this example. A scatterplot of the raw data with the connecting minimum spanning tree is shown in Figure 10.7. The edge with the largest standardized distance (0.6768) is denoted in the figure, which corresponds to a Euclidean distance of 8.7 in the original units. The test statistic is calculated as

$$G = 0.4580/0.0465 = 9.85$$

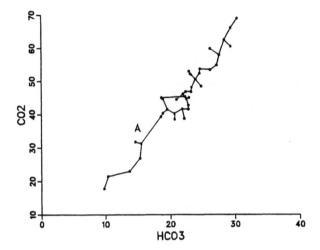

Figure 10.7 Scatterplot of HCO_3 with CO_2 for cerebrospinal fluid data (Data Set 14) with connecting minimum spanning tree. Edge marked "A" is the maximum edge.

Using the method of Greenwood and Durand (1960), the estimate of the shape parameter η is 0.73. Interpolating between entries of Table B34, it can be seen that G is greater than the 0.01 significance level critical value. Note that this edge connects a cluster of six observations with low values of both HCO_3 and CO_2 which may not have been identified by other outlier detection methods due to a masking effect.

Since this method does not evaluate individual distances from some common central value, as do distance measures, a "significant" edge can be detected between a single outlier and the remainder of the data, or between two clusters of points, as was shown in the previous example.

10.5.2 Fourier Function Plots

Andrews (1972) presented a method of plotting multidimensional data onto a single multidimensional plot. For an m-dimensional observation $\mathbf{x}_i =$

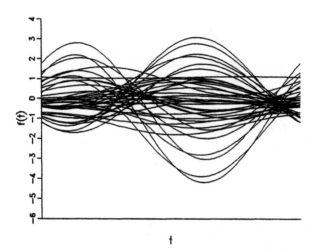

Figure 10.8 Fourier function plot of cerebrospinal fluid gas measurements on acidosis patients (Data Set 14), n = 40.

$(x_{i1}, x_{i2}, \ldots, x_{im})$, Andrews (1972) defined the function

$$f_{x_i}(t) = x_{i1}/\sqrt{2} + x_{i2}\sin(t) + x_{i3}\cos(t) + x_{i4}\sin(2t) + x_{i5}\cos(2t) + \ldots \quad (10.5)$$

which yields one dimensional projections for any specific value of t. The function (10.5) is then plotted across the range $-\pi \leq t \leq \pi$ for each observation. This results in a set of curves across the plot. For any specified value of $t = t_0$, the value of (10.5) is proportional to the length of the projection of the vector $\mathbf{x}_i$ onto the vector

$$f(t_0) = (1/\sqrt{2}, \sin(t_0), \cos(t_0), \sin(2t_0), \ldots) \quad (10.6).$$

This plot can reveal data clusters and outlier patterns and other data characteristics. One advantage of this type of plot is that a continuum (from $-\pi$ to π) of projections is available in a single plot.

Data clusters are revealed by groups of lines that band together. For observations which cluster at only some values of t rather than across the entire range, those observations are clustered in the directions defined by the corresponding vector defined by (10.6) at those values of t. Outliers can be identified by those curves which are separated (either completely or at specified values of t) from the bulk of the data.

One disadvantage of the plot is the difficulty in determining patterns when a large number of observations are plotted on a single graph. Andrews

(1972) recommended plotting the full set of data to determine the general pattern of a sample, but for assessing individual observation characteristics no more than 10 observations should be plotted at a time.

Example 10.10. Using the function

$$f_t = pH^{st}/\sqrt{2} + HCO_3^{st} \sin(t) + CO_2^{st} \cos(t)$$

all 40 observations of the cerebrospinal fluid data (Data Set 14) were plotted on $[-\pi, \pi]$ (Figure 10.8); the superscript indicates that the variables were standardized, by subtracting the median from each observation and dividing by the interquartile range. As would be expected, individual patterns are difficult to discern with so many observations on one plot, although sets of observations showing the same pattern can be discerned. No obviously different observation can be detected in this plot. Fourier plots using other functions $f(t)$ also did not show any extreme values.

The subjective nature of this method is further complicated by the fact that the function (10.5) is not invariant to variable permutations and can be affected by differences in scale. Other functions have also been suggested; for example, Tukey suggested the function

$$f_2(t_0) = (\cos(t_0), \cos(\sqrt{2}t_0), \cos(\sqrt{3}t_0, \ldots)) \qquad 0 \leq t \leq k\pi$$

for an appropriate value of k. This choice provides more complete coverage of the m dimensions and provides equal weights to the m variables (Gnanadesikan, 1977).

In order to handle a large number of cases, Gnanadesikan (1973, 1977) presented a method of reducing the Fourier plot described above. After standardizing the component variables by subtracting the median and dividing by the interquartile range, Gnanadesikan constructed a function plot, in his example using the function

$$f_3(t) = (\sin(t), \cos(t), \sin(2t), \cos(2t), \ldots).$$

At each value of t, sample quantiles (10, 25, 50, 75 and 90) were obtained and a quantile contour plot was produced. To identify outliers, the function curves for all observations that are "consistently (by some quantitative definition such as 'for more than half the values of t')" outside the deciles are superimposed on the plot.

10.5.3 A Test for Mean Slippage

Schwager and Margolin (1982) approached the outlier problem using a multivariate "mean slippage" model, wherein it is assumed that all of the observations come from a multivariate normal distribution with covariance Σ, but k of the observations have means which differ from the common mean and possibly each other. Their approach was to obtain the family of maximal invariants and their distribution. The locally best invariant test was determined to be Mardia's (1970) test for multivariate kurtosis, with the test statistic

$$b_{2,m} = n \sum_{i=1}^{n} (r_i^2)^2$$

(Section 9.4.1). Similar to Ferguson's (1961) result for kurtosis in detecting univariate outliers (Section 6.3.3), $b_{2,m}$ was shown to be locally best invariant for a fraction of outliers up to 21.13%, regardless of the dimension of the data.

In contrast to most of the other tests described in this chapter, this test is not meant to determine either the number or the identity of outliers in the sample. However, Schwager and Margolin asserted that using the multivariate kurtosis test initially to identify the presence of outliers allows subsequent identification of outliers (e.g., using sequential procedures) while still retaining the overall significance level of the test.

10.5.4 The Scale Contaminated Normal Model

Berkane and Bentler (1988) proposed a method of estimating the mixture parameters of a scale contaminated normal mixture distribution, and using the results as an informal method of identifying outliers. A scale contaminated normal distribution is defined as

$$F(x) = (1 - \epsilon)\Phi(x) + \epsilon\Phi(\frac{x}{\sigma})$$

where $\sigma > 1$ and $0 < \epsilon < 1$ is the proportion of data consisting of outliers; ϵ is usually assumed to be small, e.g., $0.01 \leq \epsilon \leq 0.1$ (Huber, 1977). Berkane and Bentler (1988) showed that the functions

$$\kappa(2) + 1 = \frac{1 + (\sigma^4 - 1)\epsilon}{(1 + (\sigma^2 - 1)\epsilon)^2}$$

and

$$\kappa(3) + 1 = \frac{1 + (\sigma^6 - 1)\epsilon}{(1 + (\sigma^2 - 1)\epsilon)^3}$$

can be estimated using

$$\hat{\kappa}(2) + 1 = \frac{b_{2,m}}{m(m+2)}$$

and

$$\hat{\kappa}(3) = \frac{b_{3,m}}{m(m+2)(m+4)}$$

respectively, where $b_{2,m}$ is Mardia's (1970) measure of multivariate kurtosis and $b_{3,m}$ is the average of the r_i^3. Solving these equations iteratively provides estimates of σ and ϵ.

Their informal procedure consisted of determining ϵ, computing the r_i^2 and deleting the $100\epsilon\%$ observations with the largest values. This procedure is then repeated using the reduced set of data until some convergence criteria is met; Berkane and Bentler proposed comparison of the correlation matrices before and after eliminating the identified outliers as a useful criterion.

10.6 Power Studies and Recommendations

Power studies for the multivariate outlier methods described in this chapter were, to our knowledge, nonexistent. This is to some extent due to the wide variety of alternatives possible. Also, many of the tests described herein are informal procedures rather than formal tests. Power of some of the general multivariate methods for detecting outliers as one of the alternatives in the power study have been conducted (see Chapter 9).

Seaman, Young and Turner (1989) conducted a robustness study in which Wilks' test was evaluated for its true significance level when the underlying distribution was heavier tailed than normal. In essence, this study showed the power of Wilks' outlier test for detecting heavy-tailed alternatives; however, the results are not relevant to power for detecting outliers. Using a multivariate form of the exponential power distribution with parameters 1.1 (near double exponential) to 1.9 (near normal), they showed that the power of Wilks' test (1) increased with sample size for fixed dimension and parameter value; (2) decreased with increased parameter value for fixed sample size and dimension; and, (3) the relationship of power to dimension for fixed sample size and parameter value is not consistent across sample sizes and parameter values. With respect to outlier detection, this indicates that a significant value of Wilks' test is not necessarily robust to departures from normality.

As with case of general multivariate normality, no single test or procedure should be relied upon to detect outliers, since there are a variety of types of outliers. It seems reasonable, therefore, to evaluate a sample with a battery of outlier detection methods. Since also, in general, the specific number of possible outliers is not known *a priori*, using tests based on the assumption of a known number of outliers may result in incorrect α levels if used as post-hoc tests or if the assumed number is incorrect.

We therefore outline the following strategy for identifying multivariate outliers. First, univariate and bivariate tests or plotting procedures can be used for the raw marginal variables and principal component variables. This could be followed by one of the Wilks' tests. When more than, say, three outliers and resultant masking may be of concern, methods to identify clusters of outliers (e.g., a robust method or the generalized gap test) could be used. Of the robust methods, due to the computational difficulty of the MVE method (Rousseeuw and van Zomeren, 1990), Campbell's (1980) method, Hadi's (1992) method or a stalactite plot is recommended. The gap test (Rohlf, 1975), a test for a normal mixtures or cluster analysis (Chapter 11) may be useful in the event that there are a large numbers of outliers. Although multivariate kurtosis (Schwager and Margolin, 1982; Mardia, 1970) is optimal for up to 21% outliers, this test is not recommended if there is need to identify which observations are outliers.

10.7 Further Reading

The first comprehensive evaluation of multivariate outlier detection methods was prepared by Gnanadesikan and Kettenring (1972); many subsequent texts and reviews of the subject (including this one) borrow heavily from that article. The chapter on outliers in Gnanadesikan (1977) is based on that article and may be more widely available for reference. Another recommended review article is Beckman and Cook (1983) which provides an overview of outlier detection techniques (univariate and multivariate).

Perhaps the most well known and comprehensive text dedicated solely to the identification of outliers is that of Barnett and Lewis (1994). In addition to addressing outliers from univariate and multivariate normal samples, many other topics are covered, including identification of outliers when sampling from other distributions, and outliers in regression. Hawkins (1980) is much more compact than the current edition of Barnett and Lewis (1994), but excludes the many outlier detection methods which have been developed since 1980. D'Agostino and Stephens (1986) includes chapters on testing for multivariate normality, but do not include much information on methods for identifying outliers from multivariate normal samples. Many other texts on general multivariate methods include descriptions of some of the methods described here; these are too numerous to mention individually.

When large numbers of observations may be separated from each other in a sample, resulting in a masking effect, normal mixtures or cluster analysis might be considered; these subjects are addressed in Chapter 11.

References

Andrews, D.F. (1972). Plots of high-dimensional data. Biometrics 28, 125-136.

Atkinson, A.C. (1993). Stalactite plots and robust estimation for the detection of multivariate outliers. In Morganthaler, S., Ronchetti, E., and Stahel, W.A., eds., **New Directions in Statistical Data Analysis and Robustness**, Birkhauser, Basel, Switzerland, 1-9.

Atkinson, A.C. (1994). Fast very robust methods for the detection of outliers. Journal of the American Statistical Association 89, 1329-1339.

Atkinson, A.C., and Mulira, H.-M. (1993). The stalactite plot for the detection of multivariate outliers. Statistics and Computing 3, 27-35.

Bacon-Shone, J., and Fung, W.K. (1987). A new graphical method for detecting single and multiple outliers in univariate and multivariate data. Applied Statistics 36, 153-162.

Barnett and Lewis (1994). **Outliers in Statistical Data**, 3rd ed. John Wiley and Sons, New York, NY.

Beckman, R.J., and Cook, R.D. (1983). Outlier.........s. Technometrics 25, 119-163.

Belsley, D.A., Kuh, E., and Welsch, R.E. (1980). **Regression Diagnostics**. John Wiley and Sons, New York, NY.

Berkane, M., and Bentler, P.M., (1988). Estimation of contamination parameters and identification of outliers in multivariate data. Sociological Methods and Research 17, 55-64.

Box, G.E.P. (1949). A general distribution theory for a class of likelihood criteria. Biometrika 36, 317-346.

Campbell, N.A. (1980). Robust procedure in Multivariate Analysis I: Robust covariance estimation. Applied Statistics 29, 231-237.

Caroni, C., and Prescott, P. (1992). Sequential application of Wilks's multivariate outlier test. Applied Statistics 41, 355-364.

Chernick, M.R. (1983). Influence functions, outlier detection, and data editing. In Wright, T., ed., **Statistical Methods and the Improvement of Data Quality**, Academic Press, Orlando, FL, 167-176.

Cochran, W.G. (1941). The distribution of the largest of a set of estimated variances as a fraction of their total. Annals of Eugenics 11, 47-52.

D'Agostino, R.B., and Stephens, M.A. (1986). **Goodness of Fit Techniques**. Marcel Dekker, Inc., New York, NY.

Devlin, S.J., Gnanadesikan, R., and Kettenring, J.R. (1975). Robust estimation and outlier detection with correlation coefficients. Biometrika 62, 531-545.

Ferguson, T.S. (1961). On the rejection of outliers. Proceedings of the 4th Berkeley Symposium on Mathematical Statistics and Probability, University of California Press, Berkeley, CA, 253-257.

Fung, W.-K. (1988). Critical values for testing in multivariate statistical outliers. Journal of Statistical Computation and Simulation 30, 195-212.

Gillespie, E.S. (1993). An application of multivariate outlier detection in assessing family characteristics for bank advertisements. The Statistician 42, 231-235.

Gnanadesikan, R. (1973). Graphical methods for informal inference in multivariate data analysis. Bulletin of the International Statistics Institute 45, 195-206.

Gnanadesikan, R. (1977). **Methods for Statistical Data Analysis of Multivariate Observations**. John Wiley and Sons, New York, NY.

Gnanadesikan, R., and Kettenring, J.R. (1972). Robust estimates, residuals, and outlier detection with multiresponse data. Biometrics 28, 81-124.

Greenwood, J.A., and Durand, D. (1960). Aids for fitting the gamma distribution by maximum likelihood. Technometrics 2, 55-65.

Hadi, A.S. (1987). The influence of a single row on the eigenstructure of a matrix. Proceedings of the Joint Statistical Meetings, Statistical Computing Section, American Statistical Association, Alexandria, VA.

Hadi, A.S. (1992). Identifying Multiple Outliers in Multivariate Data. Journal of the Royal Statistical Society B 54, 761-771.

Hampel, F.R. (1968). Contributions to the theory of robust estimation. Unpublished Ph.D. dissertation, University of California, Berkeley.

Hampel, F.R. (1973). Robust estimation: A condensed partial survey. Z. Wahrscheinlichkeitstheorie und Verw. Gebiete 27, 87-104.

Hampel, F.R. (1974). The influence curve and its role in robust estimation. Journal of the American Statistical Association 69, 383-393.

Hartigan, J.A. (1975). **Clustering Algorithms**. John Wiley and Sons, New York, NY.

Hawkins, D.M. (1980). **Identification of Outliers**. Chapman and Hall, London, U.K.

Huber, P.J. (1977). **Robust Statistical Procedures**. Society for Industrial and Applied Mathematics, Philadelphia, PA.

Jennings, L.W., and Young, D.M. (1988). Extended values of the multivariate extreme deviate test for detecting a single spurious observation. Communications in Statistics - Simulation and Computation 17, 1359-1373.

Mardia, K.V. (1970). Measures of multivariate skewness and kurtosis with applications. Biometrika 57, 519-530.

Matthews, J.N.S. (1984). Robust methods in the assessment of multivariate normality. Applied Statistics 33, 272-277.

Pearson E.S., and Hartley, H.O. (1966). **Biometrika Tables for Statisticians**, Volume I, 3rd ed. Cambridge University Press, Cambridge, U.K.

Rohlf, F.J. (1975). Generalization of the gap test for the detection of multivariate outliers. Biometrics 31, 93-101.

Rousseeuw, P.J., and Leroy, A. (1987). **Robust Regression and Outlier Detection**. John Wiley and Sons, New York, NY.

Rousseeuw, P.J., and van Zomeren, B.C. (1990). Unmasking multivariate outliers and leverage points. Journal of the American Statistical Association 85, 633-639.

Schwager, S.J., and Margolin, B.H. (1982). Detection of multivariate normal outliers. Annals of Statistics 10, 943-954.

Seaman, S.L., Young, D.M., and Turner, D.W. (1989). On the robustness of the extreme deviate test for a single multivariate outlier against heavy tailed distributions. Communications in Statistics - Theory and Methods 18, 3289-3303.

Wilk, M.B., Gnanadesikan, R., and Huyett, M.J. (1962). Estimation of parameters of the gamma distribution using order statistics. Biometrika 49, 525-545.

Wilks, S.S. (1962). **Mathematical Statistics**. John Wiley and Sons, New York, NY.

Wilks, S.S. (1963). Multivariate statistical outliers. Sankhya 25, 407- 426.

Woodruff, D.L., and Rocke, D.M. (1993). Heuristic search algorithms for the minimum volume ellipsoid. Journal of Computational and Graphical Statistics 2, 69-95.

CHAPTER 11

TESTING FOR NORMAL MIXTURES

"...detailed investigation of the analysis of finite mixture problems offers more than just a catalogue of straightforward applications of standard methods to a particular class of statistical models: our statistical approach to sand-sifting will indeed reveal a few special nuggets."

Titterington, Smith and Makov, 1985

A normal mixture occurs when a population is made up of two or more individual subpopulations (components), each of which is distributed normally, but with differing parameter values. In the context of testing for normality, testing for mixtures is somewhat distinct because often one is interested in rejecting, rather than accepting, the null hypothesis. Tests described in the previous chapters can be used to detect this type of departure from normality, since normal mixtures cover a wide range of shapes, including symmetric and skewed, and multi- and unimodal. In this chapter we will describe only those tests which have been developed specifically for testing for normal mixtures or have a specific theoretical basis which is relevant to the detection of mixtures. Since the premise of this text is in testing for normality we will focus primarily on comparisons of mixtures of two components to single normal distributions, rather than attempt to determine if there are more than two components. Many of the concepts and methods described here are directly generalizable to more than two components.

11.1 Applications of Normal Mixtures

Finite mixture distributions, and more specifically mixtures of normal distributions, have important applications in genetics and have been used in a wide variety of biomedical and other scientific fields. Finite mixture distributions arise most naturally in modeling heterogeneous populations, but they have also been used to model the occurrence of outliers and some forms of skewness. The normal mixture problem is not new, although interest has been renewed since numerical methods using computers are now available. For example, Newcomb (1886) used normal mixtures to model outliers; Pearson (1894) and Charlier (1906) attempted to solve the parameter estimation problem using the method of moments.

Some of the more well known applications of normal mixtures include the analysis of fish lengths (e.g., Bhattacharya, 1967; Hosmer, 1973) and botany e.g., Fisher's iris and Darwin's plant height data (Scott and Symons, 1971; Wolfe, 1970; O'Neill, 1978; Everitt and Hand, 1981; Box and Tiao, 1968; Aitkin and Wilson, 1980).

In genetic studies of quantitative traits, the model for a phenotype determined by a major gene plus additive polygenes and environmental factors is that the phenotype (after adjusting for known major environmental factors) has a normal mixture distribution on some scale. A trait determined by an allele which is either completely dominant (or completely recessive) would have a population distribution represented by a mixture with two components. If there is no dominance (or recessivity) then there would be three components (Elston and Stewart, 1971), each being the distribution of the phenotype for each of the three distinguishable genotypes. On the other hand, a trait determined by several unlinked genes (polygenes) each of which contributes equally to the phenotype observed or if the trait is multifactorial, i.e., caused by many genes and/or environmental factors each having small, equal and additive effects, the distribution is approximately normal (Falconer, 1960). The demonstration that phenotypes fit a mixture is therefore the first step in showing the existence of a major gene.

In the life sciences mixtures can result from a subpopulation of cells in which specific reactions take place. In the case of the MHC-DW locus responses (Mendell, Lee, Reinsmoen, Yunis, Amos and Emme, 1977; Karlof and Mendell, 1986) one wants to determine whether one has a good typing reagent and ultimately use this typing reagent to identify organ donors who would be HLA-DW locus compatible to a given recipient. Such a typing reagent would be a cell taken from an individual who is homozygous at the HLA-DW locus and for which the distribution of responses in mixed lymphocyte culture would on some scale be a normal mixture distribution. If the cell is heterozygous then the distribution of responses should be normally distributed.

11.2 Types of Normal Mixtures

A k-component normal mixture density $h(\mathbf{x})$ is defined as

$$h_m(\mathbf{x}) = \sum_{i=1}^{k} \pi_i \phi_m(\mathbf{x}; \mu_i, \Sigma_i^2) \tag{11.1}$$

where $\phi_m(\mathbf{x}; \mu, \Sigma)$ is an m-variate ($m \geq 1$) normal density with mean vector μ and covariance matrix Σ, $0 < \pi_i < 1$ for $i = 1, 2, \ldots, k$ and $\sum \pi_i = 1$. Mixtures can be defined by differences in means, covariance structure, or mixing proportions. Here we will generally be interested in testing for mixtures with $k = 2$ components against the null hypothesis of a single normal distribution.

11.2.1 Univariate Normal Mixtures

Univariate normal mixtures can be classified according to the number of components and whether the individual components have unequal means and/or variances. The two component normal mixture with unequal means and common variance (location contaminated normal mixture, LCN) is given by

$$h(x) = \frac{\pi}{\sigma} \phi \left(\frac{x - \mu_1}{\sigma} \right) + \frac{(1 - \pi)}{\sigma} \phi \left(\frac{x - \mu_2}{\sigma} \right).$$

Since all of the tests discussed here are location and scale invariant, the only parameters which are of concern in the LCN are π and $D = |\mu_2 - \mu_1|/\sigma$.

The two component normal mixture with unequal component means and variances (location and scale contaminated normal mixture, LSCN) is given by

$$h(x) = \frac{\pi}{\sigma_1} \phi \left(\frac{x - \mu_1}{\sigma_1} \right) + \frac{(1 - \pi)}{\sigma_2} \phi \left(\frac{x - \mu_2}{\sigma_2} \right)$$

where $\sigma_1 \neq \sigma_2$. Parameters which are of interest include π, the ratio σ_2/σ_1, and the distance between the component means, which, without loss of generality we will define in terms of the variance eof the first component, i.e., $D_s = |\mu_2 - \mu_1|/\sigma_1$.

The scale contaminated normal (SCN) mixture has a common mean and is given by

$$h(x) = \frac{\pi}{\sigma_1}\phi\left(\frac{x-\mu}{\sigma_1}\right) + \frac{(1-\pi)}{\sigma_2}\phi\left(\frac{x-\mu}{\sigma_2}\right).$$

The SCN mixture was Tukey's initial model for his development of robust methods (Tukey, 1960). Since the focus of the robustness research was to estimate the common mean, little work has been conducted on testing or estimating the parameters for this mixture.

One might expect mixtures to be multimodal. However, mixtures can also be unimodal and symmetric or unimodal and skewed, thereby making it difficult to distinguish between non-normal non-mixture distributions, such as beta or gamma distributions. For univariate mixtures of two components, a sufficient condition that the mixture is unimodal for any value of π is that

$$(\mu_2 - \mu_1)^2 < \frac{27\sigma_1^2\sigma_2^2}{4(\sigma_1^2 + \sigma_2^2)} \tag{11.2}$$

while a sufficient condition for bimodality is

$$(\mu_2 - \mu_1)^2 > \frac{8\sigma_1^2\sigma_2^2}{\sigma_1^2 + \sigma_2^2} \tag{11.3}$$

(Eisenberger, 1964). There is reason for concern that a test may identify a mixture when in fact the data come from a single skewed distribution. This can result in errors in interpreting the process generating the data as one which contains two or more subpopulations. For LCN mixtures, the sufficient condition for unimodality (11.2) reduces to $D < 1.84$, while for bimodality the sufficient condition (11.3) is $D > 2$. Figures 11.1a-c show the density functions of some selected LCN mixtures, including a symmetric bimodal (Figure 11.1a), skewed unimodal (Figure 11.1b) and skewed bimodal (Figure 11.1c) mixture.

By defining $\sigma_2 = c\sigma_1$, then the unimodality and bimodality conditions for an LSCN mixture are

$$D_s < 2.60\sqrt{\frac{c^2}{1+c^2}}$$

and

$$D_s > 2.83\sqrt{\frac{c^2}{1+c^2}}$$

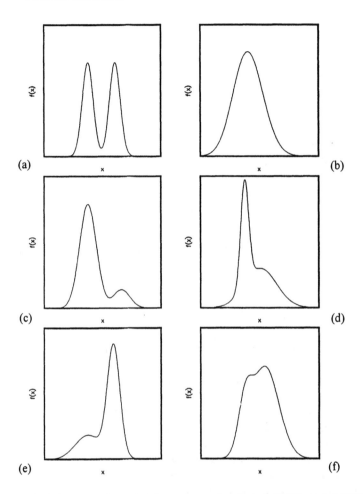

Figure 11.1. Density functions of selected LCN and LSCN mixtures: (a) LCN mixture, $\pi = 0.5$, $D = 5$; (b) LCN mixture, $\pi = 0.7$, $D = 1$; (c) LCN mixture, $\pi = 0.85$, $D = 4$; (d) LSCN mixture, $\pi = 0.4$, $D_s = 4$, $\sigma_2 = 4\sigma_1$; (e) LSCN mixture, $\pi = 0.3$, $D_s = 2$, $\sigma_2 = \sigma_1/2$; (f) LSCN $\pi = 0.3$, $D_s = 2$, $\sigma_2 = 2\sigma_1$.

for any value of π. Figures 11.1d-f are density functions of some selected LSCN mixtures, including a unimodal mixture (Figure 11.1f). SCN mixtures are always symmetric, unimodal and have fourth moment β_2 greater than 3. Figure 11.2 shows the density functions of some SCN mixtures compared to a single normal distribution.

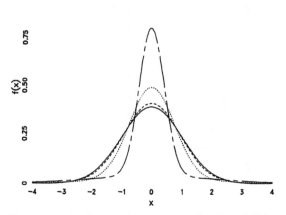

Figure 11.2. Density functions of selected SCN mixtures, each with mean 0 and variance 1: — *normal;* --- $\pi = 0.05$, $c = 2$; .... $\pi = 0.10$, $c = 3$; — $\pi = 0.20$, $c = 5$.

11.2.2 Multivariate Normal Mixtures

A two component m-variate normal mixture density is defined as

$$h_m(x) = \pi\phi_m(\mathbf{x}; \mu_1, \Sigma_1) + (1 - \pi)\phi_m(\mathbf{x}; \mu_2, \Sigma_2)$$

where ϕ_m is the multivariate normal density, $\mathbf{x}$ is an m-dimensional random variable, and $\mu_i, \Sigma_i, i = 1, 2$ are the component m-vectors of means and $m \times m$ covariance matrices, respectively. Multivariate mixtures are similar to univariate mixtures in most respects: similar methods can be used for maximum likelihood estimation and likelihood ratio testing, along with the attendant problems. The more general mixture is also categorized by differences in mean vectors and/or covariance matrices between components. Computational problems for multivariate mixtures increase as the data dimension increases, and also with increasing complexity of the covariance structure of the mixture. The simplest case would be that where the component covariance matrices were equal, diagonal and had $\sigma_{ii} = \sigma_{jj}$ for all $i, j = 1, \ldots, m$, whereas the most difficult would be the case where the component matrices were unequal with all σ_{ij} unequal for $i \leq j$.

Here, multivariate normal mixtures with unequal component means and equal component covariance matrices will be denoted as LCN mixtures, regardless of the covariance structure itself. Those with unequal means and unequal covariance matrices will be denoted as LSCN mixtures. There does not seem to be a large interest in multivariate SCN mixtures, so there is no discussion here of this type of mixture.

11.3 Tests for Mixtures

One of the outstanding problems in the field of mixtures is determining whether a finite mixture distribution is actually necessary to represent the data, and if so, how many components are necessary. The popularization of the EM algorithm by Dempster, Laird and Rubin (1977) has stimulated interest in this family of distributions since it provided a guaranteed, if slow, algorithm for obtaining the maximum likelihood estimates (MLE's) of the mixture model parameters. Despite this surge of activity and interest, major theoretical and computational problems remain unsolved. Maximum likelihood testing and estimation is the preferred method for several reasons: (1) MLE's of the parameters are more efficient than other estimates, such as moment estimates (Day, 1969; Tan and Chang, 1972); (2) the likelihood ratio test (LRT) and MLE's can be obtained simultaneously; and, (3) the LRT is one of the more powerful tests for detecting normal mixtures (Mendell, Finch and Thode, 1993).

However, the maximum likelihood approach to normal mixtures also has problematic areas which need to be overcome: global optimization of the likelihood is often difficult when components have equal variance; when the mixture components are heteroscedastic, the maximum likelihood does not exist; accurate testing criteria need to be determined; and under certain circumstances other tests may be more powerful in detecting mixtures than the LRT.

In addition to the LRT, which requires special computational techniques, other tests are available for identifying mixtures. These include some common tests for normality which have optimal properties for detecting certain types of mixtures, such as $\sqrt{b_1}$, b_2, and outlier tests, as well as tests specific to testing for mixtures. Many of the tests described here have only been developed or investigated for the univariate situation. However, they can be used for preliminary assessment of marginal distributions of multivariate data.

11.3.1 Plots

Plots of mixture data can be highly informative, but without clear separation of components graphic identification of mixtures is difficult. This is in part due to the wide variety of distributional shapes which mixture distributions can assume. In addition, due to natural variation in a sample, plots of data from unimodal (non-mixture) distributions may exhibit apparent multimodality. For mixtures with relatively small differences between components, it is difficult to distinguish mixture data from other types of non-normality using histograms or other plots. With adequate component separation, plots can indicate the presence of a mixture.

Bhattacharya (1967) suggested a plot based on a histogram of the data which can be used both for identifying a mixture and for component parameter estimation. The plot of $(\log(f_{i+1}) - \log(f_i))$ against the midpoints of the histogram intervals, where f_i is the frequency of observations within interval i, should result in a sequence of straight lines with negative slope where each line corresponds to a mixture component.

In a Q-Q plot, since each component is normal, data from each component should form an approximately linear segment with shifts in the plot based on the type and degree of difference between components. For example, an LCN mixture, involving only a shift in the mean, will result in a Q-Q plot with two straight lines with the same slope, since the slope is determined by the variance. The offset between line segments is determined by the degree of separation between components (Figure 11.3a).

Q-Q plots of LSCN mixtures will result in two line segments with different slope (Figures 11.3b, c). With a large degree of component separation and difference in variance, these characteristics will be more obvious (Figure 11.3b). For smaller differences, the results are not as clear cut (Figure 11.3c).

Q-Q plots of SCN mixtures are similar to those for long-tailed symmetric alternatives to normality, although the central portion of the plot will generally be more linear with slope roughly equal to the standard deviation of the component with smaller variance. The tails of the plot will each have slope approximately equal to the standard deviation of the component with the larger variance (Figure 11.3d). With smaller differences in variance, the linearity of the center and tails of the probability plot will not be as apparent.

Plotting procedures for multivariate mixture data should follow the recommendations given in Chapter 9. These include plotting marginal distributions and multivariate plots, e.g., beta probability plots and component vs component plots. Multivariate plots are more problematic in detecting the type of departure from normality; for example, a beta probability plot of multivariate mixture data (Figure 11.4) does not provide any visual information which would distinguish it from other types of non-normality.

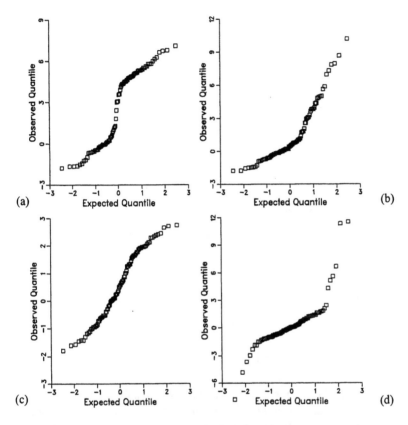

Figure 11.3.   Q-Q plots of simulated two component univariate normal mixtures, $n = 100$: (a) LCN mixture, $\pi = 0.5$, $D = 5$; (b) LSCN mixture, $\pi = 0.4$, $D_s = 4$, $\sigma_2 = 4\sigma_1$; (c) LSCN mixture, $\pi = 0.3$, $D_s = 2$, $\sigma_2 = \sigma_1/2$ (d) SCN mixture, $\pi = 0.2$, $\sigma_2 = 5\sigma_1$.

11.3.2 The Likelihood Ratio Test

The likelihood ratio test (LRT) is a general statistical procedure that can be applied to generate a test of this hypothesis. The LRT statistic is the logarithm of the ratio of the maximum likelihood of the normal mixture distribution to that of the single normal distribution given the observations, i.e.,

$$G^2 = -2(\log[L_1] - \log[L_0])$$

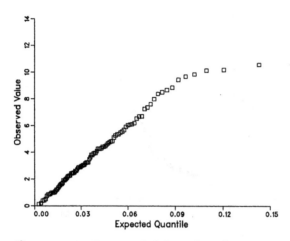

Figure 11.4. Beta probability plot of multivariate LCN mixture, $m = 4$, $\pi = 0.4$, $\Delta = 4$, $\Sigma = I$.

where

$$L_1 = \prod_{i=1}^{n} h(x_i),$$

$$L_0 = \prod_{i=1}^{n} \phi\left(\frac{x_i - \bar{x}}{s}\right),$$

$h(x)$ is given by (11.1), $\phi(y)$ is the standard normal density and $\bar{x}$ and s are the estimated mean and standard deviation, respectively. The maximum value of L_1 over all parameter estimates must be determined using iterative methods. L_1 and L_0 are similarly defined in the multivariate case.

In a general situation, for a structure such that the null distribution is nested within the alternative and with certain regularity conditions, under the null hypothesis twice the log-likelihood ratio is asymptotically distributed as a χ^2 variable with degrees of freedom equal to the difference in the number of parameters estimated in the alternative and null case. Although a $k-1$ component normal mixture is nested within a k-component mixture ($k \geq 2$), the regularity conditions do not hold (Wilks, 1938), and therefore the distribution of the LRT for this class of problems needs to be determined empirically.

For the univariate LCN mixture, Hartigan (1977) suggested that the asymptotic distribution of the LRT is between χ_1^2 and χ_2^2. This is inconsistent with the results of Wolfe (1971) who conducted a simulation study of the distribution of the LRT for univariate and multivariate LCN mixtures for samples of size 100. Based upon the results of 100 random samples, he conjectured that for a sample of size n from an m-dimensional normal sample the quantity $[(n-m-2)/n](G^2)$ is asymptotically distributed as a χ^2_{2m} random variable when testing for 2 vs 1 components. A simulation study by Everitt (1981) produced results consistent with Wolfe's approximation for the univariate problem as long as $n > 10k$. However, his findings for higher dimensions (or smaller samples) suggested an inadequacy in Wolfe's more general results.

Using simulation, Thode, Finch and Mendell (1988) obtained the null distribution of the LRT for testing for a univariate LCN mixture for sample sizes between 15 and 1000. Convergence to a χ^2 distribution with two degrees of freedom was shown to be extremely slow if indeed the null distribution converged to it at all, so that the distribution for samples of size 1000 or less may not be close to the asymptotic distribution. Their results did not rule out either $\chi^2_{1.5}$ or χ^2_2 as the asymptotic distribution, although they suggested that χ^2_2 agrees with the simulation results more closely. Based on their results, Royston (1989) suggested a fractional χ^2_m distribution, where $m = 2 + 2.5/\sqrt{n}$.

Computing MLE's for a normal mixture is difficult, as documented repeatedly (e.g., Wolfe, 1971; Everitt, 1981; Thode, Finch and Mendell, 1987). Since there is no closed form solution to the maximum of (11.1), and hence of G^2, iterative methods are necessary to calculate the test statistic. To succeed, these methods require an optimization algorithm and a vector of starting values of the parameters for input to the algorithm.

A number of optimization methods are available. Hasselblad (1966) gave a simple method which is based on the EM algorithm (Dempster, Laird and Rubin, 1977). This method is guaranteed to converge, although not necessarily to the global maximum. In addition, except for mixtures with well separated components, this algorithm can take a long time to converge (Everitt, 1984).

The Newton-Raphson method is a common optimization method, but is more difficult to implement than the EM algorithm. The Newton-Raphson approach requires the first and second derivatives of (11.1) with respect to each of the parameters. Quasi-Newton methods require only the first derivatives of (11.1), and so are much easier to program. While there is no guarantee that these methods will converge for any particular starting point, they are faster than the EM algorithm.

Different starting values may converge to different solutions for a given optimization algorithm. One common method of dealing with this problem has been to use a "large number" of starting points and take the result with the largest likelihood as the maximum. However, it is never absolutely certain that the global maximum has been found. Methods have been developed to objectively document the adequacy of a search (Finch,

Mendell and Thode, 1989).

For the univariate LCN mixture, a reliable algorithm has been determined for finding the global maximum likelihood (Thode, Finch and
Mendell, 1987). A set of five systematic starting points (SSPs) were identified which, together with a quasi-Newton optimization algorithm based
on Nash (1979), was used to obtain the MLE's and LRT. The first SSP
of π was the optimal split in the data based on the Engelman-Hartigan
statistic (Engelman and Hartigan, 1969; Section 11.3.3). The remaining
four SSPs of π were $1/n$, $1/4$, $3/4$ and $(n-1)/n$. For each SSP of π, the
starting values of μ_1 and D were calculated from the split of the ordered
data on the basis of the SSP of π; the starting value of σ was calculated
as a pooled sum of squares. It is unknown if these SSPs work for other
iterative methods.

> *Example 11.1. In the study of number of days until cancer oc-
> currence for 38 irradiated mice in a normal environment which
> develop reticulum sarcoma (Data Set 7), the value of the LRT
> under the LCN model is 12.9, which is highly significant. Based
> on the MLE's of the mixture, the 3 mice with the shortest de-
> velopment time (< 400 days) comprise one component, while the
> remainder comprise the other component of the mixture.*

There is no maximum likelihood solution to the problem of a mixture
of two normal components each having a different variance, since for any
identified mixture consisting of $n-1$ and 1 observations in the respective components the likelihood does not exist. Similarly, any pair, triplet,
quadruplet, etc., of points that are sufficiently close together will generate
a high local maximum (Everitt and Hand, 1981). Some quasi-likelihood
methods are possible for estimation of parameters, such as constraining
the variance of the smaller component, say σ_2, to be within some specified
proportion of the variance of the larger component, i.e., $c_1\sigma_1 \leq \sigma_2 \leq c_2\sigma_1$,
where $0 < c_1 < 1$ and $1 < c_2$. Alternatively, one could put a lower limit
on the variance, or require at least two observations to be in each group.
However, the routine application of these methods does not seem practical.
Everitt and Hand (1981) indicated that reasonable solutions can often be
found for reasonable sample sizes; either convergence will be to a nonsingular solution, or the largest nonsingular result can be used as the maximum.

The SCN problem is a simpler version of the LSCN mixture problem,
and its solution would be an important preparatory step in dealing with
the more general problem. Additionally, Lindsay (1983) has identified geometrical conditions on a sample which imply that the likelihood ratio test

statistic equals 0 for that sample. The ability to find the MLE's of the SCN mixture would permit an investigation of the properties of Lindsay's geometrical arguments in this case. An important question is to estimate the asymptotic probability that the likelihood ratio test statistic equals 0, an important issue in the current research on the asymptotic distribution of the LRT.

11.3.3 The Engelman-Hartigan Test

The Engelman-Hartigan test (EH) is a test based on a result from the methodology of clustering techniques (Engelman and Hartigan, 1969). For each division of the observations into two groups of size n_1 and n_2 ($n_1 + n_2 = n$) with at least one observation per group, the ratio of between to within sum of squares (SS_B/SS_W) is calculated using

$$SS_B/SS_W = \frac{n_1 n_2 (\overline{x}_1 - \overline{x}_2)^2}{(n_1 + n_2)[\sum (x_i - \overline{x}_1)^2 + \sum (x_i - \overline{x}_2)^2]}$$

where $\overline{x}_j$ is the mean of group j and the sums are over the respective two groups. The EH test statistic is calculated as

$$EH = \max[SS_B/SS_W]$$

where the maximum is taken over all possible divisions of the data into two groups. Since the maximum occurs for one of the $n - 1$ partitions of the ordered data (Fisher, 1958), only $n - 1$ values of SS_B/SS_W need to be calculated. The null hypothesis is rejected for values of EH that are too large.

Engelman and Hartigan (1969) also provided a formula for obtaining critical values for the EH test. For a test at the α level of significance the appropriate critical value c_α is defined by

$$\log(c_\alpha + 1) = -\log(1 - 2/\pi) + z_{1-\alpha}(n - 2)^{-1/2} + 2.4(n - 2)^{-1} \quad (11.4)$$

where z is the appropriate percentile of the standard normal distribution.

Example 11.2. In contrast to the results from Example 11.1, the EH test statistic calculated from the mice data (Data Set 7) is 1.48, which is not significant when compared to the 0.05 level critical value of 2.87 obtained from (11.4). The maximum value of

SS_B/SS_W is obtained by the split in the data such that the 12 lowest observations make up the first component.

11.3.4 Outlier Tests

Mixtures of two normal distributions can be viewed as distributions contaminated with outliers when $n\pi$ is small relative to n. Hence, outlier tests would intuitively be appropriate for detecting certain mixtures as well as identifying those observations which are the contaminants. Ferguson (1961) used a normal mixture model to derive outlier tests and Aitkin and Wilson (1980) used a normal mixture model to describe a sample with outliers. Grubbs' test T (Grubbs, 1950; 1969) and Dixon's $r_{10}^{(2)}$ test (Dixon 1950; 1951) were designed to detect a single outlier in an unspecified direction (see Chapter 6). Therefore, these tests should be effective at detecting mixtures where $n\pi$ is near 1. In addition, T is potentially useful as a test for detecting short-tailed symmetric distributions (Thode, 1985), which makes it a candidate for detecting symmetric LCN mixtures. Sequential outlier tests may be useful when contaminating fractions are larger.

11.3.5 Skewness and Kurtosis

The skewness test $(\sqrt{b_1})$ is a powerful test of normality over a wide range of skewed distributions, so it should be a good test for detecting asymmetric normal mixtures. It is also the locally best invariant one-sided test for up to 50% outliers or contaminating observations when all are in the same direction (Ferguson, 1961), regardless of whether the contaminating observations have identical means.

The kurtosis test (b_2) is powerful over a wide range of symmetric alternatives. Box (1953) indicated that higher order (> 2) even moment tests such as b_2 should be good for detecting short-tailed symmetric alternatives such as symmetric LCN mixtures. Ferguson (1961) determined that an upper-tailed b_2 was the locally best invariant test for an SCN mixture with $\pi \leq 0.21$.

11.3.6 Aitkin's Posterior Bayes Factor

Aitkin (1991) proposed an "average likelihood test" (ALT) as an alternative to the LRT. The solution is an adaptation of the usual Bayes factor which removes the difficulties previously encountered in its application. Formally, any two models can be compared by the usual Bayes factor. Suppose model M_j specifies that the density of x is $f_j(x|\theta_j)$, with prior density $\pi_j(\theta_j)$ for θ_j. Given the data x, the likelihood is $L_j(\theta_j)$. Then the Bayes factor for model M_1 to M_2 is the ratio $\bar{L}_1/\bar{L}_2$ of the "prior means" of the likelihoods:

$$\bar{L}_j = \int L_j(\theta_j)\pi_j(\theta_j)d\theta_j;$$

a Bayes factor of 10, 100 or 1000 constitutes mild, strong or very strong evidence in favor of model M_1 over M_2, respectively.

The Bayes factor is little used because it suffers from two related difficulties: (1) the prior mean of the likelihood is extremely sensitive to small variations in the prior, and this sensitivity does not decrease with increasing sample size; (2) if a conventional diffuse prior is used for a continuous parameter, the prior mean is not even defined if the prior is improper (e.g., if the parameter space is unbounded). These difficulties have greatly restricted the use of Bayes factors for such comparisons; the restriction to proper (e.g., conjugate) priors removes the second difficulty but not the first.

Aitkin addressed these problems by replacing the prior with the posterior in the expectation, and used the *posterior mean* of the likelihood as a direct measure of the evidence for a model. Aitkin called the ratio of two such posterior means for two models the "posterior Bayes factor". This is interpreted in the same way as the conventional Bayes factor.

The posterior mean, or average likelihood (L^A), of the likelihood for model j is defined to be

$$L_j^A = \int L_j(\theta_j)\pi_j(\theta_j|y)d\theta_j$$

with $\pi_j(\theta_j|y)$, the posterior density of θ_j:

$$\pi_j(\theta_j|y) = L_j(\theta_j)\pi_j(\theta_j)/\int L_j(\theta_j)\pi_j(\theta_j)d\theta_j.$$

Thus alternatively

$$L_j^A = \int L_j^2(\theta_j)\pi_j(\theta_j)d\theta_j/\int L_j(\theta_j)\pi_j(\theta_j)d\theta_j.$$

For the single normal distribution the average likelihood is

$$L_1^A = 2^{-n/2}\frac{\Gamma(n-\frac{1}{2})}{\Gamma(\frac{n-1}{2})}RSS^{-n/2}$$

where RSS is the sum of squares of the observations, i.e., $(n-1)s^2$. Under the LCN model the average likelihood is

$$L_2^A = \frac{\Gamma(n-1)}{2^{n/2}\Gamma(\frac{n-1}{2})} \frac{\sum_A \frac{B(2n_1+1,2n_2+1)}{\sqrt{n_1 n_2}} RSS_1^{-n+1}}{\sum_A \frac{B(n_1+1,n_2+1)}{\sqrt{n_1 n_2}} RSS_1^{-\frac{n}{2}+1}}$$

where the sums are over A, the set of all assignments of the observations into two groups of at least one observation each. B is the beta function and $RSS_1 = RSS_a + RSS_b$, the pooled estimate of the sum of squares based on the two group assignment. The ratio L_2^A/L_1^A is then interpreted as evidence for the LCN mixture.

This result may look formidable but it has a simple interpretation. Consider all $2^{n-1} - 1$ possible assignments of observations to the mixture components which have at least one observation in each component. For each assignment, we calculate the residual sums of squares RSS_a and RSS_b of the observations within each component, and construct the numerator and denominator of L_2^A from the values of the residual sums of squares for this assignment. For such an assignment, the ratio of the numerator and denominator would be the *classification likelihood* for this assignment. Then we form an equally weighted sum of averages of the numerator and denominator terms over all possible assignments. This gives the average likelihood over all assignments that the sample was drawn from a mixture. The ratio then provides the evidence that the sample does (or does not) come from a normal mixture.

The use of the ALT to provide a test for the presence of a mixture has several advantages over the conventional LRT:

(i) Since the value of the ALT is interpreted directly, a fixed value of the ratio of posterior means is used to define strong evidence against the single distribution model ("rejection" of the null hypothesis).

(ii) The computation of the ALT does not require the MLE's of the parameters, so that convergence problems and possible multiple modes of the likelihood do not affect the computation of the ALT.

(iii) The ALT has an appealing interpretation in terms of averages of components of the *classification likelihoods* obtained by formally assigning observations to the mixture components.

To calculate the ALT for even moderately sized samples an enormous amount of computational effort is required. For example, a sample of size 20 will require averaging over $2^{19} - 1 = 524287$ assignments. Under the LCN mixture model, partitions of the observations in which central observations are assigned to one component, and peripheral observations on both sides of the central observations are assigned to the other component,

have very small likelihood. A modified average likelihood test can therefore be considered as an approximation to the posterior mean in which only
contiguous partitions of the ordered observations into two components are
used. The averaging is thus carried out over only $(n-1)$ assignments of
observations rather than $2^{n-1} - 1$.

The LRT for the LSCN mixture model is often difficult to obtain because of the singularities in the likelihood. These do not affect the posterior mean of the likelihood as this involves integrating over the parameters
rather than maximizing, the singularities having measure zero in the inte-
gration. In averaging the likelihood terms over assignments of observations
to components, only those assignments in which each component has at
least two observations is considered, to ensure identifiability of the component parameters. Under the LSCN mixture model,

$$
L_2^{A'} = \frac{2^{-n/2} \sum_{A'} \frac{B(2n_1+1, 2n_2+1)}{\sqrt{n_1 n_2}} \frac{\Gamma(n_1-\frac{1}{2})\Gamma(n_2-\frac{1}{2})}{RSS_a^{n_1-\frac{1}{2}} RSS_b^{n_2-\frac{1}{2}}}}{\sum_{A'} \frac{B(n_1+1, n_2+1)}{\sqrt{n_1 n_2}} \frac{\Gamma(\frac{n_1-1}{2})\Gamma(\frac{n_2-1}{2})}{RSS_a^{\frac{n_1-1}{2}} RSS_b^{\frac{n_2-1}{2}}}}
$$

where the sums are over A', the set of all assignments into two groups with
at least two observations per group, and RSS_a and RSS_b are the respective
group sums of squares. The analytic form of the posterior mean is very
similar to that for the equal variance case, and indeed the two posterior
means can be computed concurrently.

11.3.7 Other Tests for Mixtures

As a test for LCN mixtures, Day (1969) suggested using $\hat{\Delta}$, the standardized difference between component means. Using the MLE's of the means
and covariance matrix, the estimate of the Mahalanobis distance between
the two components

$$
\Delta = [(\mu_1 - \mu_2)\Sigma^{-1}(\mu_1 - \mu_2)']^{1/2}
$$

is obtained and used as a test statistic. Everitt (1988) provided critical
values of $\hat{\Delta}$ for $m = 2(1)10$ and selected sample sizes from 20 to 500 (Table
B35). The univariate case, in which the test statistic would be the estimate
of $D = |\mu_2 - \mu_1|/\sigma$, was not included.

The possibility that the quantitative measure of interest may be a
skewed unimodal distribution rather than a mixture was considered by
MacLean, Morton, Elston and Yee (1976). Their interest was in presenting

a conservative test for identifying possible mixtures, i.e., one which would be less likely to falsely identify a skewed distribution as a mixture at the expense of failing to identify mixtures that look skewed. They suggested that one test $H_0 : X^{(q)} \sim N(\mu, \sigma^2)$ against $H_1 : X^{(r)} \sim h(x)$ where $h(x)$ is an LCN mixture, where $X^{(q)}$ and $X^{(r)}$ are scaled power transformations of the observations, given by

$$ X^{(q)} = \frac{s}{q} \left[\left(\frac{x}{s} + 1 \right)^q - 1 \right], $$

the observations x have been standardized to mean 0 and variance 1, and s is chosen so that $x/s + 1$ is positive for all observations. The powers q and r are estimated using maximum likelihood so that they are the best transformation to fit the null and alternative hypotheses, respectively.

11.4 Power of Tests for Mixtures and Recommendations

There have been few comparisons of the methods used to empirically establish the existence of mixtures. The lack of power studies is probably due in part to the relative difficulty in computing the LRT, and the fact that the testing criteria have not been accurate to this point. Elston, Namboodiri, Nino and Pollizer (1984) used $\sqrt{b_1}$ and b_2 as their first step in demonstrating that a single normal distribution did not fit the data. Lee (1975) found that b_2 was more powerful than the EH test when π was near 0 or 1, but when π was close to 0.50 EH was more powerful.

Mendell, Thode and Finch (1991) conducted power studies of the LRT for the purpose of determining the relationship between power, sample size, and mixture parameterization for LCN mixtures. It was determined that even for samples of size 50, there was reasonable power ($> 50\%$ to identify a mixture with moderately separated components, i.e., $D \geq 4$), as long as the mixing proportion was between 0.20 and 0.80; for values of π outside of this range, power dropped off dramatically.

A common misconception is that the LRT has optimal power properties compared to other tests for identifying mixtures. Without valid power comparisons, this misconception will continue. In the only large scale comparison of the LRT with other tests, for two component univariate LCN mixtures Mendell, Finch and Thode (1993) showed that, although the LRT always had high power relative to other tests, there was a limited range of parameterizations of LCN mixtures over which the LRT appeared to be the most powerful of all tests (Figure 11.5).

Two tests were consistently most powerful for some values of the mixing proportion: LRT and EH. The EH test was the most powerful test when

$0.35 < \pi < 0.65$ with the LRT being its closest competitor. In general, as the sample size or D increased for fixed π in this range, the difference in power between the two tests became smaller. Outside of this range the power of the EH test decreased.

When $0.65 \leq \pi \leq 0.85$ (or $0.15 \leq \pi \leq 0.35$), the LRT was best. For these mixing proportions, the third cumulant of the mixture has a moderate value, and the fourth cumulant is near zero, so that the standard tests of normality (e.g., skewness and kurtosis) would be at a considerable disadvantage compared to the LRT. The Anderson-Darling, Cramer-von Mises and Wilk-Shapiro tests were close competitors of the LRT in this range of mixing proportions.

When $\pi > 0.85$ or $\pi < 0.15$, $\sqrt{b_1}$, K^2 (combined $\sqrt{b_1}$ and b_2), Filliben's (1975) r, W, and the fifth moment test were about equal in power. An LCN mixture of this type is very skewed, so that the effectiveness of tests for skewness would be expected. Even though a two-tailed test was used in this study, these findings are consistent with those of Ferguson (1961). They also agree with Shapiro, Wilk, and Chen (1968) who found that W had good power for detecting skewed alternatives. The relative lack of power of the LRT for small values of π was surprising and suggests that the use of the LRT as a test for outliers (as proposed in Aitkin and Wilson, 1980) is not recommended. Although they included relatively few tests in their study, Pearson, D'Agostino and Bowman (1977) showed that $\sqrt{b_1}$ and W usually had the best power at detecting LCN mixtures for $\pi \leq 0.20$ (or $\pi \geq 0.80$). The LRT and EH tests were not included in their study.

Kurtosis was rarely competitive with the most powerful test at any specified value of π and was less powerful than the LRT in general.

The two outlier tests examined had negligible power for symmetric and near-symmetric distributions. T had slightly better power than LRT for $\pi = 0.95$ and samples of size 25 and was competitive with LRT for $n = 50$ and 100, although T was never the best test for any value of D.

When there is prior information concerning the true value of the mixing proportion, EH, LRT or $\sqrt{b_1}$ are recommended for detecting mixtures depending upon the hypothesized value of π. In the absence of prior knowledge, the LRT is recommended since it is always within a respectable distance from the best test for any value of π.

No LRT has been developed for the SCN mixture. There have been a few comparisons of other tests for normality which included SCN mixtures as alternatives. Both Pearson, D'Agostino and Bowman (1977) and Thode, Finch and Smith (1983) showed that b_2 usually had power at least as high as other tests for detecting this type of mixture. Since the maximum value of π used in these studies was 0.20, the results corroborate Ferguson's (1961) identification of b_2 as the best locally invariant test for this outlier model.

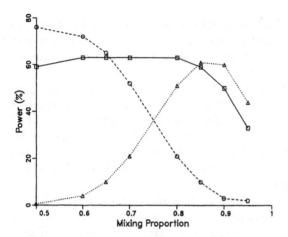

Figure 11.5. Estimated power of the likelihood ratio, Engelman-Hartigan and $\sqrt{b_1}$ tests to detect LCN mixtures with $D = 3$ for specified values of mixing proportion π for samples of size 50: $\square$ LRT; $\circ$ EH; $\triangle$ $\sqrt{b_1}$.

Most likely due to the difficulty in calculating the LRT, there have been no comparisons of tests for detecting LSCN mixtures. Power to detect multivariate LCN mixtures using the LRT have been conducted (Everitt, 1981), which indicate that for fixed values of π and D, power decreases with increasing dimensionality. Everitt (1988) also studied the power of $\hat{\Delta}$ in detecting multivariate LCN mixtures, although no other tests were compared. As might be expected, power curves from the LRT and the MLE test $\hat{\Delta}$ seem similar for the limited number of cases which can be compared between the two studies.

11.5 Further Reading

For discussion of the general mixture problem, see Everitt and Hand (1981), Titterington, Smith and Makov (1985), or McLachlan and Basford (1988), and associated references. Titterington, Smith and Makov (1985) provide an extensive bibliography of normal mixture applications in the literature over a wide range of other areas, including medicine, electrophoresis, geology, zoology and others.

References

Aitkin, M.A. (1991). Posterior Bayes factors. Journal of the Royal Statistical Society B53, 111-128.

Aitkin, M., and Wilson, G.T. (1980). Mixture models, outliers, and the EM algorithm. Technometrics 22, 325-331.

Bhattacharya, C.G. (1967). A simple method of resolution of a distribution into Gaussian components. Biometrics 23, 115-135.

Box, G.E.P. (1953). A note on regions for tests of kurtosis. Biometrika 40, 465-468.

Box, G.E.P, and Tiao, G.C. (1968). A Bayesian approach to some outlier problems. Biometrika 55, 119-129.

Charlier, C.V.L. (1906). Researches into the theory of probability. Lunds Universitets Arkskrift, Ny foljd Afd. 2.1, No. 5.

Day, N.E. (1969). Estimating the components of a mixture of normal distributions. Biometrika 56, 463-474.

Dempster, A.P., Laird, N.M., and Rubin, D.B. (1977). Maximum likelihood from incomplete data via the EM algorithm. Journal of the Royal Statistical Society B39, 1-38.

Dixon, W. (1950). Analysis of extreme values. Annals of Mathematical Statistics 21, 488-505.

Dixon, W. (1951). Ratios involving extreme values. Annals of Mathematical Statistics 22, 68-78.

Eisenberger, I. (1964). Genesis of bimodal distributions. Technometrics 6, 357-363.

Elston, R.C., Namboodiri, K.K., Nino, H.V., and Pollizer, W.S. (1984). Studies on blood and urine glucose in Seminole Indians: indications for segregation of a major gene. American Journal of Human Genetics 26, 13-34.

Elston, R.C., and Stewart, J. (1971). A general model for the genetic analysis of pedigree data. Human Heredity 21, 523-542.

Engelman, L., and Hartigan, J.A. (1969). Percentage points of a test for clusters. Journal of the American Statistical Association 64, 1647-1648.

Everitt, B.S. (1981). A Monte Carlo investigation of the likelihood ratio test for the number of components in a mixture of normal distributions, Multivariate Behavioral Research 16, 171-180.

Everitt, B.S. (1984). Maximum likelihood estimation of the parameters in a mixture of two univariate normal distributions; a comparison of different algorithms. The Statistician 33, 205-215.

Everitt, B.S. (1988). A test of multivariate normality against the alternative that the distribution is a mixture. Journal of Statistical Computation and Simulation 30, 103-115.

Everitt, B.S., and Hand, D.J. (1981). **Finite Mixture Distributions**. Chapman and Hall, New York.

Falconer, D.S. (1960). **Quantitative Genetics**. Oliver and Boyd, Ltd., Edinburgh.

Ferguson, T.S. (1961). On the rejection of outliers. Proceedings of the 4th Berkeley Symposium on Mathematical Statistics and Probability, University of California Press, Berkeley CA.

Filliben, J.J. (1975).The probability plot coefficient test for normality. Technometrics 17, 111-117.

Finch, S.J., Mendell, N.R., and Thode, Jr., H.C. (1989). Probabilistic measures of the adequacy of a numerical search for a global maximum. Journal of the American Statistical Association 84, 1020-1023.

Fisher, W.D. (1958). On grouping for maximum homogeneity. Journal of the American Statistical Association 53, 789-798.

Grubbs, F.E. (1950). Sample criteria for testing outlying observations. Annals of Mathematical Statistics 21, 27-58.

Grubbs, F. (1969). Procedures for detecting outlying observations in samples. Technometrics 11, 1-19.

Hartigan, J.A. (1977). Distribution problems in clustering. In J. van Ryzin, ed., **Classification and Clustering**, Academic Press, New York.

Hasselblad, V. (1966). Estimation of parameters for a mixture of normal distributions. Technometrics 8, 431-444.

Hosmer, D.W. (1973). A comparison of iterative maximum likelihood estimates of the parameters of a mixture of two normal distributions under three different types of sample. Biometrics 29, 761-770.

Karlof, J., and Mendell, N.R. (1986). An evaluation of typing procedures for quantitative data: results of a Monte Carlo study. Human Immunology 13, 6-18.

Lee, K.K. (1975). Some statistical tests for clusters. University of North Carolina, Institute of Statistics, Mimeo Series No. 1036.

Lindsay, B.G. (1983). The geometry of mixture likelihoods: a general theory. Annals of Statistics 11, 86-94.

MacLean, C. J., Morton, N. E., Elston, R. C., and Yee, S. (1976). Skewness in commingled distributions. Biometrics 32, 695-699.

McLachlan, G.J. (1987). On bootstrapping the likelihood ratio test statistic for the number of components in a normal mixture. Applied Statistics 36, 318-324.

McLachlan, G.J., and Basford, K.E. (1988). **Mixture Models: Inference and Applications to Clustering**. Marcel Dekker, New York.

Mendell, N.R., Finch, S.J., and Thode, Jr., H.C. (1993). Where is the likelihood ratio test powerful for detecting two component normal mixtures? Biometrics 49, 907-915.

Mendell, N.R., Lee, K.L., Reinsmoen, N., Yunis, E., Amos, D. B., and Emme, L. (1977). Statistical methods for evaluating responses in HLA-D typing. Transplantation Proceedings 9, 99-106.

Mendell, N.R., Thode, Jr., H.C., and Finch, S.J. (1991). The likelihood ratio test for the two component normal mixture problem - power and sample size analysis. Biometrics 47, 1143-1148.

Nash, J.C. (1979). **Compact Numerical Methods for Computers**. John Wiley and Sons, New York.

Newcomb, S. (1886). A generalized theory of the combination of observations so as to obtain the best result. American Journal of Mathematics 8, 343-366.

O'Neill, T.J. (1978). Normal discrimination with unclassified observations. Journal of the American Statistical Association 73, 821-826.

Pearson, K. (1894). Contributions to the mathematical theory of evolution. Philosophical Transactions of the Royal Society A 185, 71-110.

Pearson, E.S., D'Agostino, R.B., and Bowman, K.O. (1977). Tests for departure from normality: comparison of powers. Biometrika 64, 231-246.

Royston, P. (1989). Letter to the editor, 'Tests for mixtures of two normals'. Biometrics 45, 1330-1332.

Scott, A.J., and Symons, M.J. (1971). Clustering methods based on likelihood ratio criteria. Biometrics 27, 387-397.

Shapiro, S.S, Wilk, M.B., and Chen, H.J. (1968). A comparative study of various tests for normality. Journal of the American Statistical Association 63, 1343-1372.

Tan, W.Y., and Chang, W.C. (1972). Some comparisons of the method of moments and the method of maximum likelihood in estimating parameters of a mixture of two normal densities. Journal of the American Statistical Association 67, 702-708.

Thode, Jr., H.C. (1985). **Power of absolute moment tests against symmetric non-normal alternatives**. Ph.D. dissertation, University Microfilms, Ann Arbor, MI.

Thode, H.C., Jr., Finch, S.J., and Mendell, N.R. (1987). Finding the MLE in a two component normal mixture. Proceedings of the ASA Section on Statistical Computing, 1987 Joint Statistical Meetings, San Francisco, CA, August 17-20.

Thode, H.C., Jr., Finch, S.J., and Mendell, N.R. (1988). Simulated percentage points for the null distribution of the likelihood ratio test for a mixture of two normals. Biometrics 44, 1195-1201.

Thode, H.C., Jr., Smith, L.A., and Finch, S.J. (1983). Power of tests of normality for detecting scale contaminated normal samples. Communications in Statistics - Simulation and Computation 12, 675-695.

Titterington, D.M., Smith, A.F.M., and Makov, U.E. (1985). **Statistical Analysis of Finite Mixture Distributions**. John Wiley and Sons, New York.

Tukey, J.W. (1960). A survey of sampling from contaminated distributions. In I. Olkin, ed., **Contributions to Probability and Statistics I**, Stanford University Press, Stanford CA.

Wilks, S.S. (1938). The large sample distribution of the likelihood ratio for testing composite hypotheses. Annals of Mathematical Statistics 9, 60-62.

Wolfe, J.H. (1970). Pattern clustering by multivariate mixture analysis. Multivariate Behavioral Research 5, 329-350.

Wolfe, J.H. (1971). A Monte Carlo study of the sampling distribution of the likelihood ratio for mixtures of multinormal distributions. Technical Bulletin STB 72-2, San Diego U.S. Naval Personnel and Training Research Laboratory.

CHAPTER 12

ROBUST ESTIMATION OF LOCATION AND SCALE

"Robustness is a desirable but vague property ... insofar as the type of departure and the meaning of "good performance" need to be specified."

Bickel, 1988

In the preceding chapters we have been concerned with identifying whether or not a sample may have come from a normal distribution. Presumably, our concern stems from the desire to obtain efficient parameter estimates and/or conduct the most appropriate and powerful statistical test procedures with regard to the estimated parameter(s). In the event that the test(s) for normality do not reject the possibility of normality, we usually are happy to continue with appropriate parameter estimates (e.g., mean and standard deviation) and tests (e.g., t-test). However, to this point we have not addressed the issue of what could be done in the event that the hypothesis of normality is rejected. In this chapter we provide some basic methodology on obtaining robust estimates as a result of or as an alternative to testing for normality.

Since to this point we have been solely concerned with single sample tests for normality, rather than multiple samples or procedures such as regression, here we will only consider robust estimation as it relates to single (univariate and multivariate) sample parameter estimates. Specifically, we will describe robust estimation procedures for location and scale parameters. Although this chapter is not meant to provide a comprehensive exposition of robust estimation methods, we provide basic background on different types of robust estimators, the most common forms of the estimators for practical use, and references for those interested in additional in-depth information.

12.1 Background

Much of statistics is concerned with parameter estimation. Perhaps the most common estimation problems are those of location and scale estimation in single samples. The sample mean and standard deviation, which are the best location and scale estimates under the strict assumption of normality, are known to be poor robust estimates under even slight departures from the normal distribution; in fact, in their evaluation of the robustness of 68 univariate estimates of location, Andrews et al. (1972) state:

> "Which was the worst estimator in the study? If there is any clear candidate for such an overall statement, it is the arithmetic mean, long celebrated because of its many "optimality properties" and its revered use in applications. There is no contradiction: the optimality questions of mathematical theory are important for that theory and very useful, as orientation points, for applicable procedures. If taken as anything more than that, they are completely irrelevant and misleading for the broad requirements of practice."

Therefore, other estimates are desirable which are nearly as good as the mean under normality, but are better (in some sense) under departures from normality.

Departures from normality which incur the need for robust estimation include long-tailed symmetric distributions, asymmetric distributions, and contamination by outliers. In symmetric distributions, there is a natural "center" or location which can be well described by either the distribution mean or median. In contrast, in asymmetric distributions there is the question of what "center" means. Therefore, robust estimation of location is often focused on symmetric departures from normality, or contamination of a symmetric distribution with outliers. It seems obvious that for symmetric distributions these methods should provide equal weight to observations at equal distances from the center; in addition, observations that are "far" from the center should have less of an effect on the location estimate, especially if long tails and/or outliers are of concern.

Two desirable properties of a robust estimator are resistance and robustness of efficiency. An estimator is resistant when it is minimally affected by a few large errors or by small rounding and/or grouping errors. An estimator has robustness of efficiency if it has close to minimum variance for each distribution over the range of distributions of interest; further, these two properties are asymptotically equivalent (Huber, 1981). A simple measure of resistance of an estimator T_{n-1} for a given sample of size $n-1$ can be demonstrated using the sensitivity curve (SC), defined at each value of x as

$$SC(x|T_n) = n(T_n(x_1, \ldots, x_{n-1}, x) - T_{n-1}(x_1, \ldots, x_{n-1}))$$

which indicates the sensitivity of T_{n-1} to contamination by an observation x. Note the slight difference between the sensitivity curve, which indicates the sensitivity of an estimator to an additional observation for a given sample size, and the sample influence function (Section 10.3.1), which indicates the sensitivity of an estimator to each observation in a sample. The influence curve (Hampel, 1968) allows the sample size to go to infinity for a sensitivity curve, so that a single curve can be defined which shows the effect of an additional observation on an estimator for a given distribution.

12.2 Univariate Robust Estimators of Location

Probably the most commonly estimated parameter in applied statistics is that of location. A number of classes of univariate robust location estimators have been proposed. M-estimators are based on maximum likelihood estimation theory. W-estimators are weighted estimators which provide weights based on the distance of an observation from the center. L-estimators are linear combinations of order statistics with fixed weights. R-estimators are based on the ranks of combinations of paired observations. Adaptive estimators are chosen relative to characteristics of the sample. The Princeton study of robust estimates of location compared 68 robust estimates (Andrews et al., 1972); we will describe the most commonly used and some of the simpler estimates from that study, as well as some that have been developed since that study was completed.

12.2.1 Simple Estimators of Location

The most commonly used estimators of location are the sample mean and median; these are simple to calculate, well known and understood, and have desirable properties under certain assumptions. Other simple location estimators include the sample mode and midrange. Since we deal here with continuous data, we will exclude the mode in our discussion of location estimators because the number of tied observations may tend to be small unless rounding or grouping occurs.

The sample mean is the classical estimator of location. It is the most efficient estimator when a sample comes from a normal distribution, and is the MLE of the mean for that distribution. However, the sample mean gives equal weight to all observations from a sample, so that observations that are far from the center can have excessive influence on the estimate. The sensitivity curve of the sample mean is given by

$$SC(x|\bar{x}) = n(\bar{x}_n - \bar{x}_{n-1}) = \frac{n}{n-1}(x - \bar{x}_n)$$

where $\bar{x}_n$ and $\bar{x}_{n-1}$ are the sample means of a sample of size n and $n-1$, respectively. The sensitivity increases linearly as x gets farther from the mean, showing that the sample mean is not resistant, especially against outlier contamination or long-tailed distributions.

The sample median is the classical robust estimator of location. The median gives a weight of 1 to the central order statistic, and 0 to all other observations (when the sample is odd). Since it is based on only the central one or two observations, it is very robust against outliers and long tails. However, while the sample median ignores outliers, it also ignores other information in the sample, thereby having higher variance than other estimators except in certain situations. The sample median is also highly influenced by rounding or grouping of observations in the central portion of the sample.

The sample midrange

$$MR = (x_{(n)} - x_{(1)})/2$$

gives weights of 0.5 to each of the maximum and minimum observations, and 0 to all other observations. The midrange is the location estimator that is most affected by outliers or long tails and so, for the purposes of robust estimation, is not of much value.

Note that the mean, median and midrange are the location MLE's for the normal ($\beta_2 = 3$), double exponential ($\beta_2 = 6$) and uniform ($\beta_2 = 1.9$) distributions, respectively. This indicates a tendency for the "best" weights of order statistics in calculating location to flow from the extremes for short-tailed distributions to the center for long-tailed distributions. As will be seen, the sample mean and median are special cases of most of the classes of univariate robust location estimators described in this section.

12.2.2 Maximum Likelihood Estimators (M-Estimates)

The class of robust estimates known as M-estimators are so called because some of the estimators under specific circumstances are maximum likelihood estimates. M-estimates are obtained by finding the estimate which minimizes some objective function $p(x_i, t)$, where t is the estimate of location. The M-estimate T_M for an objective function p and sample $x_1, \ldots, x_n$ is that value of t which minimizes $\sum_{i=1}^{n} p(x_i, t)$. By defining

$$\psi(x_i, t) = \delta p(x_i, t)/\delta t$$

the M-estimate is the solution to the equation

$$\sum_{i=1}^{n} \psi(x_i, t) = 0.$$

In order to be location invariant, the argument (x_i, t) of p and ψ must be of the form $x - t$. In order to be scale invariant, M-estimation must also take into account the scale of the observations, and so $\psi(x - t)$ generally includes a fixed auxiliary measure of scale S_n based on the sample. A tuning constant c, used as a multiplier of S_n, is also often required to rescale the observations. The most frequently used forms of S_n are the median absolute deviation from the sample median (MAD) and the fourth-spread (d_F) (Section 12.3). Then, M-estimates are solutions to

$$\sum_{i=1}^{n} \psi(y_i) = 0 \tag{12.1}$$

where

$$y_i = \frac{x_i - t}{cS_n} \tag{12.2}$$

for appropriate choices of S_n and c, if necessary. Throughout this section we will use the notation y_i to indicate the quantity (12.2). T_M can be placed in the weighted average form

$$T_M = \left(\sum_{i=1}^{n} w_i y_i \right) / \left(\sum_{i=1}^{n} w_i \right)$$

where $w_i = |\psi(y_i)/y_i|$. M-estimators can be the most difficult to obtain of all robust estimates, as they generally require iterative procedures.

The effect of each of the observations on an estimator can be summarized by its influence curve (Hampel 1968, 1974). For an M-estimator T_M of a parameter T, the effect of the influence curve (IC) function is given by

$$T_M = T + n^{-1} \sum_{i=1}^{n} IC(y_i) + R_n$$

where the remainder R_n is negligible and $IC(y)$ is a function with the numerator $\psi(y)$ (Carroll, 1989). Thus, it can be seen that based on either influence or weight, the effect of each observation x_i on the resulting estimator is directly proportional to $\psi(y_i)$.

Special cases of M-estimators are the sample mean (MLE for the normal distribution) and median (MLE for the double exponential or Laplace distribution). Additional M-estimators that are widely used include Huber's, Tukey's biweight, Hampel's three part redescending, and Andrews' wave M-estimators. For the case where the objective function is the squared residual, i.e.,

$$p(x,t) = \left(\frac{x-t}{cS_n}\right)^2 = y^2$$

the equation to be solved, ignoring constants, is

$$\sum_{i=1}^{n}(x_i - t) = 0$$

which gives $T_M = \bar{x}$. Note that this particular estimate does not depend on cS_n. The influence for outlying observations is unbounded, since $\psi(y) = x - t$, so that the mean is not particularly robust. If the objective function is the absolute deviation from t, the M-estimate is the median, which also does not depend on cS_n.

Huber's (1964) M-estimator (T_H) is a monotone estimate ($\psi(y_i) \le \psi(y_j)$ if $y_i < y_j$), being based on the function

$$\psi(y) = \begin{cases} y & |y| \le k \\ k\,\mathrm{sgn}(y) & |y| > k \end{cases}$$

for appropriate choice of k, with the denominator of y being the normalized f-spread or normalized MAD = MAD/0.6745 (Goodall, 1983). This M-estimator was developed under the assumption of a general gross error (contaminated) model where a normal distribution is contaminated with a small proportion ϵ ($0 < \epsilon \le 0.10$) of observations from another symmetric distribution, which may or may not be normal itself. Bickel (1988) suggested a choice of k between 1 and 1.5, which corresponds to a 100ϵ of 5 to 10%. Ruppert (1988a) recommended $1 \le k \le 2$, stating that $k = 1.5$ appeared to be a good choice.

Example 12.1. The mean and median of the water alkalinity data (Data Set 10) are 36.95 and 35, respectively. The sample is long-tailed based on the value of the sample kurtosis (5.4) and some skewness appears to be present, possibly due to some outliers (see Example 6.6). Huber's M-estimate of location is 35.6, using $k = 1.339$, and is similar to the median.

Tukey's biweight, Hampel's three part redescending and Andrew's wave estimates are all "redescending" estimators, that is, $\psi(y)$ returns to 0 for large $|y|$. Tukey's biweight estimator (T_T) is the solution to (12.1) where

$$\psi(y) = \begin{cases} y(1 - y^2)^2 & |y| \le 1 \\ 0 & |y| > 1 \end{cases}$$

the denominator of (12.2) is cS_n, and S_n is the MAD. Those observations exceeding cS_n are trimmed (given a weight of 0) from the calculation. In this way, by appropriate choice of c, limits can be placed on the influence of extreme observations. Goodall (1983) proposed $6 \le c \le 12$ as reasonable values.

Hampel's three part M-estimator (T_{H3}) is another example of a redescending estimator, although in contrast to Tukey's smooth ψ curve the ψ curve is in the form of linear segments. Hampel's estimator is the solution to (12.1) with the MAD as the denominator of (12.2) and

$$\psi(y) = \begin{cases} y & |y| \le a \\ a\,\mathrm{sgn}(y) & a < |y| \le b \\ a\frac{c-|y|}{c-b}\mathrm{sgn}(y) & b < |y| \le c \\ 0 & |y| > c \end{cases}$$

for appropriate choices of constants a, b and c. Favorable choices for these constants appear to be, roughly, $1.5 \le a \le 2.5$, $2 \le b \le 4$ and $5 \le c \le 15$; Goodall (1983) also recommended that $c - b \ge 2a$.

Andrews' wave estimator (T_A) is another redescending estimator, defined using

$$\psi(y) = \begin{cases} \sin(\pi y)/\pi & |y| \le 1 \\ 0 & |y| > 1 \end{cases}$$

where the denominator of (12.2) is $c\pi$MAD. Values of c that have been explored range from about 1.5 to 2.5 (Andrews et al., 1972; Gross, 1976; Hogg, 1979).

Example 12.2. The mean of the leukemia latency data (Data Set 9) is 60.4, however, several outliers have been identified (Examples 6.4, 6.7). The M-estimates of location are considerably lower than the sample mean: $T_T = 49.8$ (c = 4.7), $T_{H3} = 51.0$ (a = 1.7, b = 3.4, c = 8.5) and $T_A = 49.8$ (c = 1.34).

12.2.3 Weighted Estimators (W-Estimates)

A robust method related to M-estimation is W- (weighted) estimation. The form of W-estimation is similar to that of (12.1) in that T_W is the solution to the equation obtained from (12.1) by substituting $yw(y)$ for $\psi(y)$, i.e.,

$$\sum_{i=1}^{n} yw(y) = 0 \qquad (12.3)$$

where $w(y)$ is a weighting function which may depend on y. Equivalence of (12.1) and (12.3) can be shown by letting

$$w(y) = \frac{\psi(y)}{y}$$

(Huber, 1977). From (12.3) we can then derive

$$T_W = \frac{\sum_{i=1}^{n} x_i w(y_i)}{\sum_{i=1}^{n} w(y_i)} \qquad (12.4).$$

Rather than requiring a more complex numerical procedure for a solution, as is required by many of the M-estimators, (12.4) suggests the use of a stepwise iterative procedure, where at step $k + 1$ the W-estimator is

$$T_W^{(k+1)} = \frac{\sum_{i=1}^{n} x_i w^{(k)}}{\sum_{i=1}^{n} w^{(k)}} \qquad (12.5)$$

where $w^{(k)} = w(y_i^{(k)})$ and

$$y_i^{(k)} = \frac{x_i - T_W^{(k)}}{cS_n}$$

is based on the estimator at the current step. The suggested starting point for W-estimation is $T_W^{(0)}$ equal to the median. If there is a unique solution

Table 12.1 Weight functions for selected W-estimators.

W-Estimator	weight function	range
mean	1	all
median	$\text{sgn}(y)/y$	all
Huber's	1 $k\text{sgn}(y)/y$	$\|y\| \leq k$ elsewhere
Tukey's biweight	$(1-y^2)^2$ 0	$\|y\| \leq 1$ elsewhere
Hampel's 3 part redescending	1 $a\text{sgn}(y)/y$ $a\frac{c-\|y\|}{c-b}\frac{\text{sgn}(y)}{\|y\|}$ 0	$\|y\| \leq a$ $a < \|y\| \leq b$ $b < \|y\| \leq c$ $\|y\| > c$
Andrews' wave	$\sin(\pi y)/\pi y$ 0	$\|y\| \leq 1$ elsewhere

to (12.3), then the W-estimator and the M-estimator converge to the same estimate.

Table 12.1 shows the weighting functions which correspond to the mean, median and each of the M-estimators discussed in the preceding section.

12.2.4 One Step m-Estimators and w-Estimators

M- and W-estimators as described in the previous sections require an iterative solution of the ψ or w function, respectively. One step m- and w-estimators (the lower case indicating a one step estimator) are those estimators obtained after using only one step of the iterative procedure. For example, using (12.5), the one step w-estimate of location is

$$T_w = \frac{\sum x_i w(y_i^0)}{\sum w(y_i^0)}$$

where

$$y_i^0 = \frac{x_i - x_M}{cS_n}$$

and w is the selected weighting function. Similarly,

$$T_m = x_M + cS_n \frac{\sum \psi(y_i^0)}{\sum \psi'(y_i^0)}$$

is the general procedure for obtaining m-estimates.

The median is the starting value of choice for one step estimation. Generally, as with W-estimation compared to M-estimation, w-estimation is simpler than m-estimation. Further, the performance of w-estimators is essentially as good as that of M-estimators (Hoaglin, Mosteller and Tukey, 1983b).

12.2.5 Linear Combinations of Order Statistics (L-Estimates)

L-estimators are defined as linear combinations of the sample order statistics. Formally, an L-estimate T_L is calculated by

$$T_L = \sum_{i=1}^{n} a_i x_{(i)}$$

where $\sum_{i=1}^{n} a_i = 1$ and $0 \le a_i \le 1$ for all i. In addition, for symmetric distributions the best linear combinations of order statistics for estimating central location employ weights such that $a_i = a_{n-i+1}$, i.e., the weights are symmetric. The robustness of an L-estimate against a particular characteristic of a sample is dependent upon the weights assigned; for example, an L-estimate with small weights for the lowest and highest order statistics would be robust against outliers and heavy-tailed distributions.

L-estimates are different from W-estimates in that the weights depend only on the order of the observations, not on their value; in other words, L-estimate weights are fixed for a given sample size, whereas W-estimate weights are iteratively estimated along with the estimator itself. Similar to Crow and Siddiqui (1967) and David (1981), the more commonly known L-estimates will be discussed here, including trimmed means, Winsorized means, linearly weighted means, and combinations of the median and two symmetric order statistics.

As with M-estimates and W-estimates, the sample mean and median are special cases of L-estimates. The sample mean has weights $a_i = 1/n$ for all i. The sample median is calculated using the weights $a_{[n+1]/2} = 1$ and all other $a_i = 0$ when n is odd, and $a_{n/2} = a_{n/2+1} = 1/2$ and all other $a_i = 0$ when n is even.

Trimmed and Winsorized Means

Let the proportion $\alpha = k/n$, for some value of $k < n/2$. Then a $100\alpha\%$-trimmed mean is defined as

$$\bar{x}_\alpha = \frac{\sum_{i=k+1}^{n-k} x_{(i)}}{n - 2k}$$

or the mean of the $n - 2k$ observations remaining after removing the k largest and k smallest observations from the sample. Outliers have little influence on the resulting L-estimator, as long as the trimming percentage excludes them, whereas the efficiency of the trimmed mean in uncontaminated normal samples is good relative to the mean (Ruppert, 1988b).

If α is prespecified and $n\alpha$ is not an integer, a trimmed mean $\bar{x}_\alpha$ can still be calculated using partial weighting of the incompletely trimmed highest and lowest observations,

$$\bar{x}_\alpha = \frac{(k - n\alpha)x_{(k)} + (k - n\alpha)x_{(n-k+1)} + \sum_{i=k+1}^{n-k} x_{(i)}}{n(1 - 2\alpha)}$$

where $k = [n\alpha] + 1$. The variance of the trimmed mean can be estimated using the sample Winsorized variance (Section 12.3).

Note that the sample mean is $\bar{x}_0$ and the median is the trimmed mean with α equal to approximately 50%. Two additional special cases of trimmed means are also somewhat popular, the midmean and the broadened median. (Note that Tukey (1960) used the term "broadened median" as an alternative to "trimmed mean", whereas elsewhere "broadened median" indicates a specific amount of trimming. Tukey (1960) also used the term α-truncated mean instead of α-trimmed mean.) Rosner (1975, 1977) used the trimmed mean in his univariate outlier test (Section 6.4.5).

The midmean is defined as $\bar{x}_{0.25}$, i.e., it is the mean of the central half of the order statistics. The broadened median has trimming proportion which is dependent on sample size, but is based on a fixed number of order statistics with nonzero weight. Rosenberger and Gasko (1983) defined the broadened median as the mean of the three central order statistics for $5 \leq n \leq 12$, and the mean of the five central order statistics for $n \geq 13$. In order to accommodate symmetric trimming, partial weights are used on the lowest and highest of the central four and six order statistics, respectively, when sample sizes are even. The broadened median is useful as an alternative to the median when robustness to grouping and rounding of central observations, as well as to outliers, is of concern.

Winsorized means are similar to trimmed means in that extreme observations are not specifically included in the calculation of the mean; instead, the high and low "trimmed" values are "censored" to the highest (lowest) value and then included in the mean calculation. Specifically, the α-Winsorized mean is defined as

$$\bar{x}_\alpha^W = \frac{\sum_{i=r+1}^{n-r} x_{(i)} + rx_{(r+1)} + rx_{(n-r)}}{n} \tag{12.6}$$

where $\alpha = r/n$, $r \le [(n-1)/2]$. When n is odd and $r = (n-1)/2$, or when n is even and $r = [(n-1)/2]$, the Winsorized mean is identically the median. The mean is obtained when $r = 0$. When a preset proportion, α, of the observations are to be Winsorized off of each tail of the data distribution and $n\alpha$ is noninteger, (12.6) is still the appropriate formula since the fractional observations are folded into $x_{(r+1)}$ and $x_{(n-r)}$ which, along with the fractional weighting of these observations, is still r (e.g., Barnett and Lewis, 1994).

Example 12.3. For the leukemia latency data (Data Set 9), the 10% trimmed and Winsorized means are obtained by trimming and Winsorizing, respectively, two observations from each tail of the ordered sample. The trimmed and Winsorized means for this sample are, respectively,

$$\bar{x}_{0.10} = 55.3$$

and

$$\bar{x}_{0.10}^W = 57.8$$

compared to the standard mean value of 60.4. The 25% trimmed mean (midmean) has a value of 55.2.

In those situations where the underlying distributions of interest are symmetric, trimming and Winsorization are conventionally performed symmetrically with a prespecified percentage of observations trimmed; asymmetric trimming or Winsorization may be applied in asymmetric cases. Anscombe-type estimators (Anscombe, 1960; Guttman and Kraft, 1980) are trimmed estimates of location in the presence of outliers. Anscombe's original estimator is the sample mean after deleting the single largest standardized residual, providing its absolute value exceeds some appropriate value c; if all standardized values are less than this cutoff value, the sample mean is used. One extension of this procedure is to exclude at most the k most extreme values if k or more standardized observations exceed the cutoff, and then take the mean of the remaining observations.

Linearly Weighted Means

The linearly weighted mean is defined as

$$\bar{x}_L = \frac{\sum_{i=1}^{n/2-r}(2i-1)(x_{(r+i)}+x_{(n-r+1-i)})}{2(\frac{n}{2}-r)^2}$$

when n is even, and

$$\bar{x}_L = \frac{\sum_{i=1}^{\frac{1}{2}(n-1)-r}(2i-1)(x_{(r+i)}+x_{(n-r+1-i)})+(n-2r)x_{((n+1)/2)}}{[\frac{1}{2}(n-1)-r]^2+[\frac{1}{2}(n+1)-r]^2}$$

when n is odd. Linearly weighted means weight observations more heavily in the center of the sample, and allow for trimming (weights of 0) of extreme observations by letting $r > 0$. This L-estimate does not have the mean as a special case, while the median is obtained by the appropriate amount of trimming.

Combinations of Three Order Statistics

A number of authors suggested using a linear combination of three order statistics as an estimator of location. In the case of a symmetric distribution, the general estimator is

$$\bar{x}(r,\gamma) = \gamma(x_{(r+1)}+x_{(n-r)})+(1-2\gamma)x_M$$

where x_M is the median, defined as appropriate for odd and even sample sizes (Mosteller 1946; see also Crow and Siddiqui, 1967, and David, 1981). The other two order statistics are symmetric (in order) about the median. Two specific estimators of this type which have been investigated include the trimean (Rosenberger and Gasko, 1983)

$$\bar{x}_{TRI} = 0.25(h_l+h_u)+0.5x_M$$

where h_l and h_u are the lower and upper hinges of the sample (Section 12.3), and Gastwirth's (1966) estimator

$$\bar{x}_G = 0.3(x_{([n/3]+1)}+x_{(n-[n/3])})+0.4x_M.$$

Example 12.4. Continuing with the leukemia latency data from the previous example, estimates of location are 54.2 and 55.0 based

on the trimean and Gastwirth's estimator, respectively. These
values are both somewhat lower than 60.4, which is the value of
the sample mean.

L-estimates have the advantage of being among the simplest robust
estimators to calculate, although they may sacrifice resistance by giving
weight to extreme observations, or may sacrifice efficiency by not giving
enough weight to tail observations that are not extreme (Goodall, 1983;
Rosenberger and Gasko, 1983).

12.2.6 Rank Estimators (R-Estimates)

Location estimators which are based on rank tests are called nonparametric
or rank (R-) estimators. For a symmetric population, any rank test for
location can be used to derive an R-estimator (Hodges and Lehmann, 1963).
The Hodges-Lehmann R-estimator

$$\bar{x}_{HL} = \operatorname*{med}_{i \leq j} M_{ij},$$

where

$$M_{ij} = \frac{x_{(i)} + x_{(j)}}{2}$$

is based on Wilcoxon's one sample rank test. For this test, the mean of
all pairs of observations are calculated and the median of those means is
used as the location estimate. This requires $n(n+1)/2$ calculations, which
can become cumbersome for even a moderate sample size. Hodges (1967)
suggested a similar estimator

$$\bar{x}_H = \operatorname*{med}_{i < j} M_{ij}$$

which does not significantly reduce the amount of computation. A third
R-estimator (Bickel and Hodges, 1967; Huber, 1972) is

$$\bar{x}_{BH} = \operatorname*{med}_{i \leq [n/2]} \frac{x_{(i)} + x_{(n-i+1)}}{2}$$

which requires the calculation of only $[n/2]$ pairwise means.

*Example 12.5. For the 15 height differences between cross- and
self- fertilized Zea mays plants (Data Set 2), the mean difference*

*was 20.93. Using R-estimation, location estimates for these data
are*

$$\bar{x}_{HL} = 24.75$$

$$\bar{x}_H = 25.5$$

$$\bar{x}_{BH} = 25.5$$

*which are all similar, and reflect the reduced influence of the two
lower outliers on the estimates.*

A general class of R-estimators of the Bickel-Hodges type are called
folded medians (Andrews et al., 1972). As with the Bickel-Hodges estimator, the mean of all pairwise ordered observations is taken,

$$\frac{x_{(i)} + x_{(n-i+1)}}{2}$$

for $i = 1, \ldots, n/2$ for n even, and $i = 1, \ldots, [n/2] + 1$ for n odd (in the
latter case, the median is averaged with itself). The resulting calculated
numbers are then sorted, and are "folded" again; at each iteration, some
observations may be trimmed. After a specified number of iterations, the
median is taken as the location estimate. In this class of R-estimators, the
Bickel-Hodges estimator is the single-iteration, no-trimming folded median.

12.2.7 Adaptive Estimators (D-Estimates)

Adaptive estimators (which we will designate D-estimators) are estimation
procedures which rely in part on the characteristics of the sample. L-
estimates are nonadaptive since they have fixed weights for a given sample
size. M-estimates with prespecified tuning constants are also nonadaptive. Trimmed and Winsorized means, where the trimming/Winsorizing α
is defined in advance of data exploration, are non-adaptive; however, by
choosing an α based on sample characteristics (e.g., some measure of tail
length or heaviness), these L-estimates become D-estimates (e.g., Hogg,
1972).

Anscombe (1960) defined a simple adaptive estimator wherein the single most extreme observation is trimmed if its standardized value is larger
than some appropriate value c. A general class of Anscombe estimators
is obtained by excluding at most a prespecified number k of the most
extreme observations whose standardized values exceed c (Guttman and
Kraft, 1980). Similar adaptive estimators against the presence of outliers

might consist of a sequential outlier test (Chapter 6), followed by calculating the mean of the observations remaining after excluding the identified outliers. For these latter estimators, although the algorithm itself is defined without respect to sample characteristics, the number of steps in the algorithm (number of deleted observations) is determined by the characteristics of the sample. A related procedure for estimation after deleting outliers is called "skipping" (see below).

Jaeckel (1971) proposed an adaptive trimming method to produce an estimate T_α, where the amount of trimming ($\alpha \le 0.25$) was selected to minimize

$$\hat{V}(\alpha) = \frac{\sum_{i=n\alpha+1}^{n-n\alpha}(x_{(i)} - T_\alpha)^2 + \alpha(x_{(n\alpha+1)} - T_\alpha)^2 + \alpha(x_{(n-n\alpha)} - T_\alpha)^2}{(1 - 2\alpha)^2}$$

which is the estimated asymptotic variance of this estimate (Andrews et al., 1972).

Hogg (1967) suggested a D-estimator based on the sample kurtosis, b_2. This estimate is a simpler trimmed mean-type estimator, with trimming proportions chosen after identifying the "tailedness" of the sample, and is calculated as

$$H = \begin{cases} \bar{x}_{0.25}^c & b_2 < 2 \\ \bar{x} & 2 \le b_2 \le 4 \\ \bar{x}_{0.25} & 4 < b_2 \le 5.5 \\ x_M & 5.5 < b_2 \end{cases}$$

where $\bar{x}_{0.25}^c$ is the mean of the lowest 25% and highest 25% order statistics (the complement of the 25% trimmed mean), and the other choices are the mean, 25% trimmed mean, and median, respectively, as sample kurtosis increases. Note that this follows the profile of location MLE's, with more weight given to extreme observations for short-tailed samples and more weight in the center of the sample for long-tailed samples. Hogg (1972) used the Q statistic (Section 4.6.2)

$$Q = \frac{\bar{U}_{0.05} - \bar{L}_{0.05}}{\bar{U}_{0.5} - \bar{L}_{0.5}}$$

as a measure of tail heaviness, and also proposed Q for use in a modification of the above estimation scheme (Hogg, 1974) rather than b_2,

$$H = \begin{cases} \bar{x}_{0.25}^c & Q < 2 \\ \bar{x} & 2 \le Q \le 2.6 \\ \bar{x}_{0.1875} & 2.6 < Q \le 3.2 \\ \bar{x}_{0.375} & 3.2 < Q \end{cases}$$

where $\bar{x}_\alpha$ is the α-trimmed mean.

Example 12.6. For the 58 water well alkalinity measurements (Data Set 10), the four choices of the location estimate for H are

$$\bar{x}^c_{0.25} = 38.41, \quad \bar{x} = 36.95, \quad \bar{x}_{0.25} = 35.48, \quad x_M = 35$$

which, since the sample kurtosis is 5.4, results in the choice of estimate $H = x_M = 35$. For the same data,

$$Q = \frac{65.1 - 23.2}{43.9 - 30.0} = 3.01$$

which suggests slightly less trimming than does the b_2 method. The resulting choice of location estimate here is $\bar{x}_{0.375} = 35.6$, which does not differ substantially from the estimate based on b_2.

When the non-normality concern is mostly a long-tailedness issue, Hogg (1974) suggested an alternate selector

$$Q_1 = \frac{\bar{U}_{0.2} - \bar{L}_{0.2}}{\bar{U}_{0.5} - \bar{L}_{0.5}}$$

and choosing trimming proportions based on bands defined by cutoff values of Q_1 and sample size. De Wet and van Wyk (1979) generalized this approach by proposing a continuous choice of the trimming proportion, α,

$$\alpha = \begin{cases} a_1 & Q_1 \le c_1 \\ a_1 + \frac{Q_1 - c_1}{c_2 - c_1}(a_2 - a_1) & c_1 \le Q_1 \le c_2 \\ a_2 & Q_1 \ge c_2 \end{cases}$$

with choices of (a_1, a_2, c_1, c_2) as (0.05, 0.4, 1.75, 1.95) or (0.1, 0.4, 1.8, 1.95).

Another form of adaptive L-estimation is that of skipping, where outlying observations are trimmed before estimating location (Tukey, 1977). In an ordered sample, the hinges and f-spread (Section 12.3) are obtained, and intervals (c_1, c_2), (t_1, t_2) and (s_1, s_2) are defined using

$$c_1 = h_l - 2d_f$$

$$c_2 = h_u + 2d_f$$

$$t_1 = h_l - 1.5d_f$$

$$t_2 = h_u + 1.5d_f$$

$$s_1 = h_l - d_f$$

$$s_2 = h_u + d_f$$

where h_l and h_u are the lower and upper hinges of the sample, respectively. Using one of the defined intervals, all observations outside of that interval are deleted or "skipped". This process may be repeated on the remaining subsample until no observations exceed the interval boundaries, at which point some L-estimator is used with the remaining observations (iterative skipping), or the estimate may be obtained after only one iteration of skipping (single skipping). Andrews et al. (1972) evaluated a number of these types of estimators, including the iteratively s-skipped trimean (iterative skipping based on (s_1, s_2) and using the trimean (Section 12.2.5) on the remaining observations), the iteratively c-skipped trimean, and the CTS skipped trimean (skipping is based on the c interval, followed by the t interval and then the s interval, and then the trimean is calculated).

Example 12.7. For the 13 erosion rate values (Data Set 11), the hinges and IQR for the full sample are

$$h_l = -1.0, \quad h_u = -0.1, \quad d_f = 0.9$$

with a corresponding s-interval of (-1.9, 0.8). This results in skipping the two observations -2.0 and -4.2; a second iteration identifies no additional skipped observations. The trimean of the remaining 11 observations is -0.44, compared to the full sample mean of -0.83.

Multiply skipped means have also been suggested (Andrews et al., 1972) where a single skip based on the t interval is performed; the number of deleted observations (k) is noted, and then some multiple of that number of observations is additionally deleted from each end of the sample before estimating location. One example is the *multiply-skipped mean, 3k deleted* where k observations are identified and deleted based on t-skipping, and then k additional observations are trimmed from each end. The mean of the remaining observations is then calculated.

The "shorth", which stands for "short half of the sample", is the mean of that subsample of half the observations with the shortest length (range). For an ordered sample of size n, the ranges

$$r_k = (x_{([n/2]+k)} - x_{(k)}), \quad k = 1, \ldots, [(n+1)/2]$$

are calculated. The mean of the sample with range r_{k^*}, where k^* indicates that range which is the minimum over all of the ranges, is the shorth mean.

Example 12.8. For the 13 erosion rate values (Data Set 11), the shorth subsample is the seven observations ranging from -0.6 to 0.1 (range = 0.7). The shorth mean is 0.27.

12.3 Univariate Robust Estimators of Scale

While the concept of a standard deviation is widely understood, the more general concept of "scale" (or "spread") estimators is vague (Mosteller and Tukey, 1977); for example, some distributions have an infinite population standard deviation (although the sample standard deviation is always finite). When we restrict ourselves to a single family of distributions, then different measures of scale can be related to each other by constant multipliers; however, as is the usual case in practice, we are never absolutely sure which distribution our data come from. Therefore, two common objectives for obtaining some precise measure of scale would be to (1) present an estimate of scale that accurately describes the data (although we can not clearly define *accurately*); and, (2) obtain an estimate of scale that works well for estimation and/or testing purposes under as wide a range of conditions as possible (i.e., an estimate that is resistant and efficient).

12.3.1 Estimates of the Standard Deviation

The sample standard deviation

$$s = \left(\frac{\sum_{i=1}^{n}(x_i - \bar{x})^2}{n-1} \right)^{\frac{1}{2}}$$

is to scale estimation what the sample mean is to location estimation: it is probably the most widely used estimate of scale, and has a number of desirable optimal properties under a very strict set of assumptions (i.e., normality). However, also like the sample mean, it is not robust against departures from those assumptions, and, in fact, outliers affect the sample standard deviation even more than they affect the mean.

Downton's (1966) estimate of σ, used in D'Agostino's D test for normality (Section 4.3.2) and based on Gini's mean difference, is unbiased and is not as influenced by outliers as the sample standard deviation. This scale estimator, however, does not seem to be used much in robust estimation.

12.3.2 Range and Quasi-Ranges

Order statistics have long been used to summarize data, and differences between two order statistics were a natural outgrowth to describe the spread of the data (Tukey, 1977; Mosteller and Tukey, 1977). The range is an obvious choice of measurement of the spread, but is highly susceptible to outliers and long tails. Quasi-ranges are differences between two order statistics which are symmetrically spaced from the median.

The fourth-spread (originally denoted the H-spread by Tukey, 1977) is defined as the difference between the upper and lower hinges (or fourths). Let the "depth" of the median (i.e., the distance or number of observations from the closest end of the ordered sample) be defined as

$$d(M) = (n + 1)/2$$

so that the median is $x_{d(M)}$ if $d(M)$ is an integer (i.e., n is odd), or $(x_{d(M)-0.5} + x_{(d(M)+0.5)})/2$ if $d(M)$ is not an integer (n is even). The lower and upper hinges are defined as the values obtained from the ordered sample with depth

$$d(H) = \frac{[d(M)] + 1}{2}$$

from each end. If $d(H)$ is an integer, the lower and upper hinges are $h_l = x_{d(H)}$ and $h_u = x_{n-d(H)+1}$, respectively; if $d(H)$ is a mixed number, then the lower fourth is defined similar to the median,

$$h_l = (x_{d(H)-1/2} + x_{d(H)+1/2})/2$$

with a similar definition for h_u. The fourth-spread (f-spread) is then calculated as

$$d_f = h_u - h_l$$

which expresses the spread as the range of the middle half of the data. The fourth spread under the normal distribution is approximately equal to 1.349σ, and the ratio d_f/s can be used as an informal determination of the approximate normality of the data.

Example 12.9. The depth of the median for the assembly times (Data Set 4) is 26, and the depth of the hinges are

$$d(H) = 27/2 = 13.5$$

so that the hinges are

$$h_l = \frac{x_{(13)} + x_{(14)}}{2} = 2.15 \qquad h_u = \frac{x_{(13)} + x_{(14)}}{2} = 3.545$$

and the f-spread is

$$d_f = 3.545 - 2.15 = 1.395.$$

The ratio d_f/s is $1.395/0.89 = 1.57$ which is not too far from the normal value.

The median splits the observations in half; the hinges split each of these halves again. Additional spread estimators are easily obtained by extension of this process, i.e., halving each of the outer splits of data (eighths, sixteenths, etc.). The depth of each proceeding pair of order statistics is

$$d(k) = \frac{[d(k-1)] + 1}{2}$$

where k defines the current step (fourths, eighths, sixteenths, ...), with averaging of adjacent observations as necessary. The limit of this process leads to the range as a measure of spread,

$$r = x_{(n)} - x_{(1)}.$$

The eighths describe the range of the middle three quarters of the data, and so on as we move farther out into the tails of the data distribution. However, d_f is fairly resistant to the shape of the data, whereas sensitivity increases as we move further out into the extremes (Velleman and Hoaglin, 1981).

12.3.3 Mean and Median Absolute Deviations

Tukey (1960) showed some of the characteristics of the mean absolute deviation from the mean

$$\bar{AD} = \frac{1}{n} \sum_{i=1}^{n} |x_i - \bar{x}| \tag{12.7}$$

relative to s for contaminated normal samples, showing an improvement in efficiency for $\bar{AD}$ for normal samples with slight contamination. Similar to Downton's estimate of σ, the ratio of this estimator to s has been used as a test for normality, being essentially Geary's test (Section 3.3). However, one of the better choices of robust scale estimates (as judged by its common usage in location estimation) appears to be the median absolute deviation from the median,

$$MAD = \operatorname*{med}_i |x_i - x_M|$$

which is almost always the scale estimate of choice for standardization. The MAD is much less sensitive to long tails than s.

Another rarely used estimate of scale is the mean absolute deviation from the median, which is calculated identically to (12.7) except the sample mean is replaced by the sample median. This is another estimate of σ that has been used as a test for normality (Section 3.4.2).

12.3.4 Trimmed and Winsorized Variances

The Winsorized variance is defined as

$$s_\alpha^2 = \frac{\sum_{i=k+1}^{n-k}(x_{(i)} - \hat{\mu})^2 + k(x_{(k)} - \hat{\mu})^2 + k(x_{(n-k+1)} - \hat{\mu})^2}{n(1 - 2\alpha)^2}$$

where $\hat{\mu}$ is either the trimmed or Winsorized mean. The Winsorized variance is the estimate of the sampling variability of the trimmed mean (Ruppert, 1988b). The ratio of $\bar{x}_\alpha$ to s_α has been suggested as a robust form of t-test, and also for the estimation of confidence intervals. The trimmed variance is similarly defined.

12.3.5 M-Estimation of Scale

Rather than obtain independent estimates of location and scale, Huber's proposal 2 (Huber, 1964) consists of simultaneously solving (12.1) and

$$\sum_{i=1}^n \chi(x_i/S_n) = 0$$

for some choice of ψ and χ. Huber proposed using

$$\psi(y) = \begin{cases} y & |y| \le k \\ k\operatorname{sgn}(y) & |y| > k \end{cases}$$

for some value of k and

$$\chi(y) = \psi^2(y) - \int \psi^2(t)dt.$$

Iglewicz (1983) gave an alternative method of calculation (from Lax, 1975), by substituting sample estimates into the formula for the asymptotic variance of the M-estimators of location. For a general estimator, this substitution yields the variance estimate

$$s_M^2 = \frac{ncMAD[\sum_{i=1}^n \psi^2(y_i)]}{|\sum_{i=1}^n \psi'(y_i)|^2}$$

for a selected M-estimator and its corresponding function, ψ. He presented two variance estimates, one for the Tukey biweight estimator

$$s_{bi}^2 = \frac{n[\sum_{|y_i|<1}(x_i - x_M)^2(1 - y_i^2)^4}{|\sum_{|y_i|<1}(1 - y_i^2)(1 - 5y_i^2)|^2} \tag{12.8}$$

and the other

$$s_{wa}^2 = \frac{ncMAD[\sum_{|y_i|<1} \sin^2(\pi y_i)]}{\pi|\sum_{|y_i|<1} \cos(\pi y_i)|^2}$$

based on Andrews' wave estimator.

For the Tukey biweight variance estimator, the y_i are approximate weights; when the y_i are small, the denominator essentially becomes n, so that (12.8) reduces to

$$s_{bi}^2 = \frac{\sum(x_i - x_M)^2}{n}$$

(Mosteller and Tukey, 1977). A modification to s_{bi}^2 which performs better is given by

$$s_{bi}^2 = \frac{n[\sum_{|y_i|<1}(x_i - x_M)^2(1 - y_i^2)^4}{[\sum_{|y_i|<1}(1 - y_i^2)(1 - 5y_i^2)][-1 + over[\sum_{|y_i|<1}(1 - y_i^2)(1 - 5y_i^2)}$$

which reduces to

$$ns_{bi}^2 = \frac{\sum(x_i - x_M)^2}{n - 1}$$

(Mosteller and Tukey, 1977).

12.4 Robust Multivariate Estimation

As with multivariate tests for normality, preliminary robust estimation of multivariate data could consist of robust estimation of the mean and spread of each component. This fails to take into account the correlational structure of the data, of course, and does not provide any estimate of the covariance matrix. Another simple alternative (when primarily concerned with slippage) would be an Anscombe-type estimator, eliminating outliers according to some rule (e.g., from some test in Chapter 10) and then calculating the mean vector and covariance matrix.

Robust estimation methods for multivariate data (Section 10.3) have the advantage of providing a test of normality simultaneously with the robust estimator. These are described in the aforementioned section.

12.5 Recommendations

As with tests of normality, there is no single "best" robust estimator under every condition. Since in practice we do not know how our data are distributed, we cannot pick the best estimator for our circumstances. Therefore we try to pick an estimator based on a number of criteria, such as: our best guess of the data distribution, robustness under as wide a variety of distributions as possible, efficiency, computational ease, etc.

As an overall recommendation, L-estimates are attractive because they are simplest to calculate; adaptive L-estimates (e.g., those proposed by Hogg, Section 12.2.7) may increase the efficiency of the L-estimates, without adding a significant amount of effort to the calculation. M-estimators are more difficult computationally, but have more theoretical advantages. R-estimates are generally less efficient than M-estimators, and more difficult to calculate than L-estimates.

Tukey (1960) suggested that for large samples it was safer to use a lightly trimmed mean when the alternative is possible small symmetric contamination of a normal distribution. Crow and Siddiqui (1967) also preferred trimmed means as well as linearly weighted means rather than Winsorized means, although they were not specific as to the amount of trimming. Of the L-estimates, Rosenberger and Gasko (1983) recommended that trimmed means with 20% to 30% trimming be used over a broad range of heavy-tailed distributions and sample sizes. Similarly, D'Agostino and Lee (1977) suggested 25% trimmed means and the Gastwirth estimate as being reasonably good estimators over a range of long-tailed distributions.

Of the M-estimators, Goodall (1983) and Carroll (1989) leaned slightly more towards the redescending estimators over Huber's M-estimator, especially if the potential for large outliers exists. Hoaglin, Mosteller and Tukey (1983b) considered w-estimates to be nearly as good as M-estimates; therefore, they suggested using a w-estimate if computational difficulty is an issue.

In their evaluation of 68 location estimators, Andrews et al. (1972) stated that there will never be a "best" estimator, which will always rely on the particular circumstances. However, they tended towards using M-estimates, seeming to favor Andrews' wave estimator. They also favored adaptive estimates, skipped-trimmed means, and trimmed means "which don't trim too little on either end" (about 25%, similar to others' suggestions).

For scale estimates, the MAD is generally preferred as at a least preliminary estimate of scale, followed by d_f. Mosteller and Tukey (1977) and Iglewicz (1983) indicated that s_{bi}^2 has good potential, and is efficient for use in testing and constructing confidence intervals.

12.6 Further Reading

Tukey's (1960) paper about the effects of small amounts of contamination on estimation is one of the classical papers which would be a good introduction to the subject of robustness. In contrast to this chapter, Hampel (1968, 1971, 1974) provides a high degree of theoretical information on robustness. Huber also tends to be theoretical in approach, and offers one of the first papers on M-estimation (Huber, 1964) as well as an early review of robustness (Huber, 1972).

Andrews et al. (1972) conducted the first large scale review of robust estimators, although they restricted themselves to location estimators for univariate samples. They also only considered alternatives to normality which were long-tailed. David (1981) gave a relatively short but practical summarization of robust estimators. Hoaglin, Mosteller and Tukey (1983a) presented a number of different topics on robustness, and several of the individual chapters from that text have been referenced in this chapter. Barnett and Lewis (1994) provided one of the more recent comprehensive reviews of robust estimation, balanced between theory and methodology, and mostly from the point of view of slippage and contamination.

Huber (1981) provided a full text on robust estimation, which also leans towards the theoretical side; a more compact version is Huber (1977, 1996). Other texts on robustness of a more recent vintage include Hampel et al. (1986) and Tiku, Tan and Balakrishnan (1986).

References

Andrews, D.F., Bickel, P.J., Hampel, F.R., Huber, P.J., Rogers, W.H., and Tukey, J.W. (1972). **Robust Estimates of Location: Survey and Advances**, Princeton University Press, Princeton, NJ.

Anscombe, F.J. (1960). Rejection of outliers. Technometrics 2, 123-147.

Barnett, V., and Lewis, T. (1994). **Outliers in Statistical Data**, 3rd ed. John Wiley and Sons, New York, NY.

Bickel, P. (1988). Robust estimation. In Kotz, S., and Johnson, N.L., eds., **Encyclopedia of Statistical Science, Vol. 8**, John Wiley and Sons, New York, 157-163.

Bickel, P.J., and Hodges, J.L. (1967). The asymptotic theory of Galton's test and a related simple estimate of location. Annals of Mathematical Statistics 38, 73-89.

Carroll, R.J. (1989). Redescending M-estimators. In Kotz, S., and Johnson, N.L., eds., **Encyclopedia of Statistical Science, Supplemental Vol.**, John Wiley and Sons, New York, 134-137.

Crow, E.L., and Siddiqui, M.M. (1967). Robust estimation of location. Journal of the American Statistical Association 62, 353-389.

D'Agostino, R.B., and Lee, A.F.S. (1977). Robustness of location estimators under changes in population kurtosis. Journal of the American Statistical Association 72, 393-396.

David, H.A. (1981). **Order Statistics**, 2nd ed., John Wiley and Sons, New York.

De Wet, T., and van Wyk, J.W.J. (1979). Efficiency and robustness of Hogg's adaptive trimmed means. Communications in Statistics A8, 117-128.

Downton, F. (1966). Linear estimates with polynomial coefficients. Biometrika 53, 129-141.

Gastwirth, J.L. (1966). On robust procedures. Journal of the American Statistical Association 61, 929-948.

Goodall, C. (1983). M-estimators of location: An outline of the theory. In Hoaglin, D.C., Mosteller, F., and Tukey, J.W., eds., **Understanding Robust and Exploratory Data Analysis**, John Wiley and Sons, New York, 339-403.

Gross, A.M. (1976). Confidence interval robustness with long-tailed symmetric distributions. Journal of the American Statistical Association 71, 409-416.

Guttman, I., and Kraft, C.H. (1980). Robustness to spurious observations of linearized Hodges-Lehmann estimators and Anscombe estimators. Technometrics 22, 55-63.

Hampel, F.R. (1968). Contributions to the theory of robustness. Ph.D. thesis, University of California, Berkeley.

Hampel, F.R. (1971). A general qualitative definition of robustness. Annals of Mathematical Statistics 42, 1887-1896.

Hampel, F.R. (1974). The influence curve and its role in robust estimation. Journal of the American Statistical Association 69, 383-393.

Hampel, F.R., Ronchetti, E.M., Rousseeuw, P.J., and Stahel, W.A. (1986). **Robust Statistics: The Approach Based on Influence Functions**. John Wiley and Sons, New York.

Hoaglin, D.C., Mosteller, F., and Tukey, J.W., eds., (1983a). **Understanding Robust and Exploratory Data Analysis**, John Wiley and Sons, New York.

Hoaglin, D.C., Mosteller, F., and Tukey, J.W. (1983b). Introduction to more refined estimators. In Hoaglin, D.C., Mosteller, F., and Tukey, J.W., eds., **Understanding Robust and Exploratory Data Analysis**, John Wiley and Sons, New York, 283-296.

Hodges, J.L. (1967). Efficiency in normal samples and tolerance for extreme values for some estimates of location. Proceedings, Fifth Berkeley Symposium on Mathematical Statistics and Probability, University of California Press, Berkeley, 163-186.

Hodges, J.L., and Lehmann, E.L. (1963). Estimates of location based on rank tests. Annals of Mathematical Statistics 34, 598-611.

Hogg, R.V. (1967). Some observations on robust estimation. Journal of the American Statistical Association 62, 1179-1186.

Hogg, R.V. (1972). More light on the kurtosis and related statistics. Journal of the American Statistical Association 67, 422-424.

Hogg, R.V. (1974). Adaptive robust procedures: a partial review and some suggestions for future applications and theory. Journal of the American Statistical Association 69, 909-927.

Hogg, R.V. (1979). Statistical robustness: one view of its use in applications today. The American Statistician 33, 108-115.

Huber, P.J. (1964). Robust estimation of a location parameter. Annals of Mathematical Statistics 35, 73-101.

Huber, P.J. (1972). Robust statistics: a review. Annals of Mathematical Statistics 43, 1041-1067.

Huber, P.J. (1977). **Robust Statistical Procedures**. Society for Industrial and Applied Mathematics, Philadelphia, PA.

Huber, P.J. (1981). **Robust Statistics**. John Wiley and Sons, New York.

Huber, P.J. (1996). **Robust Statistical Procedures, 2nd ed.**. Society for Industrial and Applied Mathematics, Philadelphia, PA.

Iglewicz, B. (1983). Robust scale estimators and confidence intervals for location. In Hoaglin, D.C., Mosteller, F., and Tukey, J.W., eds., **Understanding Robust and Exploratory Data Analysis**, John Wiley and Sons, New York, 404-431.

Jaeckel, L.A. (1971). Some flexible estimates of location. Annals of Mathematical Statistics 42, 1540-1552.

Lax, D. A. (1975). An interim report of a Monte Carlo study of robust estimators of width. Technical Report No. 93 (Series 2), Dept. of Statistics, Princeton University.

Mosteller, F. (1946). On some useful "inefficient" statistics. Annals of Mathematical Statistics 17, 377-408.

Mosteller, F., and Tukey, J.W. (1977). **Data Analysis and Regression**. Addison-Wesley, Reading, MA.

Rosenberger, J.L., and Gasko, M. (1983). Comparing location estimators: Trimmed means, medians, and trimean. In Hoaglin, D.C., Mosteller, F., and Tukey, J.W., eds., **Understanding Robust and Exploratory Data Analysis**, John Wiley and Sons, New York, 297-338.

Rosner, B. (1975). On the detection of many outliers. Technometrics 17, 221-227.

Rosner, B. (1977). Percentage points for the RST many outlier procedure. Technometrics 19, 307-312.

Ruppert, D. (1988a). M-Estimators. In Kotz, S., and Johnson, N.L., eds., **Encyclopedia of Statistical Science, Vol. 5**, John Wiley and Sons, New York, 443-449.

Ruppert, D. (1988b). Trimming and Winsorization. In Kotz, S., and Johnson, N.L., eds., **Encyclopedia of Statistical Science, Vol. 9**, John Wiley and Sons, New York, 348-353.

Tiku, M.L., Tan, Y.U., and Balakrishnan, N. (1986). **Robust Inference**. Marcel Dekker, New York.

Tukey, J.W. (1960). A survey of sampling from contaminated distributions. In Olkin, I., Ghurye, S.G., Hoeffding, W., Madow, W.G., and Mann,

H.B., eds. **Contributions to Probability and Statistics, Essays in Honor of Harold Hotelling**, Stanford University Press, Stanford CA, 448-485.

Tukey, J.W. (1977). **Exploratory Data Analysis.** Addison-Wesley, Reading, MA.

Velleman, P.F., and Hoaglin, D.C. (1981).**Applications, Basics, and Computing of Exploratory Data Analysis**, Duxbury Press, Boston MA.

CHAPTER 13

COMPUTATIONAL ISSUES

"...a major purpose of conducting a distributional test is to decide whether or not it is meaningful to use the standard statistical procedures which are based on the assumed model."

Shapiro, 1980

In the previous chapters we have defined a multitude of test procedures for determining whether a sample may or may not have come from a normal distribution. These procedures have varied widely in a number of ways, including objectivity of the method; optimal conditions for use; power; and computational simplicity and, hence, availability for use. While ideally the first three considerations should determine which procedure(s) should be used, if they cannot be easily accessed then they will not be used.

We addressed the computational simplicity issue to some degree in Chapter 7 for full sample univariate tests, and in the Recommendations sections of the chapters on censored data, multivariate tests for normality and tests for outliers, and normal mixture tests. Many of the procedures described, especially for the univariate case, can be performed using a basic scientific hand calculator. A larger set can easily be calculated using simple computer programming languages from the formulas provided. Others, particularly the multivariate and normal mixture tests, require more sophisticated computational procedures.

In this chapter we provide details on some of the more common generic computational methods which are needed to calculate some of the test statistics used in this book. This will allow those with more sophisticated programming abilities and capabilities to compute the more complex tests, which may not be available at all in commercial software packages. We also provide some background and examples of procedures which are contained in a commercially available statistical software package.

13.1 Selected Computational Methods

In this section we describe in more detail some of the computational methods which are necessary to conduct certain tests presented in the preceding chapters, but which we considered too much of a tangential issue to provide in the main body of the text. While the methods were selected because they are necessary to perform certain tests for normality, they generally have more universal applications also.

13.1.1 Iterative Methods for Function Minimization

Optimization of some objective function for a set of observations is often required in statistical computation, for those functions of r unknown variables (parameters), $r \geq 1$, which do not have closed form solutions. Common statistical problems of this type include obtaining nonlinear least squares or maximum likelihoods, in which the parameter estimates for the maximum or minimum value of the objective function are often of interest. Methods that require iterative or numerical methods for obtaining solutions are found throughout this book, e.g., for normal mixture estimation and testing (Chapter 11) or multivariate normal tests (Chapter 9). We will describe the method generically so that it can be used for any computational problem, not just those relevant to the specific methods described in previous chapters. This section is a summarization of relevant sections of Nash (1979).

Function optimization (obtaining the minimum or maximum of a function) is most often described in terms of a minimization problem; this is easily managed, since maxima (such as maximum likelihoods) can be obtained by minimizing the negative of the function to be maximized. We will only discuss the case of unconstrained parameters. Often in the case where parameters are constrained, transformation of the parameter will eliminate the problem. For example, in the case of an unknown parameter b of a function which has the constraint that $b \geq 0$, define

$$b = e^{\beta} \tag{13.1}$$

and substituting the right side of (13.1) into the function to be minimized. The parameter β is unconstrained, and the methods described below can be used to obtain β, which gives b. An example where this situation might arise is the case where a standard deviation or variance is one of the unknowns.

There are a number of strategies which can be used to attain function minimization using iterative methods. These include direct search methods and descent methods (Nash, 1979). We have found that a form of the latter method, a variable metric procedure, works very well for all of the optimization problems described in previous chapters. We will give a general outline of how descent methods work, then focus specifically on the method we prefer. A number of issues that we raise below, including starting points and finding the global minimum, pertain to all of the iterative methods.

Descent Methods

Let $\mathbf{L}(\mathbf{b}, \mathbf{X})$ be an objective function to be minimized with unknown parameter vector $\mathbf{b}$ which has entries $b_i, i = 1, \ldots, r$, and dependent on the $n \times m$ data matrix $\mathbf{X}$. To obtain the minimum of $\mathbf{L}(\mathbf{b}, \mathbf{X})$, the partial derivative functions

$$g_i = g_i(\mathbf{b}) = \delta \mathbf{L} / \delta b_i = 0$$

must be solved for all parameters b_i. In the above notation, for simplicity we will use $\mathbf{L}$ for $\mathbf{L}(\mathbf{b}, \mathbf{X})$ and g_i to indicate the partial derivative of $\mathbf{L}$ with respect to parameter element b_i.

For the vector $\mathbf{g}$ with entries g_i, a Taylor expansion can be written as

$$\mathbf{g} = \mathbf{g}(\mathbf{b_0}) + \mathbf{H}(\mathbf{b_o})(\mathbf{b} - \mathbf{b_0}) = 0 \qquad (13.2)$$

where $\mathbf{b_0}$ is a vector of "starting points" for $\mathbf{b}$ and $\mathbf{H}$ is the matrix of second derivatives (Hessian matrix) of $\mathbf{L}$ evaluated at $\mathbf{b_0}$. Rearranging (13.2) gives the estimate

$$\mathbf{b} = \mathbf{b_0} - \mathbf{H}^{-1}(\mathbf{b_0})\mathbf{g}(\mathbf{b_0}) \qquad (13.3).$$

Descent methods of function optimization use a general iterative process based on (13.3), i.e., the estimates of the parameters at step i, $\mathbf{b}^{(i)}$, are obtained by

$$\mathbf{b}^{(i)} = \mathbf{b}^{(i-1)} - c^{(i-1)}\mathbf{B}^{(i-1)}\mathbf{g}^{(i-1)} \qquad (13.4)$$

where the superscripts indicate that the estimates of $\mathbf{b}$ at step i are based on the results of the previous $(i - 1)$ step. Here, $\mathbf{B}$ is some $r \times r$ matrix and c is a scalar search direction. The differences among descent methods center on the choice and updating methods of c and $\mathbf{B}$, which are chosen so that $\mathbf{L}$ is reduced with each step of the process. In particular, the method of steepest descents uses the identity matrix for $\mathbf{B}$ and c is any appropriate step length that results in a reduction of the objective function at each step.

Note that $\mathbf{B}$, c and/or $\mathbf{g}$ may be updated at each step of the iterative process, with $\mathbf{B}$ and $\mathbf{g}$ usually being dependent on the current step estimates $\mathbf{b}^{(i-1)}$. The Newton-Raphson method is another descent algorithm, and reduces to (13.2) directly from (13.4), by using $c = 1$ and

$$\mathbf{B} = \mathbf{H}^{-1}$$

which, at step i, is the inverse of the Hessian matrix evaluated at the point $\mathbf{b}^{i-1}$, requiring updates at each step of $\mathbf{B}$ and $\mathbf{g}$ (but not c).

The Variable Metric Method

Variable metric methods are also called quasi-Newton methods because $\mathbf{H}^{-1}$ is approximated by the matrix $\mathbf{B}$ based on the first derivative information. These methods use the iterative process (13.4) to move towards the final set of parameter estimates. Variable metric methods differ based on the method of updating $\mathbf{B}$ and c at each step. The product $\mathbf{Bg}$ is the *search direction* at each step and c is the *step length*. The basic problem to be solved here is, at each step, to generate new estimates of $\mathbf{b}$ such that

$$\Delta \mathbf{L} = \mathbf{L}^{(i)} - \mathbf{L}^{(i-1)} < 0 \qquad (13.5),$$

i.e., at each step the objective function is moving towards a minimum. Let the search step be defined as

$$\mathbf{t}^{(i-1)} = \mathbf{b}^{(i)} - \mathbf{b}^{(i-1)} = -c\mathbf{B}^{(i-1)}\mathbf{g}^{(i-1)}$$

(from 13.4) with $c = 1$ at the initial step. Noting that the Taylor expansion of the objective function is

$$\mathbf{L}(\mathbf{b}) = \mathbf{L}(\mathbf{b_0}) + \mathbf{g}(\mathbf{b_0})(\mathbf{b} - \mathbf{b_0})$$

then (13.5) is given by $\mathbf{g}^T\mathbf{t}$, which is negative when $\mathbf{t}$ is in a downhill direction. By choosing $c = w^i, i = 0, 1, \ldots$ at each step for some $0 < w < 1$, a $\mathbf{t}$ can be found such that

$$\frac{\Delta \mathbf{L}}{\mathbf{g}^T\mathbf{t}} > q > 0, \qquad 0 < q <<< 1$$

for some chosen tolerance q unless the minimum of $\mathbf{L}$ has been reached.

For the variable metric method we utilized, as suggested by Nash (1979), the tolerance q is set at 0.0001, $w = 0.2$, and the initial matrix $\mathbf{B}$ is the identity matrix. The update of $\mathbf{B}$ at each step is given by

$$\mathbf{B}^{(i)} = \mathbf{B}^{(i-1)} + \mathbf{A}^{(i)}$$

where

$$\mathbf{A}^{(i)} = d_2 \mathbf{t}\mathbf{t}^T - \frac{\mathbf{t}(\mathbf{By})^T + (\mathbf{By})\mathbf{t}^T}{d_1}$$

with

$$\mathbf{y} = \mathbf{g}(\mathbf{b}^i) - \mathbf{g}(\mathbf{b}^{i-1})$$

and constants d_1 and d_2 calculated as $d_1 = \mathbf{t}^T \mathbf{y}$ and

$$d_2 = \frac{1 + \mathbf{y}^T \mathbf{By}/d_1}{d_1}.$$

For more details on the general iterative problme or on the specific algorithm described above, see Nash (1979). A copy of a FORTRAN program which performs the algorithm is provided in Appendix C.

Note on Reaching the Global Minimum

The reader should be aware that nonlinear functions may contain more than one locally minimum value. When minimizing such a function, regardless of the algorithm, there is the potential that the algorithm will get caught in a "valley" and descend to the local minimum. Therefore, it is recommended that a minimization problem be solved multiple times, each time with a new set of starting values for the $\mathbf{b}$, with the smallest minimum value being accepted as the global (true) minimum of the function. The certainty with which one accepts this global minimum is dependent on the strategy for choosing the starting points. The strategy may consist of trying a large number of random starting points, or some logical sequence. For example, Thode, Finch and Mendell (1987, 1988) found that only five starting points were required to find the maximum likelihood for a univariate normal mixture problem (Chapter 11).

13.1.2 Approximating Expected Normal Order Statistics

In this section, we provide a method for estimating expected values and covariances of normal order statistics. This is based on the method described in David (1981).

The method uses Kimball's (1960) plotting position

$$p_k = \frac{k}{n+1} = E(U_{(k)})$$

where $U_{(k)}$ is the kth uniform order statistic. Let $q_k = 1 - p_k$. Then for a (general) continuous probability distribution defined by $P(X_{(k)}) = U_{(k)}$ with inverse $Q(U_{(k)}) = X_{(k)}$, and order statistics given by $X_{(k)}$, a Taylor expansion of $X_{(k)}$ and taking the expected value gives (to order $(n+2)^{-2}$)

$$E(X_{(k)}) = Q(p_k) + \frac{p_k q_k}{2(n+2)} Q''(p_k)$$
$$+ \frac{p_k q_k}{(n+2)^2} \left[\frac{(q_k - p_k)}{3} Q'''(p_k) + \frac{p_k q_k}{8} Q''''(p_k) \right]$$

where $Q', Q'', \ldots$ are the first derivative, second derivative, etc., of Q. For the normal distribution, these functions are given by

$$Q(p) = \Phi^{-1}(p)$$

$$Q'(p) = \frac{1}{\phi(Q)}$$

$$Q''(p) = \frac{Q}{\phi^2(Q)}$$

$$Q'''(p) = \frac{1 + 2Q^2}{\phi^3(Q)}$$

and

$$Q''''(p) = \frac{Q(7 + 6Q^2)}{\phi^4(Q)}$$

(using the notation Q for $Q(p)$ in the derivative equations for simplicity). Similar methods lead to the expression for the covariances

$$cov(X_{(j)}, X_{(k)}) = \frac{p_j q_k}{n+2} Q'(p_j) Q'(p_k) + \frac{p_j q_k}{(n+2)^2} [(q_j - p_j) Q''(p_j) Q'(p_k)$$
$$+ (q_k - p_k) Q'(p_j) Q''(p_k) + \frac{p_j q_j}{2} Q'''(p_j) Q'(p_k)$$
$$+ \frac{p_k q_k}{2} Q'(p_j) Q'''(p_k) + \frac{p_j q_k}{2} Q''(p_j) Q''(p_k)]$$

with the variances of the order statistics being obtained by calculating $cov(X_{(j)}, X_{(j)})$ using the above equation. While the expected value estimates should be good for samples of size 50 or greater, the covariance estimates should be good for any sample size (LaRiccia, 1986).

13.1.3 Minimum Spanning Trees

Rohlf (1975) uses a minimum spanning tree (MST) to visually identify and to formally test for multivariate outliers (Section 10.5.1). An MST (or shortest simply connected graph) is defined as a graph which connects points (vertices) in m-space by lines (edges) such that (1) all points are connected to adjacent (nearest neighbor) points; (2) there is one and only one path along the edges from any point to any other point; and, (3) the sum of the edges in the graph is the minimum possible. Algorithms used to obtain an MST can be found in Kruskal (1956) and Rohlf (1975). Here we provide a straightforward algorithm for obtaining an MST.

Let an m-variate set of n points, in no particular order, be defined such that the m components are appropriately scaled, if necessary (see Section 10.5.1 for the scaling used for the generalized gap test). Define an interpoint distance, d_{ij}, which is some measure of the distance between two points, such as Euclidean distance. Now form the $n \times n$ matrix of distances, where the entry (i, j) is the distance between x_i and x_j. This will be a symmetric matrix with zeros down the main diagonal.

The algorithm begins by connecting an arbitrary point to its nearest neighbor; thereafter, all unconnected points are searched to find that point with the smallest distance to one of the points which is already connected. This is repeated until all points are connected. Following is a verbal "flow chart" of how this algorithm works:

1. Form the $n \times n$ matrix of interpoint distances as described above. Create two vectors of length n which will be column and row indicators, that will be used to identify those points which have already been connected. All entries of these vectors should be initialized to 0.

2. Set the first entry in both the row and column indicator vectors to 1, i.e., arbitrarily selecting x_1 as the starting point. Then, in row 1 of the distance matrix, find the smallest distance between x_1 and all other points $x_i, i > 1$.

3. Let x_m be the observation with the smallest distance to x_1. Then the MST will have an edge connecting these two points. Set the row and column indicator vector entries corresponding to the observation indexed by x_m to 1 (the index of x_1).

4. The process now becomes iterative. Conduct a search for the minimum distance through all rows with a row index *not equal to* 0 of all columns with a column index *equal to* 0. If the smallest observed distance is between (unconnected point) x_k and (connected point) x_j, set the kth entry of the row and column vectors to j. Continue this procedure until all row indicator vector entries are greater than 0.

5. The row (or column) indicator vector entries will indicate which points are connected to each other, e.g., if $r(7) = 14$, then there is an edge connecting $\mathbf{x_7}$ and $\mathbf{x_{14}}$. Using this information, the edge lengths in the MST can be identified from the distance matrix.

In the event, at some step in the process, that two or more distances are tied for the minimum value, the choice of edge to select at that iteration is arbitrary and does not affect the results. For a relatively small number of points, identifying the MST manually is fairly simple once the matrix of interpoint distances is calculated.

13.1.4 Estimating Gamma Parameters

Parameter estimates for the gamma distribution are required for conducting the generalized gap test (Section 10.5.1). There are several methods available for estimating the parameters of a two parameter gamma distribution, given by

$$f(x) = \frac{\exp^{-(x/\lambda)}(x/\lambda)^{\eta-1}}{\lambda\Gamma(\eta)}$$

with parameters λ and η. For a sample of size n, the maximum likelihood estimators can be obtained by iteratively solving the equation

$$\log(\overline{x}) - \log(\overline{x_g}) = \log(\hat{\eta}) - \psi(\hat{\eta}) \qquad (13.6)$$

where $\overline{x_g}$ is the geometric mean and ψ is the digamma function; the second parameter λ can then be estimated by $\overline{x}/\hat{\eta}$.

Here, we will describe a simple tabular method of estimating the gamma parameters used by Greenwood and Durand (1960). We will use notation which agrees with Rohlf (1975), who used η to denote the shape parameter, whereas Greenwood and Durand use ρ to denote the shape parameter and η to denote the left side of (13.6). Note that to perform the generalized gap test (Section 10.5.1), only the shape parameter η needs to be estimated except for in large samples. The second parameter was denoted by λ and a by Rohlf (1975) and Greenwood and Durand (1960), respectively.

Let z denote the left side of (13.6), i.e.,

$$\hat{z} = \log(\overline{x}) - \log(\overline{x_g}).$$

The required calculations are simple, consisting of obtaining the mean and geometric mean of the observations, which are used to estimate of the right side of (13.6). For ease of application, the mean of the $\log(x_i)$

can be used to directly obtain the second term of $\hat{z}$. Greenwood and Durand (1960) provided tables of z with corresponding values of $z\eta$ for $z = 0(0.01)1.40(0.20)18.0$. Interpolation is used for values of z intermediate to those tabulated, and the value of $z\eta$ is obtained for $\hat{z}$. The shape parameter can then be estimated from these two values,

$$\hat{\eta} = z\eta/\hat{z}$$

and the estimate of λ calculated as $\bar{x}/\hat{\eta}$. The table of $(z, z\eta)$ is reproduced in Table B36.

Example 13.1. The squared distances from Example 10.9 have a mean of 0.0465 and mean of the logarithms of -3.8911. This gives

$$\hat{z} = \log(0.0465) - (-3.8911) = 0.8228.$$

From Table B36, interpolating between $z = 0.82$ and $z = 0.83$ gives a value of $z\eta = 0.600820$, and

$$\hat{\eta} = 0.600820/0.8228 = 0.730$$

and

$$\hat{\lambda} = 0.0465/0.730 = 0.637$$

are the parameter estimates. The estimate of η above is used in Example 10.9 for the generalized gap test.

Wilk, Gnanadesikan and Huyett (1962) showed an alternative method of parameter estimation using order statistics.

13.2 Testing for Normality Using SPSS

The wide availability of personal computers and statistical software has the advantage of providing powerful statistical capabilities in the hands of many. Researchers can run off literally hundreds of t-tests, regression, ANOVAs, and other procedures with little or no regard for the assumptions behind those procedures. Unfortunately, we have found that only a limited

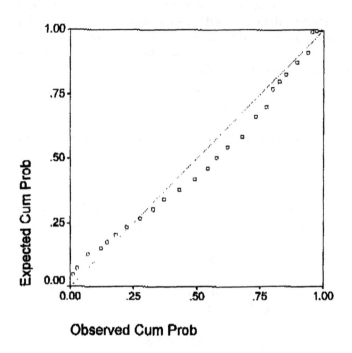

Figure 13.1 Normal probability P-P plot for water well alkalinity data (Data Set 10) using the **Graphs:P-P Plots** *procedure.*

number of the more basic procedures have been incorporated into software packages.

Virtually all software packages have some basic graphic and univariate normality tests. Histograms and box plots are usually available for univariate inspection, as are one or more type of probability plot. Two, and sometimes three, dimensional scatterplots are also basic components of statistical packages. However, the only procedures available for testing univariate normality in most packages are skewness and kurtosis, and multivariate procedures are essentially nonexistent. Some robust estimation procedures are also available, although only for univariate samples.

In this section we provide an overview of the procedures available in the SPSS software package for personal computers (SPSS, Inc). Only those procedures relevant to the topics in this book are described below. The SPSS procedures are performed using a menu bar with drop down menus and submenus. In order to identify the location of the software proce-

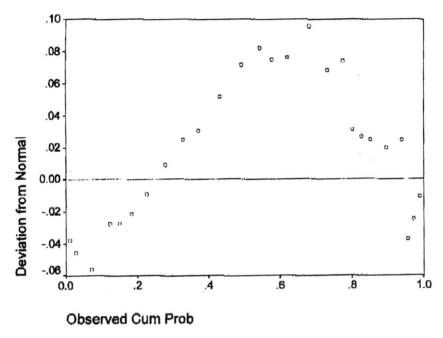

Figure 13.2 Detrended normal probability P-P plot for water well alkalinity data (Data Set 10) using the **Graphs:P-P Plots** *procedure.*

dures, we will use the notation **Menu:Submenu:Submenu**, for example, **Statistics:Descriptives:Explore** will indicate that we are using the Explore procedure in the Descriptives submenu from the Statistics menu on the menu bar.

13.2.1 Univariate Plots

SPSS includes a number of different plots for inspecting the distribution of a data sample, including simple plots (i.e., histograms, stem-and-leaf and box plots) and probability plots. These plots are available as stand alone procedures and/or as options under other procedures. For all plots, output can be modified with respect to axis labeling, colors, shading patterns, plot symbols (size, shape), titling, line types and text.

VAR1 Stem-and-Leaf Plot

```
Frequency      Stem &  Leaf

     1.00        0 .  7
    10.00        1 .  1366778899
    17.00        2 .  00111234447788889
    10.00        3 .  0111235679
    13.00        4 .  0111222226677
     2.00        5 .  15
     4.00        6 .  4459
     3.00        7 .  244
     1.00        8 .  1
     4.00  Extremes     (>=88)

Stem width:       10.00
Each leaf:         1 case(s)
```

Figure 13.3 Stem-and-leaf plot for total suspended particulate data (Data Set 1) using the **Statistics:Summarize:Explore** *procedure.*

Probability Plots

SPSS contains procedures for univariate probability plots, including both *P-P* and *Q-Q* plots and detrended *P-P* and *Q-Q* plots. All four of these probability plots are available in the **Graph** menu, while the *Q-Q* plots are also available in the **Statistics:Summarize:Explore** procedure as an option.

P-P and *Q-Q* plots (including detrended plots) are available in the **Graphs:P-P Plots** and **Graphs:Q-Q Plots**, respectively. These procedures include a number of different options, including four different plotting positions (see Chapter 2), twelve null distributions in addition to the normal (e.g., exponential, lognormal, uniform), and the option of entering *a priori* fixed parameters or using parameter estimates for the plot setup. Under the **Statistics:Summarize:Explore** procedure, however, the *Q-Q* plots are limited to the normal distribution with estimated parameters and the default plotting position. Figures 13.1 and 13.2 show a sample chart output of a *P-P* plot and a detrended *P-P* plot, respectively, for the well water alkalinity data (Data Set 10).

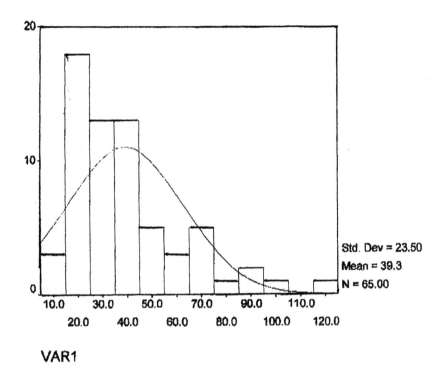

VAR1

*Figure 13.4 Histogram for total suspended particulate data (Data Set 1)
using the* **Graphs:Histogram** *procedure.*

Data Plots

Raw data plots which can be generated in SPSS include histograms,
stem-and-leaf plots and box plots (Section 2.1); scatterplots are also available for comparing variables two at a time in the multivariate case. Stem-
and-leaf plots are only available in the **Statistics:Summarize:Explore**
procedure using the Plots option. Figure 13.3 contains the stem-and-leaf
plot for the total suspended particulate data (Data Set 1).

Figure 13.4 contains the output from the plotting procedure for a histogram, from the **Graphs:Histogram** procedure. The data in Figure 13.4
are also the suspended particulate data (Data Set 1). The normal probability distribution function curve for the normal distribution with parameters
given by the sample mean and variance can be inscribed on the histogram
in this procedure. However, this procedure (unlike the probability plots)
does not provide options for this overlay from other distributions. Basic

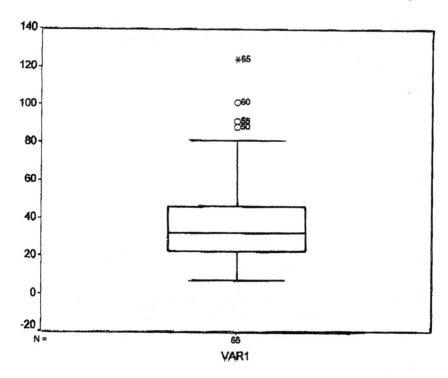

Figure 13.5 Box plot for total suspended particulate data (Data Set 1) using the **Graphs:Boxplot** *procedure.*

statistics (mean, standard deviation and sample size) can be included in the output, as shown in the figure. A histogram is also provided in the Plots option of the **Statistics:Summarize:Explore** procedure, but the normal curve overlay is not available.

The box plot for the TSP data is shown in Figure 13.5, from the SPSS procedure **Graphs:Boxplot**. Note that outliers and far outliers are identified using different plotting symbols. Box plots are also available under the Plots option of **Statistics:Summarize:Explore**.

13.2.2 Univariate Tests

Several univariate tests for normality and goodness of fit are available in SPSS, although some of them are of limited use for use with a normal null distribution.

Descriptives

			Statistic	Std. Error
Assembly Time	Mean		2.8400	.1250
	95% Confidence	Lower Bound	2.5890	
	Interval for Mean	Upper Bound	3.0910	
	5% Trimmed Mean		2.8034	
	Median		2.6900	
	Variance		.797	
	Std. Deviation		.8926	
	Minimum		1.32	
	Maximum		4.96	
	Range		3.64	
	Interquartile Range		1.4500	
	Skewness		.538	.333
	Kurtosis		-.350	.656

Figure 13.6 Tests of skewness and kurtosis for assembly times of mechanical parts data (Data Set 4) using the **Statistics:Summarize:Explore** *procedure.*

Skewness and Kurtosis

SPSS procedures for basic descriptive statistics include measures of skewness and kurtosis and their estimated standard errors (Sections 3.2, 6.3). These are available from the **Statistics:Summarize:Explore** procedure (by default) or in the **Statistics:Summarize:Descriptives** procedure as an option. The output from the **Statistics:Summarize:Explore** procedure is contained in Figure 13.6. Note that the values for skewness and kurtosis are slightly different than those given in Examples 3.1 and 3.3, since moments about the mean in SPSS are calculated recursively using a

	Kolmogorov-Smirnov[a]			Shapiro-Wilk		
	Statistic	df	Sig.	Statistic	df	Sig.
VAR1	.143	32	.095	.919	32	.029

a. Lilliefors Significance Correction

One-Sample Kolmogorov-Smirnov Test

		VAR1
N		32
Normal Parameters[a,b]	Mean	111.7500
	Std. Deviation	18.1943
Most Extreme	Absolute	.143
Differences	Positive	.101
	Negative	-.143
Kolmogorov-Smirnov Z		.809
Asymp. Sig. (2-tailed)		.530

a. Test distribution is Normal.

b. Calculated from data.

Figure 13.7 Wilk-Shapiro and Kolmogorov-Smirnov tests for the newborn birthweights (Data Set 3) using the **Statistics:Summarize:Explore** *procedure.*

provisional means algorithm (Spicer, 1972; Bliss, 1967). Kurtosis in SPSS has a normal value of 0 rather than 3.

Although no formal tests for outliers are available other than skewness and kurtosis, this procedure also has an option for listing the five highest and five lowest "extreme values" (Figure 13.6).

Other Univariate Tests

Other univariate tests for normality which are available from SPSS include the Wilk-Shapiro test (Section 2.3.1), the Kolmogorov-Smirnov test (Section 5.1) and the χ^2 test (Section 5.2).

The **Statistics:Summarize:Explore** procedure can be used to generate the Kolmogorov-Smirnov test with Lilliefor's correction (Section 5.1) and the Wilk-Shapiro tests, using the Plots option. The Kolmogorov-Smirnov test will always be provided under this option, while the Wilk-Shapiro test is only available when the sample size is less than 50. The output for the 32 newborn birthweights (Data Set 3) is given in Figure 13.7. A more general Kolmogorov-Smirnov test is available in the **Nonparametric Tests:1-Sample K-S** procedure, which offers four different null distributions to test against (normal, Poisson, uniform and exponential). However, this procedure bases the test on the estimated parameters while the significance level is not based on Lilliefor's correction. Using this

CHI

	Observed N	Expected N	Residual
1.00	9	11.6	-2.6
2.00	18	11.6	6.4
3.00	14	11.6	2.4
4.00	8	11.6	-3.6
5.00	9	11.6	-2.6
Total	58		

Test Statistics

	CHI
Chi-Square[a]	6.310
df	4
Asymp. Sig.	.177

a. 0 cells (.0%) have expected frequencies less than 5.
The minimum expected cell frequency is 11.6.

Figure 13.8 Robust estimates of location for the leukemia latency data (Data Set 9) using the **Statistics:Summarize:Explore** *procedure.*

procedure for the birthweight data results in a significance level of 0.53, as compared to the corrected version, which gives a significance level of 0.095 (Figure 13.7).

The **Statistics:Nonparametric Tests:Chi-Square** procedure can be used to perform the χ^2 test (Section 5.2). The number of cells used in this test is user-selected; however, the user must also enter the expected cell counts based on the null distribution, and the data must be recoded to the categories identified by the cell ranges, making this an awkward test to use. The generality of this procedure permits any null distribution to be implicitly specified through the use of the expected cell counts, so that, in the absence of other tests of goodness of fit, the χ^2 test procedure can be used for distributions other than the normal. The significance level of the test is based on the χ^2_{k-1} distribution, however, making it valid only for the simple hypothesis case.

13.2.3 Robust Estimates of Location (Univariate)

SPSS offers five univariate robust estimates of location in the **Statistics:Summarize:Explore** procedure, one included in the default descriptive statistics and the other four which are chosen in the Statistics option of the procedure. This provides a choice of location estimate in addition to the mean and median. The only limitation is that the parameters used to perform the calculations are fixed default values. The robust estimators include four M-estimates and one L-estimate.

The L-estimate is the trimmed mean (Section 12.2.5), with the trimming percentage fixed at 5%. The M-estimators of location (Section 12.2.2) include the Tukey's biweight (with parameter $c = 4.685$), Andrews' wave (with parameter $c = 1.339\pi$), Huber's (with parameter $c = 1.339$) and Hampel's 3 part redescending (with parameters $a = 1.7, b = 3.4, c = 8.5$) estimates. Figure 13.8 shows the five robust estimators for the leukemia latency data (Data Set 9). The three redescending M-estimates from SPSS were used in Example 12.2, while a 10% trimmed mean was used in example 12.3 for these data, which differs only slightly from the 5% trimmed mean shown here (57.8 vs 57.2).

13.2.4 Robust Estimates of Scale (Univariate)

The interquartile range (f-spread) and range (Section 12.3.2) are given in the **Statistics:Summarize:Explore** procedure (Figure 13.6, Figure 13.8), with the values of Tukey's hinges provided as an option in this procedure.

13.2.5 Multivariate Tests

The only tests for multivariate normality that are available in SPSS are obtained by calculating the leverage values from the hat matrix (Section 10.4.1) and/or the Mahalanobis distances (Section 9.1.2). This requires an indirect approach, since a linear regression needs to be performed in order to obtain these statistics. For the example here, an extra variable was obtained by generating random numbers (using the RV.NORMAL function

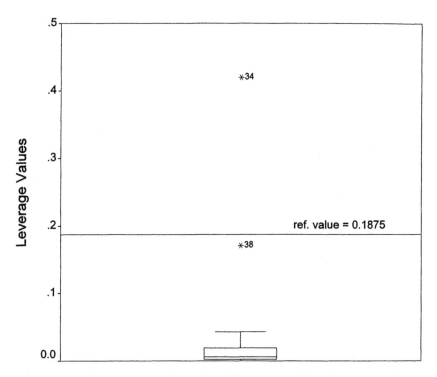

Figure 13.9 Box plot of leverage values for the cerebrospinal fluid data (Data Set 14) using the **Statistics:Regression:Linear** *procedure.*

from the **Transform:Compute** menu). The three cerebrospinal fluid variables were then used as independent variables with the generated variable as the dependent variable in the **Statistics:Regression:Linear** procedure. The leverage values and Mahalanobis distances are then saved as new variables using the Save option of the procedure.

The leverages and distances are not tests of normality, since there is a value for each case in a sample. Both of these measures are, however, equivalent to Wilks' test for multivariate outliers (Section 10.2.2). The cutoff value for the leverage values, $2.5m/n$, can be used as a test fro multivariate outliers (Section 10.4.1). The Mahalanobis distances (squared radii, Section 9.1.2) can also be manipulated to perform tests of multivariate normality (see Chapter 9), but since these manipulations are not performed directly by the SPSS software we will not discuss them here.

A box plot of the leverage values for the cerebrospinal fluid data is contained in Figure 13.9. The cutoff value of 0.1875 is shown as a reference line for identifying outliers; the outlying cases are labeled by case number, with "large" values identified by circles and outliers by filled circles (Section 2.1). The identification of the two outliers using the leverage values agrees with the results for Wilks' multivariate outlier test (Example 10.4).

References

Bliss, C.I. (1967). **Statistics in Biology**, Volume 1. McGraw Hill, New York, NY.

David, H.A. (1981). **Order Statistics**, 2nd ed. John Wiley and Sons, New York, NY.

Greenwood, J.A., and Durand, D. (1960). Aids for fitting the gamma distribution by maximum likelihood. Technometrics 2, 55-65.

Kimball, B.F. (1960). On the choice of plotting positions on probability paper. Journal of the American Statistical Association 55, 546-560.

Kruskal, J.B. (1956). On the shortest spanning subtree of a graph and the traveling salesman problem. Proceedings of the American Mathematical Society 7, 48-50.

LaRiccia, V.N. (1986). Asymptotically chi-squared distributed tests of normality for Type II censored samples. Journal of the American Statistical Association 81, 1026-1031.

Nash, J.C. (1979). **Compact Numerical Methods for Computers**. John Wiley and Sons, New York.

Rohlf, F.J. (1975). Hierarchical clustering using the minimum spanning tree. Computer Journal 16, 93-95.

Shapiro, S.S. (1980). **How to Test Normality and Other Distributional Assumptions**. American Society for Quality Control, Milwaukee, WI.

Spicer, C.C. (1972). Algorithm AS52: Calculation of power sums of deviations about the mean. Applied Statistics 21, 226-227.

SPSS, Inc. SPSS for Windows V8.0, Chicago, IL.

Thode, H.C., Jr., Finch, S.J., and Mendell, N.R. (1987). Finding the MLE in a two component normal mixture. Proceedings of the ASA Section on Statistical Computing, 1987 Joint Statistical Meetings, San Francisco, CA, August 17-20.

Thode, H.C., Jr., Finch, S.J., and Mendell, N.R. (1988). Simulated percentage points for the null distribution of the likelihood ratio test for a mixture of two normals. Biometrics 44, 1195-1201.

Wilk, M.B., Gnanadesikan, R., and Huyett, M.J. (1962). Estimation of parameters of the gamma distribution using order statistics. Biometrika 49, 525-545.

APPENDIX A

DATA SETS USED IN EXAMPLES

A1. Total Suspended Particulates from an Air Sampling Station (n = 65)
A2. Heights and Height Differences in *Zea Mays* Plants (n = 15)
A3. Newborn Birthweights (n = 32)
A4. Mechanical Part Assembly Times (n = 51)
A5. Sodium Content of Margarine (n = 28)
A6. Pulse Rates (n = 40)
A7. Reticulum Cell Sarcoma Latency Period in Mice (n = 38)
A8. July 1985 Average Daily Wind Speed (n = 31)
A9. Leukemia Latency Period in Chemotherapy Patients (n = 20)
A10. Alkalinity in Public Water Wells (n = 58)
A11. Shoreline Erosion Rates (n = 13)
A12. *Iris Setosa* Plant Characteristics (n = 50)
A13. Simulated bivariate normal data with correlation $\rho_{xy} = 0.45$ (n = 50)
A14. Cerebrospinal fluid gas measurements in acidosis patients (n = 40).

Data Set 1. Total suspended particulate measurements ($\mu g/m^3$) at an air sampling station near the Navajo Generating Station, 1974, n = 65.

Examples: [2.2], [4.3], [4.11]

Figures: [2.1], [2.2], [2.3], [4.6], [13.3], [13.4], [13.5]

7	21	28	41	55
11	21	29	41	64
13	21	30	41	64
16	22	31	42	65
16	23	31	42	69
17	24	31	42	72
17	24	32	42	74
18	24	33	42	74
18	27	35	46	81
19	27	36	46	88
19	28	37	47	91
20	28	39	47	101
20	28	40	51	124

$$\overline{x} = 39.28 \qquad \sqrt{b_1} = 1.39$$
$$s^2 = 552.30 \qquad b_2 = 4.84$$

Data Set 2. Heights and height differences in eighths of an inch between cross- and self-fertilized Zea Mays plants, n = 15 (from Fisher, 1971).

Examples: [2.1], [2.4], [4.9], [6.2], [6.3]

Figures: [2.8], [4.4]

Block (Pot)	Cross-fert. (a)	Self-fert. (b)	Difference (c)
I	188	138	50
	96	163	-67
	168	160	8
II	176	160	16
	153	147	6
	172	149	23
III	177	149	28
	163	122	41
	146	132	14
	173	144	29
	186	130	56
IV	168	144	24
	177	102	75
	184	124	60
	96	144	-48

(a)	$\bar{x} = 161.53$	$\sqrt{b_1} = -1.55$
	$s^2 = 837.27$	$b_2 = 4.18$
(b)	$\bar{x} = 140.60$	$\sqrt{b_1} = -0.72$
	$s^2 = 269.40$	$b_2 = 3.14$
(c)	$\bar{x} = 20.93$	$\sqrt{b_1} = -0.99$
	$s^2 = 1424.64$	$b_2 = 3.61$

Data Set 3. Newborn baby birthweights in ounces, n = 32 (from Armitage and Berry, 1987; copyright 1987 Blackwell Publishers. Used with permission).

Examples: [2.5], [5.1], [5.2]

Figures: [2.9], [5.1], [13.7]

72	80	123	127	133
112	81	116	86	106
111	84	125	142	103
107	115	126	132	118
119	118	122	87	114
92	128	126	123	94
126	128			

$$\bar{x} = 111.75 \qquad \sqrt{b_1} = -0.64$$
$$s^2 = 331.03 \qquad b_2 = 2.33$$

Data Set 4. Time in minutes required to assemble a mechanical part, n = 51 (from Shapiro, 1970; copyright 1970. Reprinted with permission of ASQ Quality Press, Milwaukee, WI).

Examples: [3.1], [3.2], [3.3], [3.4], [3.5]

Figures: [3.3], [13.6]

3.67	4.09	3.93	2.69	1.95
3.73	1.32	2.94	1.92	4.96
3.43	3.01	2.37	2.80	3.07
2.57	2.40	1.62	2.71	4.08
2.29	3.33	3.20	2.37	2.03
1.83	3.55	4.75	4.04	1.55
1.87	3.12	2.44	2.20	2.10
1.84	1.71	1.77	2.57	2.96
2.69	4.83	3.77	3.58	2.09
2.51	3.54	2.28	2.49	3.65
				2.63

$$\bar{x} = 2.84 \qquad \sqrt{b_1} = 0.52$$
$$s^2 = 0.80 \qquad b_2 = 2.57$$

Data Set 5. Sodium content of margarine[1] (mg/tbsp), n = 28 (from Consumer Reports, 1989; copyright 1989 by Consumers Union of U.S., Inc., Yonkers, NY 10703-1057, a nonprofit organization. Reprinted with permission from the Sept. 1989 issue of Consumer Reports for educational purposes only. No commercial use or photocopying permitted. To learn more about Consumers Union, log onto www.consumerreports.org.).

Examples: [3.6], [4.1], [4.6]

Figures:

I Can't Believe It's Not		Promise (tub)	90
Butter (stick)	95	Shedd's Spread Country	
Land O Lakes	115	Crock Classic	110
Lady Lee Vegetable Oil	95	Acme Corn Oil	110
Fleischman's Squeeze	95	Saffola	95
I Can't Believe It's Not		Blue Bonnet	95
Butter (tub)	95	Promise (stick)	90
Land O Lakes Soft	115	A&P Soft	100
Kroger Corn Oil	115	Chiffon Soft	105
Albertsons Corn Oil	120	Mazola Premium	100
Lucerne Vegetable Oil	95	Parkay Squeeze	100
Blue Bonnet Spread	100	Parkay Spread	110
Lucerne 100% Corn Oil	95	Fleischman's	95
Blue Bonnet Soft	95	Parkay	115
Shedd's Spread Country		Shoprite 100% Corn Oil	100
Crock	100	Parkey Soft	115

$$\bar{x} = 101.96 \qquad \sqrt{b_1} = 0.51$$
$$s^2 = 83.96 \qquad b_2 = 1.88$$

[1] Excludes sweet and diet margarines.

Data Set 6. Pulse rates of subjects after completion of standard physical fitness test, n = 40 (from Langley, 1971; used with permission of Dover Publications, Inc.).

Examples: [4.2], [4.4]

Figures: [4.1]

136	121	77	129	81
120	146	90	114	110
108	107	128	168	125
137	92	108	112	129
110	108	99	97	128
115	138	100	137	105
121	108	93	144	111
100	101	125	116	115

$$\bar{x} = 115.22 \qquad \sqrt{b_1} = 0.38$$
$$s^2 = 345.51 \qquad b_2 = 3.35$$

Data Set 7. Number of days until occurrence of cancer (reticulum cell sarcoma) in mice, n = 38 (from Hoel, 1972; copyright 1972. Used with permission of the International Biometric Society.).

Examples: [4.5], [4.10]

Figures: [4.2], [4.5], [8.1]

317	554	612	649	700
318	557	621	661	705
399	558	628	663	712
495	571	631	666	713
525	586	636	670	738
536	594	643	695	748
549	596	647	697	753
552	605	648		

$$\bar{x} = 609.16 \qquad \sqrt{b_1} = -1.236$$
$$s^2 = 10480.68 \qquad b_2 = 4.634$$

Data Set 8. *Average daily wind speed at Long Island MacArthur Airport (mph), July 1985, n = 31 (from NOAA, 1985).*

Examples: [4.7], [4.8], [5.4], [5.6], [6.5], [6.7]

Figures: [4.3]

7.7	11.1	7.8	9.5	5.9
8.5	8.8	11.5	5.6	10.7
6.9	8.9	10.2	6.2	7.7
11.1	9.0	8.7	10.4	5.2
17.1	11.2	10.7	12.5	3.8
13.3	6.2	8.8	8.1	7.4
8.9				

$$\bar{x} = 9.01 \qquad \sqrt{b_1} = 0.65$$
$$s^2 = 7.24 \qquad b_2 = 4.10$$

Data Set 9. *Induced acute leukemia latency period (in months) following chemotherapy, n = 20 (from Kapadia, Krause, Ellis, Pan and Wald, 1980; copyright 1980 American Cancer Society. Reprinted by permission of Wiley-Liss, Inc., a subsidiary of John Wiley and Sons, Inc.).*

Examples: [5.3], [5.7], [6.4], [6.8]

Figures: [5.2], [6.2], [13.8]

16	72	54	52	62
12	21	44	56	32
60	60	168	66	50
11	132	48	120	72

$$\bar{x} = 60.40 \qquad \sqrt{b_1} = 1.20$$
$$s^2 = 1596.57 \qquad b_2 = 4.16$$

Data Set 10. Average alkalinity (mg/l) of public water wells, 1990, n = 58 (from Newsday, 1991).

Examples: [5.5], [6.6]

Figures: [6.3], [13.1], [13.2]

29	48	34	41	39
34	48	41	29	28
30	35	39	38	34
36	46	36	60	26
48	32	72	41	26
32	27	38	29	30
42	45	39	44	26
36	45	63	50	37
48	23	35	33	34
38	35	31	33	31
42	31	21	38	33
31	27	26		

$$\overline{x} = 36.95 \qquad \sqrt{b_1} = 1.30$$
$$s^2 = 92.19 \qquad b_2 = 5.40$$

Data Set 11. Average annual erosion rates (m/year) for thirteen states on the East Coast of the U.S., n = 13 (from Dean, 1988, from data originally published in May, Dolan and Hayden, 1983. Copyright 1983 American Geophysical Union. Reproduced/modified by permission of American Geophysical Union).

Examples: [6.1]

Figures: [6.1]

Maine	-0.4	Maryland	-1.5
New Hampshire	-0.5	Virginia	-4.2
Massachusetts	-0.9	North Carolina	-0.6
Rhode Island	-0.5	South Carolina	-2.0
New York	0.1	Georgia	0.7
New Jersey	-1.0	Florida	-0.1
Delaware	0.1		

$$\bar{x} = -0.83 \qquad\qquad \sqrt{b_1} = -1.59$$
$$s^2 = 1.52 \qquad\qquad b_2 = 5.39$$

Data Set 12. Characteristics of Iris Setosa plants, n = 50 (from Fisher, 1936; copyright 1936. Used with permission of the Cambridge University Press.).

Examples: [9.1], [9.3], [10.2]

Figures: [9.2]

Sepal Length (a)	Sepal Width (b)	Petal Length (c)	Petal Width (d)
5.1	3.5	1.4	0.2
4.9	3.0	1.4	0.2
4.7	3.2	1.3	0.2
4.6	3.1	1.5	0.2
5.0	3.6	1.4	0.2
5.4	3.9	1.7	0.4
4.6	3.4	1.4	0.3
5.0	3.4	1.5	0.2
4.4	2.9	1.4	0.2
4.9	3.1	1.5	0.1
5.4	3.7	1.5	0.2
4.8	3.4	1.6	0.2
4.8	3.0	1.4	0.1
4.3	3.0	1.1	0.1
5.8	4.0	1.2	0.2
5.7	4.4	1.5	0.4
5.4	3.9	1.3	0.4
5.1	3.5	1.4	0.3
5.7	3.8	1.7	0.3
5.1	3.8	1.5	0.3
5.4	3.4	1.7	0.2
5.1	3.7	1.5	0.4
4.6	3.6	1.0	0.2
5.1	3.3	1.7	0.5
4.8	3.4	1.9	0.2
5.0	3.0	1.6	0.2
5.0	3.4	1.6	0.4
5.2	3.5	1.5	0.2
5.2	3.4	1.4	0.2
4.7	3.2	1.6	0.2

Data Set 12. Characteristics of Iris Setosa plants, n = 50 (Cont'd).

Sepal Length (a)	Sepal Width (b)	Petal Length (c)	Petal Width (d)
4.8	3.1	1.6	0.2
5.4	3.4	1.5	0.4
5.2	4.1	1.5	0.1
5.5	4.2	1.4	0.2
4.9	3.1	1.5	0.2
5.0	3.2	1.2	0.2
5.5	3.5	1.3	0.2
4.9	3.6	1.4	0.1
4.4	3.0	1.3	0.2
5.1	3.4	1.5	0.2
5.0	3.5	1.3	0.3
4.5	2.3	1.3	0.3
4.4	3.2	1.3	0.2
5.0	3.5	1.6	0.6
5.1	3.8	1.9	0.4
4.8	3.0	1.4	0.3
5.1	3.8	1.6	0.2
4.6	3.2	1.4	0.2
5.3	3.7	1.5	0.2
5.0	3.3	1.4	0.2

(a)	$\overline{x} = 5.01$	$\sqrt{b_1} = 0.12$
	$s^2 = 0.12$	$b_2 = 2.75$
(b)	$\overline{x} = 3.43$	$\sqrt{b_1} = 0.04$
	$s^2 = 0.14$	$b_2 = 3.95$
(c)	$\overline{x} = 1.46$	$\sqrt{b_1} = 0.11$
	$s^2 = 0.03$	$b_2 = 4.02$
(d)	$\overline{x} = 0.25$	$\sqrt{b_1} = 1.25$
	$s^2 = 0.01$	$b_2 = 4.72$

Data Set 13. Simulated bivariate normal data with correlation $\rho_{xy} = 0.45$, n = 50.

Examples: [9.2], [9.4]

Figures: [9.1], [9.3]

x	y	x	y	x	y
-0.065	-0.802	-0.705	-1.546	0.069	-0.913
0.322	0.309	0.686	-1.108	0.440	-1.075
-0.478	0.576	1.074	2.311	0.330	-2.160
0.214	0.312	-0.523	-0.768	0.928	0.441
-1.047	-2.811	-0.495	-1.576	-0.432	-0.398
-1.048	-1.222	-0.208	0.706	0.007	-0.199
1.163	-0.921	0.956	2.000	-0.302	-0.268
0.775	0.243	1.155	-0.269	0.586	0.835
-0.599	-1.693	0.109	-0.220	-0.526	0.202
-0.928	-0.885	0.742	-0.244	1.982	1.123
-0.278	2.100	1.371	2.172	0.909	0.672
-0.693	-0.591	0.223	1.489	-1.648	-2.915
-1.001	1.608	1.139	-0.303	-1.056	-1.405
-0.477	-1.534	0.093	-1.071	-0.478	1.372
-0.174	-1.931	-0.039	0.295	0.252	0.679
0.566	1.775	1.406	1.106	1.694	0.651
-0.059	0.248	0.970	0.572		

$$(x) \qquad \bar{x} = 0.14 \qquad \sqrt{b_1} = 0.15$$
$$s^2 = 0.66 \qquad b_2 = 2.44$$
$$(y) \qquad \bar{y} = -0.10 \qquad \sqrt{b_1} = -0.07$$
$$s^2 = 1.65 \qquad b_2 = 2.51$$

Data Set 14. Cerebrospinal fluid gas measurements in acidosis patients, n = 40 (from Hartigan, 1975; used with permission of the author).

Examples: [10.1], [10.3], [10.4], [10.5], [10.6], [10.7], [10.8], [10.9], [10.10], [13.1]

Figures: [10.1], [10.2], [10.3], [10.4], [10.5], [10.6], [10.7], [10.8], [13.9]

pH nanomol/ liter	HCO_3 millemol/ liter	CO_2 mm mercury	pH nanomol/ liter	HCO_3 millemol/ liter	CO_2 mm mercury
39.8	22.2	38.8	50.9	23.3	52.0
53.7	18.7	45.1	50.0	24.6	53.8
47.3	23.3	48.2	49.0	24.5	52.4
41.7	22.8	41.6	49.4	22.9	53.1
44.7	24.8	48.5	47.2	27.2	54.9
47.9	22.0	46.2	47.7	26.2	53.5
48.4	21.0	44.5	49.0	27.6	58.0
48.4	23.9	50.6	53.1	26.2	59.9
48.4	18.6	39.4	52.5	29.4	66.2
41.7	9.8	17.8	51.3	28.4	62.6
46.2	15.5	31.3	52.7	30.4	69.0
48.4	19.6	41.6	48.2	29.4	60.6
49.6	14.6	31.8	42.7	20.7	38.6
47.3	10.4	21.5	44.2	20.7	40.3
42.7	15.3	26.9	43.6	21.9	41.7
38.5	13.7	23.0	49.0	22.4	46.9
46.2	23.2	46.9	54.9	22.9	45.1
51.3	23.1	52.3	46.6	22.5	44.8
49.0	18.9	40.4	47.5	22.3	45.4
46.0	18.9	44.8	44.3	22.8	42.8

(pH)	$\bar{x} = 21.91$	$\sqrt{b_1} = -0.65$	
	$s^2 = 22.93$	$b_2 = 3.54$	
(HCO_3)	$\bar{x} = 45.57$	$\sqrt{b_1} = -0.39$	
	$s^2 = 132.01$	$b_2 = 3.37$	
(CO_2)	$\bar{x} = 47.52$	$\sqrt{b_1} = -0.37$	
	$s^2 = 14.07$	$b_2 = 3.00$	

References

Armitage, P., and Berry, G. (1987). **Statistical Methods in Medical Research.** Blackwell Scientific Publications, Oxford, U.K.

Consumer Reports (1989). Butter vs. margarine. September 1989 issue, 551-556, Consumers Union of United States, Inc., Mount Vernon, New York.

Dean, R.G. (1988). Managing sand and preserving shorelines. Oceanus 31, 49-55.

Fisher, R.A. (1936). The use of multiple measurements in taxonomic problems. Annals of Eugenics 7, 179-188.

Fisher, R.A. (1971). **The Design of Experiments.** Hafner Press, New York, NY.

Hartigan, J.A. (1975). **Clustering Algorithms.** John Wiley and Sons, New York, NY.

Hoel, D.G. (1972). A representation of mortality data by competing risks. Biometrics 28, 475-488.

Kapadia, S.B., Krause, J.R., Ellis, L.D., Pan, S.F., and Wald, N. (1980). Induced acute non-lymphocytic leukemia following long term chemotherapy. Cancer 45, 1315-1321.

Langley, R. (1971). **Practical Statistics.** Dover Publications, New York, NY.

May, S.K., Dolan, R., and Hayden, B.P. (1983). Erosion of U.S. shorelines. EOS 64, 551-553.

National Oceanic and Atmospheric Administration, U.S. Dept. of Commerce (1985). Local climatological data monthly summary, ISSN 0742-8707.

Newsday, 1990 annual water quality statement, Suffolk County Water Authority, March 28, 1991, 127-135.

Shapiro, S.S. (1980). **How to Test Normality and Other Distributional Assumptions.** American Society for Quality Control, Milwaukee, WI.

APPENDIX B

PARAMETER AND CRITICAL VALUES

B1. Coefficients a_i for computing the Wilk-Shapiro W test for normality

B2. Lower percentage points of the Wilk-Shapiro W test for normality

B3. Coefficients a_i and b_i for computing LaBreque's nonlinearity tests for normality

B4. Upper percentage points of LaBreque's nonlinearity tests for normality

B5. Upper and lower percentage points of the skewness test for normality, $\sqrt{b_1}$

B6. Upper and lower percentage points of the kurtosis test for normality, b_2

B7. Upper and lower percentage points of absolute moment tests, a(c)

B8. Upper and lower percentage points of Grubbs' extreme outlier tests T_1, T_n and $T = max(T_1, T_n)$

B9. Upper and lower percentage points of the deviations from the median test, U

B10. Upper and lower percentage points of the range test

B11. Percentage points of the most powerful location and scale invariant tests T'_s (upper percentage points) and S_s (lower percentage points)

B12. Upper and lower percentage points of the T^* test based on a U statistic

B13. Upper and lower percentage points of the standardized version, Y, of DAgostino's D

B14. Lower 5% percentage points of the sample entropy tests, K_{mn}

Table B1. *Coefficients a_i for computing the Wilk-Shapiro test for normality, W (from Shapiro and Wilk, 1965; used with permission of the Biometrika Trustees)*

				sample size			
i	2	3	4	5	6	7	8
1	0.7071	0.7071	0.6872	0.6646	0.6431	0.6233	0.6052
2	-	0.0000	0.1677	0.2413	0.2806	0.3031	0.3164
3	-	-	-	0.0000	0.0875	0.1401	0.1743
4	-	-	-	-	-	0.0000	0.0561

i	9	10	11	12	13	14	15	16
1	0.5888	0.5739	0.5601	0.5475	0.5359	0.5251	0.5150	0.5056
2	0.3244	0.3291	0.3315	0.3325	0.3325	0.3318	0.3306	0.3290
3	0.1976	0.2141	0.2260	0.2347	0.2412	0.2460	0.2495	0.2521
4	0.0947	0.1224	0.1429	0.1586	0.1707	0.1802	0.1878	0.1939
5	0.0000	0.0399	0.0695	0.0922	0.1099	0.1240	0.1353	0.1447
6	-	-	0.0000	0.0303	0.0539	0.0727	0.0880	0.1005
7	-	-	-	-	0.0000	0.0240	0.0433	0.0593
8	-	-	-	-	-	-	0.0000	0.0196

i	17	18	19	20	21	22	23	24
1	0.4968	0.4886	0.4808	0.4734	0.4643	0.4590	0.4542	0.4493
2	0.3273	0.3253	0.3232	0.3211	0.3185	0.3156	0.3126	0.3098
3	0.2540	0.2553	0.2561	0.2565	0.2578	0.2571	0.2563	0.2554
4	0.1988	0.2027	0.2059	0.2085	0.2119	0.2131	0.2139	0.2145
5	0.1524	0.1587	0.1641	0.1686	0.1736	0.1764	0.1787	0.1807
6	0.1109	0.1197	0.1271	0.1334	0.1399	0.1443	0.1480	0.1512
7	0.0725	0.0837	0.0932	0.1013	0.1092	0.1150	0.1201	0.1245
8	0.0359	0.0496	0.0612	0.0711	0.0804	0.0878	0.0941	0.0997
9	0.0000	0.0163	0.0303	0.0422	0.0530	0.0618	0.0696	0.0764
10	-	-	0.0000	0.0140	0.0263	0.0368	0.0459	0.0539
11	-	-	-	-	0.0000	0.0122	0.0228	0.0321
12	-	-	-	-	-	-	0.0000	0.0107

Table B1 (cont'd). Coefficients a_i for computing the Wilk-Shapiro W test for normality

i	25	26	27	28	29	30	31	32
1	0.4450	0.4407	0.4366	0.4328	0.4291	0.4254	0.4220	0.4188
2	0.3069	0.3043	0.3018	0.2992	0.2968	0.2944	0.2921	0.2898
3	0.2543	0.2533	0.2522	0.2510	0.2499	0.2487	0.2475	0.2463
4	0.2148	0.2151	0.2152	0.2151	0.2150	0.2148	0.2145	0.2141
5	0.1822	0.1836	0.1848	0.1857	0.1864	0.1870	0.1874	0.1878
6	0.1539	0.1563	0.1584	0.1601	0.1616	0.1630	0.1641	0.1651
7	0.1283	0.1316	0.1346	0.1372	0.1395	0.1415	0.1433	0.1449
8	0.1046	0.1089	0.1128	0.1162	0.1192	0.1219	0.1243	0.1265
9	0.0823	0.0876	0.0923	0.0965	0.1002	0.1036	0.1066	0.1093
10	0.0610	0.0672	0.0728	0.0778	0.0822	0.0862	0.0899	0.0931
11	0.0403	0.0476	0.0540	0.0598	0.0650	0.0697	0.0739	0.0777
12	0.0200	0.0284	0.0358	0.0424	0.0483	0.0537	0.0585	0.0629
13	0.0000	0.0094	0.0178	0.0253	0.0320	0.0381	0.0435	0.0485
14	-	-	0.0000	0.0084	0.0159	0.0227	0.0289	0.0344
15	-	-	-	-	0.0000	0.0076	0.0144	0.0206
16	-	-	-	-	-	-	0.0000	0.0068

Table B1 (cont'd). Coefficients a_i for computing the Wilk-Shapiro W test for normality

i	33	34	35	36	37	38	39
1	0.4156	0.4127	0.4096	0.4068	0.4040	0.4015	0.3989
2	0.2876	0.2854	0.2834	0.2813	0.2794	0.2774	0.2755
3	0.2451	0.2439	0.2427	0.2415	0.2403	0.2391	0.2380
4	0.2137	0.2132	0.2127	0.2121	0.2116	0.2110	0.2104
5	0.1880	0.1882	0.1883	0.1883	0.1883	0.1881	0.1880
6	0.1660	0.1667	0.1673	0.1678	0.1683	0.1686	0.1689
7	0.1463	0.1475	0.1487	0.1496	0.1505	0.1513	0.1520
8	0.1284	0.1301	0.1317	0.1331	0.1344	0.1356	0.1366
9	0.1118	0.1140	0.1160	0.1179	0.1196	0.1211	0.1225
10	0.0961	0.0988	0.1013	0.1036	0.1056	0.1075	0.1092
11	0.0812	0.0844	0.0873	0.0900	0.0924	0.0947	0.0967
12	0.0669	0.0706	0.0739	0.0770	0.0798	0.0824	0.0848
13	0.0530	0.0572	0.0610	0.0645	0.0677	0.0706	0.0733
14	0.0395	0.0441	0.0484	0.0523	0.0559	0.0592	0.0622
15	0.0262	0.0314	0.0361	0.0404	0.0444	0.0481	0.0515
16	0.0131	0.0187	0.0239	0.0287	0.0331	0.0372	0.0409
17	0.0000	0.0062	0.0119	0.0172	0.0220	0.0264	0.0305
18	-	-	0.0000	0.0057	0.0110	0.0158	0.0203
19	-	-	-	-	0.0000	0.0053	0.0101
20	-	-	-	-	-	-	0.0000

Table B1 (cont'd). Coefficients a_i for computing the Wilk-Shapiro W test for normality

i	40	41	42	43	44	45	46
1	0.3964	0.3940	0.3917	0.3894	0.3872	0.3850	0.3830
2	0.2737	0.2719	0.2701	0.2684	0.2667	0.2651	0.2635
3	0.2368	0.2357	0.2345	0.2334	0.2323	0.2313	0.2302
4	0.2098	0.2091	0.2085	0.2078	0.2072	0.2065	0.2058
5	0.1878	0.1876	0.1874	0.1871	0.1868	0.1865	0.1862
6	0.1691	0.1693	0.1694	0.1695	0.1695	0.1695	0.1695
7	0.1526	0.1531	0.1535	0.1539	0.1542	0.1545	0.1548
8	0.1376	0.1384	0.1392	0.1398	0.1405	0.1410	0.1415
9	0.1237	0.1249	0.1259	0.1269	0.1278	0.1286	0.1293
10	0.1108	0.1123	0.1136	0.1149	0.1160	0.1170	0.1180
11	0.0986	0.1004	0.1020	0.1035	0.1049	0.1062	0.1073
12	0.0870	0.0891	0.0909	0.0927	0.0943	0.0959	0.0972
13	0.0759	0.0782	0.0804	0.0824	0.0842	0.0860	0.0876
14	0.0651	0.0677	0.0701	0.0724	0.0745	0.0765	0.0783
15	0.0546	0.0575	0.0602	0.0628	0.0651	0.0673	0.0694
16	0.0444	0.0476	0.0506	0.0534	0.0560	0.0584	0.0607
17	0.0343	0.0379	0.0411	0.0442	0.0471	0.0497	0.0522
18	0.0244	0.0283	0.0318	0.0352	0.0383	0.0412	0.0439
19	0.0146	0.0188	0.0227	0.0263	0.0296	0.0328	0.0357
20	0.0049	0.0094	0.0136	0.0175	0.0211	0.0245	0.0277
21	-	0.0000	0.0045	0.0087	0.0126	0.0163	0.0197
22	-	-	-	0.0000	0.0042	0.0081	0.0118
23	-	-	-	-	-	0.0000	0.0039

Table B1 (cont'd). Coefficients a_i for computing the Wilk-Shapiro W test for normality

i	47	48	49	50
1	0.3808	0.3789	0.3770	0.3751
2	0.2620	0.2604	0.2589	0.2574
3	0.2291	0.2281	0.2271	0.2260
4	0.2052	0.2045	0.2038	0.2032
5	0.1859	0.1855	0.1851	0.1847
6	0.1695	0.1693	0.1692	0.1691
7	0.1550	0.1551	0.1553	0.1554
8	0.1420	0.1423	0.1427	0.1430
9	0.1300	0.1306	0.1312	0.1317
10	0.1189	0.1197	0.1205	0.1212
11	0.1085	0.1095	0.1105	0.1113
12	0.0986	0.0998	0.1010	0.1020
13	0.0892	0.0906	0.0919	0.0932
14	0.0801	0.0817	0.0832	0.0846
15	0.0713	0.0731	0.0748	0.0764
16	0.0628	0.0648	0.0667	0.0685
17	0.0546	0.0568	0.0588	0.0608
18	0.0465	0.0489	0.0511	0.0532
19	0.0385	0.0411	0.0436	0.0459
20	0.0307	0.0335	0.0361	0.0386
21	0.0229	0.0259	0.0288	0.0314
22	0.0153	0.0185	0.0215	0.0244
23	0.0076	0.0111	0.0143	0.0174
24	0.0000	0.0037	0.0071	0.0104
25	-	-	0.0000	0.0035

Table B2. *Lower percentage points of the Wilk-Shapiro W test for normality (from Shapiro and Wilk, 1965; used with permission of the Biometrika Trustees)*

		percentage point		
n	1%	2%	5%	10%
3	0.753	0.756	0.767	0.789
4	0.687	0.707	0.748	0.792
5	0.686	0.715	0.762	0.806
6	0.713	0.743	0.788	0.826
7	0.730	0.760	0.803	0.838
8	0.749	0.778	0.818	0.851
9	0.764	0.791	0.829	0.859
10	0.781	0.806	0.842	0.869
11	0.792	0.817	0.850	0.876
12	0.805	0.828	0.859	0.883
13	0.814	0.837	0.866	0.889
14	0.825	0.846	0.874	0.895
15	0.835	0.855	0.881	0.901
16	0.844	0.863	0.887	0.906
17	0.851	0.869	0.892	0.910
18	0.858	0.874	0.897	0.914
19	0.863	0.879	0.901	0.917
20	0.868	0.884	0.905	0.920
21	0.873	0.888	0.908	0.923
22	0.878	0.892	0.911	0.926
23	0.881	0.895	0.914	0.928
24	0.884	0.898	0.916	0.930
25	0.888	0.901	0.918	0.931
26	0.891	0.904	0.920	0.933
27	0.894	0.906	0.923	0.935
28	0.896	0.908	0.924	0.936
29	0.898	0.910	0.926	0.937
30	0.900	0.912	0.927	0.939

Table B2 (cont'd). Lower percentage points of the Wilk-Shapiro W test for normality

n	1%	2%	5%	10%
31	0.902	0.914	0.929	0.940
32	0.904	0.915	0.930	0.941
33	0.906	0.917	0.931	0.942
34	0.908	0.919	0.933	0.943
35	0.910	0.920	0.934	0.944
36	0.912	0.922	0.935	0.945
37	0.914	0.924	0.936	0.946
38	0.916	0.925	0.938	0.947
39	0.917	0.927	0.939	0.948
40	0.919	0.928	0.940	0.949
41	0.920	0.929	0.941	0.950
42	0.922	0.930	0.942	0.951
43	0.923	0.932	0.943	0.951
44	0.924	0.933	0.944	0.952
45	0.926	0.934	0.945	0.953
46	0.927	0.935	0.945	0.953
47	0.928	0.936	0.946	0.954
48	0.929	0.937	0.947	0.954
49	0.929	0.937	0.947	0.955
50	0.930	0.938	0.947	0.955

Table B3. Coefficients a_i and b_i for computing LaBreque's nonlinearity tests for normality (from LaBreque, 1977; reproduced with permission from Technometrics. Copyright 1977 by the American Statistical Association. All rights reserved.)

	n=4		n=5		n=6	
i	a	b	a	b	a	b
1	1.1387	-0.6261	1.2002	-0.8403	1.2266	-0.9701
2	-1.1387	2.1699	-0.7157	1.9742	-0.4545	1.7195
3	-	-	-0.9689	-0.0000	-0.7721	0.6243

	n=7		n=8		n=9	
i	a	b	a	b	a	b
1	1.2367	-1.0574	1.2385	-1.1196	1.2355	-1.1658
2	-0.2768	1.4962	-0.1483	1.3092	-0.0512	1.1527
3	-0.6095	0.8410	-0.4796	0.9090	-0.3753	0.9145
4	-0.7009	0.0000	-0.6106	0.3174	-0.5245	0.4864
5	-	-	-	-	-0.5691	0.0000

	n=10		n=11		n=12	
i	a	b	a	b	a	b
1	1.2298	-1.2010	1.2225	-1.2284	1.2143	-1.2501
2	0.0245	1.0205	0.0852	0.9075	0.1346	0.8100
3	-0.2902	0.8919	-0.2198	0.8564	-0.1606	0.8157
4	-0.4478	0.5771	-0.3807	0.6236	-0.3222	0.6441
5	-0.5164	0.1979	-0.4629	0.3239	-0.4125	0.4048
6	-	-	-0.4886	0.0000	-0.4536	0.1375

	n=13		n=14		n=15	
i	a	b	a	b	a	b
1	1.2056	-1.2674	1.1966	-1.2814	1.1875	-1.2927
2	0.1756	0.7249	0.2100	0.6499	0.2393	0.5834
3	-0.1104	0.7733	-0.0673	0.7313	-0.0299	0.6906
4	-0.2710	0.6489	-0.2261	0.6438	-0.1865	0.6327
5	-0.3660	0.4566	-0.3237	0.4888	-0.2854	0.5079
6	-0.4171	0.2342	-0.3813	0.3027	-0.3473	0.3512
7	-0.4333	0.0000	-0.4082	0.1023	-0.3815	0.1789
8	-	-	-	-	-0.3925	0.0000

Table B3 (cont'd). Coefficients a_i and b_i for computing LaBreque's non-linearity tests for normality

	n=16		n=17		n=18	
i	a	b	a	b	a	b
1	1.1785	-1.3020	1.1695	-1.3095	1.1607	-1.3156
2	0.2645	0.5240	0.2863	0.4705	0.3053	0.4222
3	0.0027	0.6517	0.0315	0.6147	0.0571	0.5797
4	-0.1513	0.6178	-0.1199	0.6006	-0.0917	0.5822
5	-0.2507	0.5177	-0.2192	0.5210	-0.1905	0.5198
6	-0.3154	0.3854	-0.2858	0.4092	-0.2583	0.4252
7	-0.3548	0.2366	-0.3289	0.2802	-0.3040	0.3130
8	-0.3736	0.0797	-0.3532	0.1421	-0.3324	0.1911
9	-	-	-0.3611	0.0000	-0.3461	0.0642

	n=19		n=20		n=21	
i	a	b	a	b	a	b
1	1.1521	-1.3206	1.1436	-1.3246	1.1363	-1.3283
2	0.3220	0.3782	0.3367	0.3380	0.3447	0.3041
3	0.0798	0.5467	0.1002	0.5156	0.1223	0.4834
4	-0.0663	0.5630	-0.0433	0.5437	-0.0216	0.5242
5	-0.1644	0.5153	-0.1406	0.5085	-0.1184	0.5002
6	-0.2329	0.4354	-0.2093	0.4413	-0.1876	0.4440
7	-0.2805	0.3377	-0.2583	0.3560	-0.2376	0.3695
8	-0.3119	0.2298	-0.2920	0.2603	-0.2730	0.2844
9	-0.3300	0.1162	-0.3134	0.1583	-0.2969	0.1926
10	-0.3358	0.0000	-0.3237	0.0531	-0.3107	0.0972
11	-	-	-	-	-0.3152	0.0000

Table B3 (cont'd). Coefficients a_i and b_i for computing LaBreque's non-linearity tests for normality

	n=22		n=23		n=24	
i	a	b	a	b	a	b
1	1.1283	-1.3309	1.1205	-1.3329	1.1129	-1.3345
2	0.3561	0.2705	0.3663	0.2395	0.3754	0.2107
3	0.1390	0.4556	0.1543	0.4292	0.1681	0.4044
4	-0.0023	0.5051	0.0151	0.4864	0.0312	0.4682
5	-0.0984	0.4906	-0.0798	0.4801	-0.0626	0.4692
6	-0.1673	0.4439	-0.1485	0.4419	-0.1310	0.4384
7	-0.2179	0.3791	-0.1995	0.3854	-0.1822	0.3893
8	-0.2548	0.3031	-0.2373	0.3178	-0.2208	0.3291
9	-0.2805	0.2207	-0.2648	0.2432	-0.2494	0.2618
10	-0.2971	0.1336	-0.2833	0.1644	-0.2698	0.1896
11	-0.3051	0.0451	-0.2942	0.0827	-0.2828	0.1148
12	-	-	-0.2977	0.0000	-0.2891	0.0386

	n=25		n=26		n=27	
i	a	b	a	b	a	b
1	1.1056	-1.3356	1.0984	-1.3363	1.0914	-1.3367
2	0.3836	0.1840	0.3910	0.1591	0.3977	0.1358
3	0.1808	0.3809	0.1925	0.3586	0.2031	0.3376
4	0.0460	0.4505	0.0596	0.4334	0.0723	0.4168
5	-0.0466	0.4579	-0.0318	0.4464	-0.0181	0.4349
6	-0.1147	0.4337	-0.0995	0.4282	-0.0851	0.4220
7	-0.1661	0.3913	-0.1506	0.3918	-0.1365	0.3908
8	-0.2050	0.3377	-0.1904	0.3438	-0.1762	0.3483
9	-0.2347	0.2766	-0.2205	0.2888	-0.2071	0.2984
10	-0.2563	0.2111	-0.2434	0.2286	-0.2307	0.2434
11	-0.2713	0.1420	-0.2597	0.1653	-0.2483	0.1849
12	-0.2798	0.0715	-0.2702	0.1000	-0.2603	0.1246
13	-0.2827	0.0000	-0.2754	0.0335	-0.2674	0.0625
14	-	-	-	-	-0.2697	0.0000

Table B3 (cont'd). Coefficients a_i and b_i for computing LaBreque's nonlinearity tests for normality

	n=28		n=29		n=30	
i	a	b	a	b	a	b
1	1.0845	-1.3369	1.0779	-1.3367	1.0714	-1.3364
2	0.4038	0.1140	0.4093	0.0935	0.4144	0.0743
3	0.2130	0.3176	0.2221	0.2985	0.2305	0.2805
4	0.0841	0.4007	0.0949	0.3853	0.1052	0.3703
5	-0.0054	0.4235	0.0068	0.4120	0.0180	0.4008
6	-0.0717	0.4152	-0.0593	0.4081	-0.0474	0.4008
7	-0.1229	0.3888	-0.1102	0.3861	-0.0980	0.3827
8	-0.1628	0.3512	-0.1501	0.3528	-0.1382	0.3533
9	-0.1941	0.3060	-0.1820	0.3117	-0.1701	0.3163
10	-0.2186	0.2556	-0.2067	0.2659	-0.1955	0.2741
11	-0.2371	0.2017	-0.2263	0.2159	-0.2156	0.2261
12	-0.2503	0.1459	-0.2406	0.1639	-0.2308	0.1799
13	-0.2591	0.0878	-0.2505	0.1103	-0.2419	0.1294
14	-0.2633	0.0295	-0.2564	0.0553	-0.2491	0.0783
15	-	-	-0.2583	0.0000	-0.2527	0.0261

Table B3 (cont'd). Coefficients a_i and b_i for computing LaBreque's non-linearity tests for normality

	n=31		n=32		n=33	
i	a	b	a	b	a	b
1	1.0650	-1.3359	1.0589	-1.3352	1.0528	-1.3344
2	0.4190	0.0561	0.4232	0.0390	0.4270	0.0229
3	0.2383	0.2632	0.2456	0.2468	0.2524	0.2311
4	0.1147	0.3559	0.1237	0.3420	0.1321	0.3285
5	0.0286	0.3897	0.0384	0.3789	0.0475	0.3682
6	-0.0365	0.3932	-0.0259	0.3855	-0.0159	0.3778
7	-0.0865	0.3788	-0.0759	0.3744	-0.0656	0.3697
8	-0.1268	0.3529	-0.1158	0.3519	-0.1057	0.3502
9	-0.1590	0.3195	-0.1483	0.3217	-0.1379	0.3231
10	-0.1847	0.2809	-0.1743	0.2863	-0.1642	0.2907
11	-0.2052	0.2385	-0.1953	0.2471	-0.1858	0.2542
12	-0.2213	0.1933	-0.2120	0.2051	-0.2029	0.2153
13	-0.2333	0.1464	-0.2247	0.1614	-0.2163	0.1743
14	-0.2416	0.0984	-0.2341	0.1161	-0.2265	0.1320
15	-0.2466	0.0495	-0.2402	0.0701	-0.2336	0.0881
16	-0.2482	0.0000	-0.2432	0.0231	-0.2377	0.0445
17	-	-	-	-	-0.2391	0.0000

Table B3 (cont'd). Coefficients a_i and b_i for computing LaBreque's nonlinearity tests for normality

	n=34		n=35		n=36	
i	a	b	a	b	a	b
1	1.0469	-1.3335	1.0412	-1.3324	1.0356	-1.3313
2	0.4305	0.0075	0.4338	-0.0071	0.4367	-0.0209
3	0.2587	0.2162	0.2646	0.2019	0.2702	0.1881
4	0.1399	0.3156	0.1473	0.3031	0.1543	0.2910
5	0.0565	0.3577	0.0648	0.3475	0.0724	0.3376
6	-0.0066	0.3701	0.0022	0.3623	0.0108	0.3545
7	-0.0560	0.3647	-0.0469	0.3595	-0.0382	0.3541
8	-0.0957	0.3481	-0.0864	0.3454	-0.0775	0.3424
9	-0.1284	0.3236	-0.1190	0.3235	-0.1099	0.3230
10	-0.1547	0.2940	-0.1454	0.2966	-0.1368	0.2982
11	-0.1765	0.2604	-0.1676	0.2653	-0.1589	0.2696
12	-0.1941	0.2241	-0.1855	0.2316	-0.1772	0.2379
13	-0.2082	0.1855	-0.2001	0.1956	-0.1921	0.2043
14	-0.2189	0.1459	-0.2115	0.1581	-0.2042	0.1689
15	-0.2269	0.1048	-0.2202	0.1194	-0.2134	0.1326
16	-0.2321	0.0632	-0.2262	0.0801	-0.2202	0.0953
17	-0.2347	0.0210	-0.2298	0.0403	-0.2247	0.0573
18	-	-	-0.2310	0.0000	-0.2269	0.0193

Table B3 (cont'd). Coefficients a_i and b_i for computing LaBreque's non-linearity tests for normality

	n=37		n=38		n=39	
i	a	b	a	b	a	b
1	1.0301	-1.3300	1.0248	-1.3287	1.0195	-1.3273
2	0.4395	-0.0340	0.4420	-0.0465	0.4443	-0.0585
3	0.2754	0.1750	0.2803	0.1624	0.2848	0.1503
4	0.1607	0.2794	0.1669	0.2681	0.1729	0.2572
5	0.0801	0.3277	0.0870	0.3182	0.0936	0.3090
6	0.0186	0.3469	0.0264	0.3393	0.0335	0.3318
7	-0.0299	0.3486	-0.0219	0.3431	-0.0143	0.3375
8	-0.0691	0.3391	-0.0609	0.3356	-0.0532	0.3319
9	-0.1013	0.3220	-0.0934	0.3204	-0.0854	0.3187
10	-0.1284	0.2993	-0.1199	0.3000	-0.1123	0.3000
11	-0.1505	0.2729	-0.1425	0.2755	-0.1347	0.2776
12	-0.1691	0.2434	-0.1615	0.2479	-0.1540	0.2518
13	-0.1845	0.2118	-0.1769	0.2185	-0.1697	0.2242
14	-0.1969	0.1787	-0.1897	0.1874	-0.1828	0.1948
15	-0.2068	0.1442	-0.2003	0.1543	-0.1937	0.1638
16	-0.2143	0.1088	-0.2081	0.1213	-0.2022	0.1320
17	-0.2195	0.0728	-0.2142	0.0868	-0.2087	0.0998
18	-0.2226	0.0365	-0.2180	0.0524	-0.2132	0.0669
19	-0.2236	0.0000	-0.2199	0.0178	-0.2160	0.0333
20	-	-	-	-	-0.2168	0.0000

Table B3 (cont'd). Coefficients a_i and b_i for computing LaBreque's non-linearity tests for normality

i	n=40 a	b	n=41 a	b	n=42 a	b
1	1.0144	-1.3259	1.0094	-1.3244	1.0045	-1.3229
2	0.4465	-0.0699	0.4484	-0.0808	0.4501	-0.0911
3	0.2890	0.1388	0.2933	0.1275	0.2971	0.1167
4	0.1784	0.2466	0.1835	0.2365	0.1885	0.2266
5	0.1001	0.2999	0.1061	0.2911	0.1118	0.2825
6	0.0402	0.3245	0.0468	0.3172	0.0530	0.3100
7	-0.0071	0.3318	-0.0002	0.3262	0.0064	0.3205
8	-0.0457	0.3280	-0.0387	0.3240	-0.0319	0.3199
9	-0.0780	0.3166	-0.0706	0.3143	-0.0638	0.3117
10	-0.1046	0.2998	-0.0976	0.2991	-0.0905	0.2981
11	-0.1276	0.2788	-0.1203	0.2798	-0.1134	0.2804
12	-0.1463	0.2553	-0.1394	0.2578	-0.1326	0.2599
13	-0.1629	0.2288	-0.1559	0.2332	-0.1492	0.2369
14	-0.1760	0.2015	-0.1695	0.2072	-0.1632	0.2121
15	-0.1873	0.1721	-0.1810	0.1795	-0.1746	0.1865
16	-0.1962	0.1422	-0.1903	0.1511	-0.1847	0.1588
17	-0.2033	0.1110	-0.1978	0.1217	-0.1923	0.1315
18	-0.2084	0.0800	-0.2035	0.0916	-0.1986	0.1025
19	-0.2118	0.0480	-0.2075	0.0614	-0.2031	0.0736
20	-0.2135	0.0164	-0.2099	0.0310	-0.2061	0.0442
21	-	-	-0.2107	0.0000	-0.2076	0.0150

Table B3 (cont'd). Coefficients a_i and b_i for computing LaBreque's non-linearity tests for normality

i	n=43 a	b	n=44 a	b	n=45 a	b
1	0.9996	-1.3213	0.9949	-1.3197	0.9903	-1.3180
2	0.4519	-0.1011	0.4534	-0.1107	0.4548	-0.1199
3	0.3006	0.1064	0.3040	0.0965	0.3072	0.0869
4	0.1932	0.2171	0.1978	0.2078	0.2020	0.1989
5	0.1174	0.2740	0.1224	0.2660	0.1274	0.2579
6	0.0590	0.3030	0.0648	0.2960	0.0703	0.2892
7	0.0126	0.3149	0.0186	0.3093	0.0243	0.3037
8	-0.0253	0.3157	-0.0189	0.3115	-0.0129	0.3072
9	-0.0570	0.3090	-0.0509	0.3061	-0.0447	0.3031
10	-0.0837	0.2969	-0.0774	0.2953	-0.0712	0.2936
11	-0.1069	0.2805	-0.1001	0.2805	-0.0938	0.2802
12	-0.1259	0.2616	-0.1195	0.2629	-0.1135	0.2636
13	-0.1427	0.2400	-0.1367	0.2424	-0.1306	0.2445
14	-0.1569	0.2167	-0.1506	0.2208	-0.1446	0.2242
15	-0.1689	0.1919	-0.1631	0.1971	-0.1571	0.2020
16	-0.1787	0.1664	-0.1731	0.1730	-0.1677	0.1788
17	-0.1870	0.1399	-0.1817	0.1476	-0.1766	0.1544
18	-0.1937	0.1124	-0.1887	0.1215	-0.1838	0.1297
19	-0.1986	0.0851	-0.1941	0.0951	-0.1896	0.1046
20	-0.2022	0.0563	-0.1982	0.0681	-0.1941	0.0787
21	-0.2043	0.0286	-0.2008	0.0409	-0.1972	0.0525
22	-0.2050	0.0000	-0.2021	0.0141	-0.1991	0.0264
23	-	-	-	-	-0.1997	0.0000

Table B3 (cont'd). Coefficients a_i and b_i for computing LaBreque's non-linearity tests for normality

	n=46		n=47		n=48	
i	a	b	a	b	a	b
1	0.9858	-1.3163	0.9814	-1.3146	0.9770	-1.3129
2	0.4560	-0.1286	0.4572	-0.1371	0.4583	-0.1451
3	0.3103	0.0775	0.3131	0.0687	0.3160	0.0599
4	0.2062	0.1901	0.2100	0.1817	0.2137	0.1735
5	0.1321	0.2502	0.1367	0.2426	0.1410	0.2352
6	0.0754	0.2825	0.0803	0.2760	0.0854	0.2695
7	0.0301	0.2981	0.0352	0.2927	0.0403	0.2873
8	-0.0074	0.3029	-0.0016	0.2985	0.0035	0.2942
9	-0.0387	0.3000	-0.0332	0.2968	-0.0275	0.2935
10	-0.0651	0.2918	-0.0594	0.2897	-0.0537	0.2875
11	-0.0881	0.2794	-0.0821	0.2786	-0.0767	0.2774
12	-0.1074	0.2642	-0.1017	0.2644	-0.0962	0.2643
13	-0.1245	0.2464	-0.1190	0.2476	-0.1129	0.2491
14	-0.1390	0.2270	-0.1331	0.2297	-0.1281	0.2315
15	-0.1516	0.2061	-0.1462	0.2096	-0.1408	0.2128
16	-0.1624	0.1838	-0.1571	0.1885	-0.1520	0.1926
17	-0.1713	0.1610	-0.1662	0.1669	-0.1611	0.1724
18	-0.1789	0.1373	-0.1742	0.1441	-0.1695	0.1502
19	-0.1852	0.1128	-0.1806	0.1208	-0.1762	0.1280
20	-0.1899	0.0885	-0.1858	0.0973	-0.1817	0.1053
21	-0.1935	0.0633	-0.1898	0.0730	-0.1859	0.0825
22	-0.1959	0.0381	-0.1926	0.0491	-0.1891	0.0594
23	-0.1971	0.0127	-0.1943	0.0245	-0.1914	0.0352
24	-	-	-0.1948	0.0000	-0.1924	0.0115

Table B3 (cont'd). Coefficients a_i and b_i for computing LaBreque's nonlinearity tests for normality

	n=49		n=50		n=51	
i	a	b	a	b	a	b
1	0.9728	-1.3112	0.9686	-1.3094	0.9645	-1.3076
2	0.4593	-0.1530	0.4602	-0.1605	0.4610	-0.1677
3	0.3185	0.0515	0.3209	0.0435	0.3233	0.0357
4	0.2171	0.1656	0.2207	0.1577	0.2239	0.1503
5	0.1452	0.2279	0.1490	0.2211	0.1529	0.2142
6	0.0898	0.2632	0.0943	0.2570	0.0985	0.2509
7	0.0453	0.2819	0.0500	0.2767	0.0545	0.2715
8	0.0087	0.2898	0.0138	0.2855	0.0184	0.2811
9	-0.0224	0.2902	-0.0173	0.2869	-0.0120	0.2834
10	-0.0482	0.2852	-0.0431	0.2828	-0.0385	0.2803
11	-0.0714	0.2761	-0.0660	0.2747	-0.0605	0.2732
12	-0.0906	0.2642	-0.0853	0.2637	-0.0807	0.2629
13	-0.1078	0.2496	-0.1027	0.2501	-0.0972	0.2506
14	-0.1228	0.2332	-0.1175	0.2347	-0.1126	0.2359
15	-0.1353	0.2158	-0.1303	0.2182	-0.1254	0.2202
16	-0.1469	0.1965	-0.1419	0.1999	-0.1371	0.2028
17	-0.1564	0.1768	-0.1517	0.1809	-0.1470	0.1848
18	-0.1648	0.1559	-0.1601	0.1613	-0.1555	0.1661
19	-0.1717	0.1347	-0.1673	0.1409	-0.1632	0.1461
20	-0.1774	0.1132	-0.1734	0.1200	-0.1693	0.1264
21	-0.1822	0.0907	-0.1784	0.0985	-0.1745	0.1059
22	-0.1857	0.0683	-0.1823	0.0767	-0.1788	0.0847
23	-0.1883	0.0456	-0.1851	0.0555	-0.1818	0.0645
24	-0.1897	0.0230	-0.1870	0.0331	-0.1843	0.0423
25	-0.1903	0.0000	-0.1880	0.0112	-0.1855	0.0217
26	-	-	-	-	-0.1860	0.0000

Table B3 (cont'd). Coefficients a_i and b_i for computing LaBreque's nonlinearity tests for normality

i	n=52 a	b	n=53 a	b	n=54 a	b
1	0.9605	-1.3059	0.9565	-1.3041	0.9526	-1.3023
2	0.4617	-0.1747	0.4625	-0.1815	0.4631	-0.1880
3	0.3255	0.0280	0.3276	0.0207	0.3296	0.0136
4	0.2269	0.1430	0.2297	0.1360	0.2326	0.1290
5	0.1565	0.2075	0.1602	0.2008	0.1635	0.1945
6	0.1024	0.2450	0.1063	0.2392	0.1102	0.2334
7	0.0592	0.2662	0.0633	0.2611	0.0673	0.2562
8	0.0229	0.2769	0.0274	0.2726	0.0317	0.2684
9	-0.0073	0.2800	-0.0027	0.2765	0.0016	0.2731
10	-0.0336	0.2777	-0.0286	0.2752	-0.0243	0.2724
11	-0.0557	0.2715	-0.0513	0.2697	-0.0464	0.2678
12	-0.0753	0.2622	-0.0704	0.2613	-0.0658	0.2602
13	-0.0927	0.2504	-0.0879	0.2503	-0.0833	0.2500
14	-0.1075	0.2370	-0.1027	0.2377	-0.0980	0.2382
15	-0.1209	0.2217	-0.1162	0.2232	-0.1116	0.2245
16	-0.1320	0.2059	-0.1277	0.2078	-0.1231	0.2100
17	-0.1426	0.1880	-0.1379	0.1913	-0.1336	0.1940
18	-0.1512	0.1702	-0.1468	0.1741	-0.1426	0.1776
19	-0.1587	0.1517	-0.1545	0.1564	-0.1503	0.1607
20	-0.1654	0.1319	-0.1615	0.1372	-0.1575	0.1424
21	-0.1707	0.1127	-0.1667	0.1192	-0.1632	0.1244
22	-0.1751	0.0928	-0.1718	0.0990	-0.1681	0.1056
23	-0.1788	0.0719	-0.1753	0.0802	-0.1720	0.0874
24	-0.1813	0.0518	-0.1784	0.0597	-0.1754	0.0674
25	-0.1830	0.0310	-0.1803	0.0402	-0.1776	0.0491
26	-0.1839	0.0102	-0.1816	0.0203	-0.1793	0.0287
27	-	-	-0.1820	0.0000	-0.1800	0.0102

Table B3 (cont'd). Coefficients a_i and b_i for computing LaBreque's non-linearity tests for normality

	n=55		n=56		n=57	
i	a	b	a	b	a	b
1	0.9488	-1.3005	0.9450	-1.2987	0.9414	-1.2969
2	0.4636	-0.1943	0.4641	-0.2003	0.4646	-0.2062
3	0.3315	0.0066	0.3333	-0.0001	0.3351	-0.0066
4	0.2354	0.1224	0.2378	0.1159	0.2403	0.1095
5	0.1665	0.1884	0.1700	0.1821	0.1729	0.1762
6	0.1140	0.2277	0.1172	0.2224	0.1206	0.2170
7	0.0710	0.2513	0.0749	0.2464	0.0788	0.2415
8	0.0359	0.2641	0.0401	0.2599	0.0434	0.2559
9	0.0061	0.2696	0.0099	0.2662	0.0143	0.2626
10	-0.0199	0.2697	-0.0157	0.2670	-0.0113	0.2642
11	-0.0419	0.2659	-0.0375	0.2638	-0.0335	0.2617
12	-0.0617	0.2589	-0.0570	0.2577	-0.0529	0.2562
13	-0.0784	0.2497	-0.0746	0.2488	-0.0700	0.2483
14	-0.0938	0.2383	-0.0889	0.2386	-0.0853	0.2383
15	-0.1068	0.2258	-0.1027	0.2264	-0.0982	0.2271
16	-0.1191	0.2114	-0.1148	0.2129	-0.1103	0.2144
17	-0.1290	0.1969	-0.1247	0.1992	-0.1208	0.2010
18	-0.1385	0.1807	-0.1345	0.1835	-0.1304	0.1862
19	-0.1465	0.1643	-0.1423	0.1681	-0.1387	0.1709
20	-0.1534	0.1473	-0.1498	0.1512	-0.1456	0.1557
21	-0.1596	0.1294	-0.1557	0.1348	-0.1523	0.1389
22	-0.1645	0.1120	-0.1611	0.1174	-0.1576	0.1226
23	-0.1688	0.0936	-0.1656	0.0996	-0.1623	0.1056
24	-0.1723	0.0752	-0.1691	0.0824	-0.1662	0.0882
25	-0.1749	0.0569	-0.1722	0.0635	-0.1692	0.0714
26	-0.1768	0.0373	-0.1742	0.0461	-0.1718	0.0528
27	-0.1779	0.0191	-0.1757	0.0277	-0.1733	0.0361
28	-0.1783	0.0000	-0.1764	0.0084	-0.1744	0.0176
29	-	-	-	-	-0.1747	0.0000

Table B3 (cont'd). Coefficients a_i and b_i for computing LaBreque's non-linearity tests for normality

	n=58		n=59		n=60	
i	a	b	a	b	a	b
1	0.9377	-1.2951	0.9342	-1.2933	0.9306	-1.2915
2	0.4650	-0.2118	0.4654	-0.2173	0.4657	-0.2227
3	0.3367	-0.0129	0.3382	-0.0190	0.3398	-0.0250
4	0.2428	0.1032	0.2451	0.0972	0.2471	0.0915
5	0.1756	0.1705	0.1785	0.1648	0.1814	0.1591
6	0.1239	0.2116	0.1272	0.2064	0.1301	0.2014
7	0.0823	0.2369	0.0853	0.2324	0.0887	0.2278
8	0.0471	0.2518	0.0514	0.2475	0.0547	0.2436
9	0.0183	0.2592	0.0215	0.2558	0.0252	0.2524
10	-0.0076	0.2614	-0.0036	0.2586	0.0004	0.2557
11	-0.0294	0.2595	-0.0254	0.2574	-0.0218	0.2551
12	-0.0487	0.2547	-0.0446	0.2531	-0.0406	0.2515
13	-0.0657	0.2475	-0.0618	0.2465	-0.0578	0.2455
14	-0.0810	0.2382	-0.0770	0.2378	-0.0732	0.2373
15	-0.0944	0.2274	-0.0902	0.2278	-0.0861	0.2281
16	-0.1063	0.2155	-0.1023	0.2164	-0.0985	0.2171
17	-0.1169	0.2024	-0.1127	0.2042	-0.1092	0.2053
18	-0.1261	0.1889	-0.1227	0.1905	-0.1186	0.1927
19	-0.1348	0.1740	-0.1309	0.1768	-0.1272	0.1793
20	-0.1422	0.1588	-0.1382	0.1626	-0.1350	0.1651
21	-0.1485	0.1434	-0.1453	0.1468	-0.1414	0.1509
22	-0.1542	0.1274	-0.1508	0.1317	-0.1475	0.1357
23	-0.1590	0.1110	-0.1557	0.1160	-0.1525	0.1206
24	-0.1632	0.0939	-0.1600	0.1000	-0.1569	0.1052
25	-0.1663	0.0778	-0.1635	0.0836	-0.1606	0.0895
26	-0.1691	0.0603	-0.1665	0.0668	-0.1638	0.0728
27	-0.1710	0.0434	-0.1686	0.0505	-0.1660	0.0577
28	-0.1723	0.0260	-0.1701	0.0338	-0.1680	0.0406
29	-0.1730	0.0087	-0.1711	0.0167	-0.1691	0.0251
30	-	-	-0.1714	0.0000	-0.1697	0.0075

Table B3 (cont'd). Coefficients a_i and b_i for computing LaBreque's non-linearity tests for normality

	n=61		n=62		n=63	
i	a	b	a	b	a	b
1	0.9272	-1.2897	0.9238	-1.2879	0.9204	-1.2861
2	0.4660	-0.2279	0.4662	-0.2328	0.4664	-0.2377
3	0.3411	-0.0308	0.3426	-0.0365	0.3437	-0.0418
4	0.2494	0.0856	0.2513	0.0801	0.2534	0.0745
5	0.1838	0.1538	0.1864	0.1484	0.1886	0.1433
6	0.1330	0.1964	0.1358	0.1915	0.1387	0.1866
7	0.0922	0.2232	0.0950	0.2188	0.0981	0.2144
8	0.0577	0.2398	0.0614	0.2357	0.0644	0.2319
9	0.0291	0.2489	0.0322	0.2456	0.0356	0.2422
10	0.0037	0.2529	0.0075	0.2500	0.0108	0.2472
11	-0.0178	0.2528	-0.0142	0.2505	-0.0107	0.2482
12	-0.0368	0.2498	-0.0334	0.2480	-0.0297	0.2462
13	-0.0544	0.2442	-0.0503	0.2431	-0.0466	0.2418
14	-0.0690	0.2369	-0.0651	0.2363	-0.0620	0.2354
15	-0.0824	0.2280	-0.0791	0.2277	-0.0750	0.2276
16	-0.0948	0.2176	-0.0908	0.2181	-0.0870	0.2186
17	-0.1052	0.2066	-0.1018	0.2073	-0.0984	0.2080
18	-0.1152	0.1941	-0.1111	0.1961	-0.1078	0.1971
19	-0.1235	0.1816	-0.1202	0.1833	-0.1165	0.1854
20	-0.1311	0.1683	-0.1278	0.1705	-0.1244	0.1729
21	-0.1384	0.1536	-0.1345	0.1575	-0.1313	0.1600
22	-0.1438	0.1402	-0.1412	0.1427	-0.1379	0.1461
23	-0.1494	0.1249	-0.1459	0.1296	-0.1427	0.1335
24	-0.1541	0.1096	-0.1511	0.1140	-0.1482	0.1181
25	-0.1577	0.0948	-0.1549	0.0998	-0.1519	0.1049
26	-0.1610	0.0793	-0.1583	0.0848	-0.1558	0.0894
27	-0.1637	0.0632	-0.1612	0.0693	-0.1585	0.0755
28	-0.1656	0.0483	-0.1633	0.0545	-0.1611	0.0602
29	-0.1672	0.0313	-0.1651	0.0386	-0.1629	0.0453
30	-0.1679	0.0164	-0.1661	0.0238	-0.1642	0.0300
31	-0.1683	0.0000	-0.1667	0.0068	-0.1650	0.0153
32	-	-	-	-	-0.1653	0.0000

Table B3 (cont'd). Coefficients a_i and b_i for computing LaBreque's non-linearity tests for normality

	n=64	
i	a	b
1	0.9172	-1.2843
2	0.4666	-0.2424
3	0.3450	-0.0472
4	0.2551	0.0693
5	0.1912	0.1381
6	0.1410	0.1821
7	0.1013	0.2100
8	0.0671	0.2282
9	0.0392	0.2388
10	0.0141	0.2444
11	-0.0073	0.2459
12	-0.0259	0.2444
13	-0.0436	0.2404
14	-0.0577	0.2347
15	-0.0719	0.2271
16	-0.0836	0.2186
17	-0.0946	0.2088
18	-0.1044	0.1982
19	-0.1131	0.1870
20	-0.1210	0.1749
21	-0.1280	0.1626
22	-0.1344	0.1497
23	-0.1400	0.1364
24	-0.1450	0.1227
25	-0.1493	0.1086
26	-0.1530	0.0947
27	-0.1560	0.0808
28	-0.1587	0.0659
29	-0.1608	0.0511
30	-0.1622	0.0374
31	-0.1633	0.0215
32	-0.1637	0.0080

Table B4. Upper percentage points of LaBreque's nonlinearity tests for normality (from LaBreque, 1977; reproduced with permission from Technometrics. Copyright 1977 by the American Statistical Association. All rights reserved.)

		F1	
n	90%	95%	99%
4	2.029	2.405	2.939
5	2.080	2.652	3.731
6	2.109	2.769	4.165
7	2.128	2.829	4.424
8	2.131	2.865	4.995
9	2.142	2.872	4.974
10	2.152	2.879	4.955
11	2.160	2.884	4.940
12	2.167	2.889	4.296

		F2	
n	90%	95%	99%
4	2.730	3.419	4.379
5	2.739	3.562	5.008
6	2.739	3.638	5.367
7	2.736	3.684	5.594
8	2.733	3.713	5.748
9	2.730	3.732	5.859
10	2.726	3.745	5.941
11	2.723	3.755	6.005
12	2.724	3.770	6.049

		F3	
n	90%	95%	99%
4	2.274	3.373	5.215
5	2.490	3.425	5.247
6	2.581	3.475	5.430
7	2.621	3.518	5.635
8	2.638	3.554	5.825
9	2.644	3.583	5.990
10	2.644	3.608	6.130
11	2.641	3.628	6.249
12	2.612	3.673	6.579

Table B5. Upper and lower percentage points of the skewness test for normality, $\sqrt{b_1}$ (from Pearson and Hartley, 1976; used with permission of the Biometrika Trustees)

	percentage points			
n	95%	97.5%	99%	99.5%
20	0.772	0.940	1.150	1.304
25	0.711	0.866	1.059	1.200
30	0.662	0.806	0.986	1.117
35	0.621	0.756	0.923	1.044
40	0.588	0.714	0.871	0.985
45	0.559	0.679	0.826	0.934
50	0.534	0.647	0.788	0.889
60	0.492	0.596	0.724	0.816
70	0.459	0.556	0.673	0.758
80	0.432	0.522	0.632	0.710
90	0.409	0.494	0.597	0.670
100	0.390	0.470	0.567	0.636
125	0.351	0.422	0.508	0.569
150	0.322	0.387	0.465	0.519
175	0.299	0.359	0.430	0.481
200	0.280	0.336	0.403	0.449
250	0.251	0.301	0.361	0.402
300	0.230	0.275	0.329	0.366
350	0.213	0.255	0.305	0.339
400	0.200	0.239	0.285	0.317
450	0.188	0.225	0.269	0.299
500	0.179	0.214	0.255	0.283

Note: Since the distribution of $\sqrt{b_1}$ is symmetric with mean 0, the lower percentage points are the negative of the upper percentage points.

Table B6. Upper and lower percentage points of the kurtosis test for normality, b_2 (from Pearson and Hartley, 1976; used with permission of the Biometrika Trustees)

				percentage points				
n	0.5%	1%	2.5%	5%	95%	97.5%	99%	99.5%
20	1.58	1.64	1.73	1.83	4.18	4.68	5.38	5.91
30	1.73	1.79	1.89	1.98	4.12	4.57	5.20	5.69
40	1.83	1.89	1.99	2.07	4.06	4.46	5.04	5.48
50	1.91	1.95	2.06	2.15	4.00	4.36	4.88	5.28
75	2.05	2.08	2.19	2.27	3.87	4.17	4.59	4.90
100	2.13	2.18	2.27	2.35	3.77	4.03	4.39	4.66
125	2.19	2.24	2.32	2.40	3.70	3.93	4.24	4.48
150	2.24	2.29	2.37	2.45	3.65	3.86	4.13	4.34
175	2.28	2.34	2.41	2.48	3.61	3.79	4.04	4.23
200	2.32	2.37	2.44	2.51	3.57	3.75	3.98	4.16
250	-	2.42	-	2.55	3.52	-	3.87	-
300	-	2.46	-	2.59	3.47	-	3.79	-
400	-	2.52	-	2.64	3.41	-	3.67	-
500	-	2.57	-	2.67	3.37	-	3.60	-
600	-	2.60	-	2.70	3.34	-	3.54	-
700	-	2.62	-	2.72	3.31	-	3.50	-
800	-	2.65	-	2.74	3.29	-	3.46	-
900	-	2.66	-	2.75	3.28	-	3.43	-
1000	-	2.68	-	2.76	3.26	-	3.41	-
2000	-	2.77	-	2.83	3.18	-	3.28	-

Table B7. Upper and lower percentage points of absolute moment tests,
a(c) (based on simulation of 20,000 samples)

		percentage point			
c	n	2.5%	5%	95%	97.5%
0.5	12	0.722	0.743	0.918	0.930
1	-	0.694	0.716	0.901	0.915
1.5	-	0.787	0.805	0.932	0.941
2.5	-	1.088	1.101	1.336	1.373
3	-	1.207	1.238	1.876	1.992
4	-	1.548	1.632	4.088	4.653
5	-	2.058	2.253	9.644	11.638
6	-	2.795	3.188	23.467	29.695
0.5	17	0.734	0.753	0.901	0.911
1	-	0.704	0.725	0.883	0.896
1.5	-	0.795	0.809	0.920	0.929
2.5	-	1.107	1.119	1.330	1.363
3	-	1.250	1.284	1.871	1.969
4	-	1.681	1.779	4.139	4.625
5	-	2.357	2.573	10.057	11.982
0.5	25	0.751	0.765	0.889	0.898
1	-	0.721	0.737	0.870	0.881
1.5	-	0.805	0.817	0.911	0.917
2.5	-	1.126	1.137	1.322	1.349
3	-	1.299	1.327	1.856	1.943
4	-	1.820	1.912	4.118	4.603
5	-	2.685	2.903	10.206	12.268
6	-	4.088	4.580	26.985	34.709
0.5	33	0.762	0.772	0.879	0.888
1	-	0.731	0.743	0.860	0.869
1.5	-	0.811	0.820	0.904	0.910
2.5	-	1.137	1.148	1.317	1.339
3	-	1.329	1.357	1.841	1.920
4	-	1.929	2.017	4.121	4.552
5	-	2.965	3.167	10.295	12.399
6	-	4.722	5.191	27.554	36.148

Table B7 (cont'd). Upper and lower percentage points of absolute moment tests, a(c)

c	n	2.5%	5%	95%	97.5%
0.5	50	0.770	0.779	0.867	0.875
1	-	0.742	0.752	0.848	0.856
1.5	-	0.819	0.828	0.896	0.901
2.5	-	1.153	1.163	1.303	1.320
3	-	1.371	1.397	1.806	1.862
4	-	2.072	2.152	3.971	4.301
5	-	3.335	3.547	9.992	11.456
6	-	5.555	6.133	27.564	33.312
0.5	100	0.785	0.792	0.855	0.860
1	-	0.758	0.765	0.834	0.839
1.5	-	0.831	0.836	0.886	0.890
2.5	-	1.174	1.181	1.285	1.298
3	-	1.425	1.445	1.752	1.791
4	-	2.272	2.341	3.763	3.994
5	-	3.872	4.083	9.335	10.542
6	-	6.969	7.531	25.582	30.898
0.5	250	0.798	0.802	0.843	0.846
1	-	0.771	0.776	0.820	0.824
1.5	-	0.840	0.844	0.876	0.879
2.5	-	1.195	1.200	1.268	1.276
3	-	1.483	1.498	1.700	1.727
4	-	2.493	2.553	3.522	3.642
5	-	4.572	4.743	8.472	9.097
6	-	8.928	9.422	22.790	25.814
0.5	500	0.805	0.807	0.837	0.839
1	-	0.779	0.782	0.814	0.817
1.5	-	0.846	0.848	0.872	0.874
2.5	-	1.206	1.209	1.258	1.264
3	-	1.514	1.525	1.671	1.690
4	-	2.626	2.672	3.382	3.474
5	-	5.004	5.170	7.935	8.370
6	-	10.221	10.710	21.020	22.956

Table B8. Upper and lower percentage points of Grubbs' extreme outlier tests T_1, T_n and $T = max(T_1, T_n)$ (partially from Grubbs and Beck, 1972; reproduced with permission from Technometrics. Copyright 1972 by the American Statistical Association. All rights reserved. Remainder of table based on simulation of 20,000 samples)

		percentage points of T_1, T_n			
n	0.5%	1%	2.5%	5%	10%
10	0.880	0.926	1.008	1.082	1.169
15	1.053	1.100	1.176	1.248	1.337
20	1.156	1.215	1.293	1.368	1.459
25	1.249	1.301	1.375	1.449	1.542
30	1.319	1.362	1.450	1.524	1.615
35	1.381	1.426	1.513	1.590	1.678
40	1.431	1.481	1.560	1.632	1.724
45	1.477	1.531	1.609	1.678	1.775
50	1.524	1.567	1.648	1.722	1.815
60	1.584	1.640	1.717	1.790	1.881
70	1.640	1.691	1.775	1.850	1.940
80	1.703	1.753	1.830	1.903	1.993
90	1.750	1.798	1.877	1.946	2.038
100	1.790	1.840	1.915	1.989	2.081
250	2.113	2.158	2.233	2.304	2.397
500	2.348	2.401	2.478	2.542	2.627

Table B8 (cont'd). Upper and lower percentage points of Grubbs' extreme outlier tests T_1, T_n and $T = max(T_1, T_n)$

	percentage points of T_1, T_n				
n	90%	95%	97.5%	99%	99.5%
3	1.148	1.153	1.155	1.155	1.155
4	1.425	1.463	1.481	1.492	1.496
5	1.602	1.672	1.715	1.749	1.764
6	1.729	1.822	1.887	1.944	1.973
7	1.828	1.938	2.020	2.097	2.139
8	1.909	2.032	2.126	2.221	2.274
9	1.977	2.110	2.215	2.323	2.387
10	2.036	2.176	2.290	2.410	2.482
11	2.088	2.234	2.355	2.485	2.564
12	2.134	2.285	2.412	2.550	2.636
13	2.175	2.331	2.462	2.607	2.699
14	2.213	2.371	2.507	2.659	2.755
15	2.247	2.409	2.549	2.705	2.806
16	2.279	2.443	2.585	2.747	2.852
17	2.309	2.475	2.620	2.785	2.894
18	2.335	2.504	2.651	2.821	2.932
19	2.361	2.532	2.681	2.854	2.968
20	2.385	2.557	2.709	2.884	3.001
21	2.408	2.580	2.733	2.912	3.031
22	2.429	2.603	2.758	2.939	3.060
23	2.448	2.624	2.781	2.963	3.087
24	2.467	2.644	2.802	2.987	3.112
25	2.486	2.663	2.822	3.009	3.135
26	2.502	2.681	2.841	3.029	3.157
27	2.519	2.698	2.859	3.049	3.178
28	2.534	2.714	2.876	3.068	3.199
29	2.549	2.730	2.893	3.085	3.218
30	2.563	2.745	2.908	3.103	3.236

Table B8 (cont'd). Upper and lower percentage points of Grubbs' extreme outlier tests T_1, T_n and $T = max(T_1, T_n)$

		percentage points of T_1, T_n			
n	90%	95%	97.5%	99%	99.5%
31	2.577	2.759	2.924	3.119	3.253
32	2.591	2.773	2.938	3.135	3.270
33	2.604	2.786	2.952	3.150	3.286
34	2.616	2.799	2.965	3.164	3.301
35	2.628	2.811	2.979	3.178	3.316
36	2.639	2.823	2.991	3.191	3.330
37	2.650	2.835	3.003	3.204	3.343
38	2.661	2.846	3.014	3.216	3.356
39	2.671	2.857	3.025	3.228	3.369
40	2.682	2.866	3.036	3.240	3.381
41	2.692	2.877	3.046	3.251	3.393
42	2.700	2.887	3.057	3.261	3.404
43	2.710	2.896	3.067	3.271	3.415
44	2.719	2.905	3.075	3.282	3.425
45	2.727	2.914	3.085	3.292	3.435
46	2.736	2.923	3.094	3.302	3.445
47	2.744	2.931	3.103	3.310	3.455
48	2.753	2.940	3.111	3.319	3.464
49	2.760	2.948	3.120	3.329	3.474
50	2.768	2.956	3.128	3.336	3.483
51	2.775	2.964	3.136	3.345	3.491
52	2.783	2.971	3.143	3.353	3.500
53	2.790	2.978	3.151	3.361	3.507
54	2.798	2.986	3.158	3.368	3.516
55	2.804	2.992	3.166	3.376	3.524
56	2.811	3.000	3.172	3.383	3.531
57	2.818	3.006	3.180	3.391	3.539
58	2.824	3.013	3.186	3.397	3.546
59	2.831	3.019	3.193	3.405	3.553
60	2.837	3.025	3.199	3.411	3.560

Table B8 (cont'd). Upper and lower percentage points of Grubbs' extreme outlier tests T_1, T_n and $T = max(T_1, T_n)$

percentage points of T_1, T_n

n	90%	95%	97.5%	99%	99.5%
61	2.842	3.032	3.205	3.418	3.566
62	2.849	3.037	3.212	3.424	3.573
63	2.854	3.044	3.218	3.430	3.579
64	2.860	3.049	3.224	3.437	3.586
65	2.866	3.055	3.230	3.442	3.592
66	2.871	3.061	3.235	3.449	3.598
67	2.877	3.066	3.241	3.454	3.605
68	2.883	3.071	3.246	3.460	3.610
69	2.888	3.076	3.252	3.466	3.617
70	2.893	3.082	3.257	3.471	3.622
71	2.897	3.087	3.262	3.476	3.627
72	2.903	3.092	3.267	3.482	3.633
73	2.908	3.098	3.272	3.487	3.638
74	2.912	3.102	3.278	3.492	3.643
75	2.917	3.107	3.282	3.496	3.648
76	2.922	3.111	3.287	3.502	3.654
77	2.927	3.117	3.291	3.507	3.658
78	2.931	3.121	3.297	3.511	3.663
79	2.935	3.125	3.301	3.516	3.669
80	2.940	3.130	3.305	3.521	3.673
81	2.945	3.134	3.309	3.525	3.677
82	2.949	3.139	3.315	3.529	3.682
83	2.953	3.143	3.319	3.534	3.687
84	2.957	3.147	3.323	3.539	3.691
85	2.961	3.151	3.327	3.543	3.695
86	2.966	3.155	3.331	3.547	3.699
87	2.970	3.160	3.335	3.551	3.704
88	2.973	3.163	3.339	3.555	3.708
89	2.977	3.167	3.343	3.559	3.712
90	2.981	3.171	3.347	3.563	3.716

Table B8 (cont'd). Upper and lower percentage points of Grubbs' extreme outlier tests T_1, T_n and $T = max(T_1, T_n)$

percentage points of T_1, T_n

n	90%	95%	97.5%	99%	99.5%
91	2.984	3.174	3.350	3.567	3.720
92	2.989	3.179	3.355	3.570	3.725
93	2.993	3.182	3.358	3.575	3.728
94	2.996	3.186	3.362	3.579	3.732
95	3.000	3.189	3.365	3.582	3.736
96	3.003	3.193	3.369	3.586	3.739
97	3.006	3.196	3.372	3.589	3.744
98	3.011	3.201	3.377	3.593	3.747
99	3.014	3.204	3.380	3.597	3.750
100	3.017	3.207	3.383	3.600	3.754
101	3.021	3.210	3.386	3.603	3.757
102	3.024	3.214	3.390	3.607	3.760
103	3.027	3.217	3.393	3.610	3.765
104	3.030	3.220	3.397	3.614	3.768
105	3.033	3.224	3.400	3.617	3.771
106	3.037	3.227	3.403	3.620	3.774
107	3.040	3.230	3.406	3.623	3.777
108	3.043	3.233	3.409	3.626	3.780
109	3.046	3.236	3.412	3.629	3.784
110	3.049	3.239	3.415	3.632	3.787
111	3.052	3.242	3.418	3.636	3.790
112	3.055	3.245	3.422	3.639	3.793
113	3.058	3.248	3.424	3.642	3.796
114	3.061	3.251	3.427	3.645	3.799
115	3.064	3.254	3.430	3.647	3.802
116	3.067	3.257	3.433	3.650	3.805
117	3.070	3.259	3.435	3.653	3.808
118	3.073	3.262	3.438	3.656	3.811
119	3.075	3.265	3.441	3.659	3.814
120	3.078	3.267	3.444	3.662	3.817

Table B8 (cont'd). Upper and lower percentage points of Grubbs' extreme
outlier tests T_1, T_n and $T = max(T_1, T_n)$

		percentage points of T_1, T_n			
n	90%	95%	97.5%	99%	99.5%
121	3.081	3.270	3.447	3.665	3.819
122	3.083	3.274	3.450	3.667	3.822
123	3.086	3.276	3.452	3.670	3.824
124	3.089	3.279	3.455	3.672	3.827
125	3.092	3.281	3.457	3.675	3.831
126	3.095	3.284	3.460	3.677	3.833
127	3.097	3.286	3.462	3.680	3.836
128	3.100	3.289	3.465	3.683	3.838
129	3.102	3.291	3.467	3.686	3.840
130	3.104	3.294	3.470	3.688	3.843
131	3.107	3.296	3.473	3.690	3.845
132	3.109	3.298	3.475	3.693	3.848
133	3.112	3.302	3.478	3.695	3.850
134	3.114	3.304	3.480	3.697	3.853
135	3.116	3.306	3.482	3.700	3.856
136	3.119	3.309	3.484	3.702	3.858
137	3.122	3.311	3.487	3.704	3.860
138	3.124	3.313	3.489	3.707	3.863
139	3.126	3.315	3.491	3.710	3.865
140	3.129	3.318	3.493	3.712	3.867
141	3.131	3.320	3.497	3.714	3.869
142	3.133	3.322	3.499	3.716	3.871
143	3.135	3.324	3.501	3.719	3.874
144	3.138	3.326	3.503	3.721	3.876
145	3.140	3.328	3.505	3.723	3.879
146	3.142	3.331	3.507	3.725	3.881
147	3.144	3.334	3.509	3.727	3.883
250	3.305	3.497	3.667	3.893	4.032
500	3.505	3.691	3.865	4.088	4.248

Table B8 (cont'd). Upper and lower percentage points of Grubbs' extreme outlier tests T_1, T_n and $T = max(T_1, T_n)$

percentage points of $T = max(T_1, T_n)$

n	0.5%	1%	2.5%	5%	10%
10	1.298	1.329	1.379	1.429	1.496
15	1.434	1.477	1.532	1.587	1.659
20	1.528	1.574	1.638	1.696	1.767
25	1.611	1.648	1.711	1.769	1.847
30	1.674	1.713	1.778	1.837	1.916
35	1.721	1.762	1.830	1.897	1.975
40	1.767	1.809	1.882	1.941	2.018
45	1.813	1.856	1.924	1.989	2.063
50	1.846	1.895	1.959	2.023	2.101
60	1.917	1.955	2.021	2.086	2.163
70	1.971	2.009	2.075	2.139	2.215
80	2.004	2.051	2.117	2.182	2.262
90	2.046	2.089	2.161	2.227	2.304
100	2.077	2.123	2.189	2.262	2.339
250	2.393	2.434	2.500	2.559	2.643
500	2.611	2.649	2.714	2.774	2.853

Table B8 (cont'd). Upper and lower percentage points of Grubbs' extreme outlier tests T_1, T_n and $T = max(T_1, T_n)$

	percentage points of $T = max(T_1, T_n)$				
n	90%	95%	97.5%	99%	99.5%
10	2.179	2.291	2.387	2.482	2.549
15	2.413	2.555	2.681	2.825	2.921
20	2.559	2.719	2.843	3.002	3.097
25	2.659	2.809	2.948	3.123	3.229
30	2.740	2.899	3.054	3.227	3.350
35	2.803	2.966	3.122	3.292	3.397
40	2.864	3.037	3.188	3.383	3.510
45	2.907	3.068	3.227	3.398	3.536
50	2.969	3.131	3.293	3.488	3.613
60	3.032	3.208	3.361	3.558	3.694
70	3.072	3.243	3.407	3.604	3.765
80	3.119	3.301	3.469	3.659	3.792
90	3.168	3.342	3.518	3.716	3.876
100	3.206	3.381	3.543	3.751	3.905
250	3.497	3.667	3.830	4.037	4.185
500	3.684	3.861	4.017	4.231	4.396

Table B9. Upper and lower percentage points of the deviations from the median test, U (based on simulation of 20,000 samples)

n	0.5%	1%	2.5%	5%	10%
10	0.607	0.633	0.663	0.689	0.716
15	0.639	0.659	0.684	0.705	0.726
20	0.663	0.677	0.700	0.718	0.738
25	0.676	0.690	0.709	0.726	0.744
30	0.695	0.703	0.720	0.733	0.748
35	0.698	0.709	0.724	0.737	0.751
40	0.704	0.714	0.728	0.741	0.754
45	0.710	0.719	0.733	0.745	0.757
50	0.714	0.721	0.736	0.746	0.759
60	0.723	0.730	0.742	0.751	0.763
70	0.729	0.736	0.746	0.755	0.765
80	0.732	0.739	0.750	0.758	0.768
90	0.735	0.743	0.753	0.761	0.769
100	0.740	0.745	0.754	0.762	0.771
250	0.762	0.766	0.771	0.775	0.780
500	0.773	0.775	0.779	0.782	0.786

n	90%	95%	97.5%	99%	99.5%
10	0.880	0.900	0.917	0.935	0.946
15	0.859	0.874	0.885	0.898	0.906
20	0.855	0.869	0.880	0.893	0.901
25	0.847	0.860	0.870	0.881	0.888
30	0.844	0.857	0.866	0.877	0.884
35	0.841	0.852	0.860	0.871	0.878
40	0.839	0.850	0.859	0.869	0.874
45	0.836	0.845	0.854	0.863	0.870
50	0.835	0.845	0.853	0.861	0.868
60	0.832	0.840	0.848	0.857	0.862
70	0.829	0.837	0.844	0.852	0.858
80	0.827	0.835	0.841	0.850	0.855
90	0.826	0.833	0.840	0.847	0.850
100	0.825	0.832	0.838	0.845	0.849
250	0.815	0.819	0.823	0.827	0.831
500	0.810	0.813	0.816	0.819	0.821

Table B10. *Upper and lower percentage points of the range test (from David, Hartley and Pearson, 1954; used with permission of the Biometrika Trustees)*

	percentage points				
n	0.5%	1%	2.5%	5%	10%
3	-	-	-	-	-
4	-	-	-	-	-
5	-	-	-	-	-
6	-	-	-	-	-
7	-	-	-	-	-
8	-	-	-	-	-
9	-	-	-	-	-
10	2.47	2.51	2.59	2.67	2.77
11	2.53	2.58	2.66	2.74	2.84
12	2.59	2.65	2.73	2.80	2.91
13	2.65	2.70	2.78	2.86	2.97
14	2.70	2.75	2.83	2.91	3.02
15	2.75	2.80	2.88	2.96	3.07
16	2.80	2.85	2.93	3.01	3.13
17	2.84	2.90	2.98	3.06	3.17
18	2.88	2.94	3.02	3.10	3.21
19	2.92	2.98	3.06	3.14	3.25
20	2.95	3.01	3.10	3.18	3.29
30	3.22	3.27	3.37	3.46	3.58
40	3.41	3.46	3.57	3.66	3.79
50	3.57	3.61	3.72	3.82	3.94
60	3.69	3.74	3.85	3.95	4.07
80	3.88	3.93	4.05	4.15	4.27
100	4.02	4.09	4.20	4.31	4.44
150	4.30	4.36	4.47	4.59	4.72
200	4.50	4.56	4.67	4.78	4.90
500	5.06	5.13	5.25	5.37	5.49
1000	5.50	5.57	5.68	5.79	5.92

Table B10 (cont'd). Upper and lower percentage points of the range test

			percentage points		
n	90%	95%	97.5%	99%	99.5%
3	2.00	2.00	2.00	2.00	2.00
4	2.41	2.43	2.44	2.44	2.45
5	2.71	2.75	2.78	2.80	2.81
6	2.95	3.01	3.06	3.10	3.12
7	3.14	3.22	3.28	3.34	3.37
8	3.31	3.40	3.47	3.54	3.58
9	3.45	3.55	3.63	3.72	3.77
10	3.57	3.68	3.78	3.88	3.94
11	3.68	3.80	3.90	4.01	4.08
12	3.78	3.91	4.01	4.13	4.21
13	3.87	4.00	4.11	4.24	4.32
14	3.95	4.09	4.21	4.34	4.43
15	4.02	4.17	4.29	4.43	4.53
16	4.09	4.24	4.37	4.51	4.62
17	4.15	4.31	4.44	4.59	4.69
18	4.21	4.38	4.51	4.66	4.77
19	4.27	4.43	4.57	4.73	4.84
20	4.32	4.49	4.63	4.79	4.91
30	4.70	4.89	5.06	5.25	5.39
40	4.96	5.15	5.34	5.54	5.69
50	5.15	5.35	5.54	5.77	5.91
60	5.29	5.50	5.70	5.93	6.09
80	5.51	5.73	5.93	6.18	6.35
100	5.68	5.90	6.11	6.36	6.54
150	5.96	6.18	6.39	6.64	6.84
200	6.15	6.38	6.59	6.85	7.03
500	6.72	6.94	7.15	7.42	7.60
1000	7.11	7.33	7.54	7.80	7.99

Table B11. Percentage points of the most powerful location and scale invariant tests T_s' (upper percentage points) and S_s (lower percentage points) (from Spiegelhalter, 1977, 1980; used with permission of the Biometrika Trustees)

percentage points by test

	T_s'		S_s	
n	90%	95%	5%	10%
5	1.512	1.532	0.124	0.146
10	1.417	1.453	0.089	0.128
15	1.387	1.423	-	-
20	1.369	1.403	0.096	0.160
30	-	-	0.120	0.205
40	-	-	0.142	0.250
50	1.317	1.337	0.206	0.343
100	1.294	1.308	0.485	0.740

Table B12. Upper and lower percentage points of the T* test based on a U statistic (Locke and Spurrier, 1981; reprinted by courtesy of Marcel Dekker, Inc.)

				percentage points				
n	1%	2.5%	5%	10%	90%	95%	97.5%	99%
10	0.329	0.373	0.412	0.458	0.754	0.786	0.810	0.835
12	0.364	0.403	0.437	0.477	0.730	0.758	0.779	0.801
15	0.398	0.432	0.461	0.495	0.708	0.731	0.750	0.770
17	0.414	0.446	0.472	0.503	0.698	0.719	0.737	0.756
20	0.432	0.461	0.485	0.513	0.686	0.706	0.722	0.740
25	0.454	0.478	0.499	0.523	0.673	0.691	0.705	0.721
30	0.468	0.490	0.509	0.530	0.665	0.681	0.694	0.708
35	0.480	0.500	0.516	0.536	0.658	0.673	0.685	0.699
40	0.488	0.506	0.522	0.540	0.653	0.667	0.678	0.691
45	0.495	0.512	0.527	0.543	0.649	0.662	0.673	0.685
50	0.507	0.517	0.531	0.546	0.646	0.658	0.669	0.680
60	0.510	0.524	0.537	0.551	0.641	0.652	0.662	0.672
70	0.517	0.530	0.541	0.554	0.637	0.647	0.656	0.666
80	0.522	0.534	0.545	0.557	0.634	0.644	0.652	0.661
90	0.526	0.538	0.548	0.559	0.631	0.641	0.648	0.657
100	0.530	0.541	0.550	0.561	0.629	0.638	0.646	0.654
150	0.543	0.551	0.559	0.567	0.622	0.630	0.636	0.643
200	0.550	0.557	0.564	0.571	0.618	0.625	0.630	0.636
250	0.555	0.561	0.567	0.573	0.616	0.621	0.626	0.632
300	0.558	0.564	0.569	0.575	0.614	0.619	0.624	0.629
350	0.561	0.566	0.571	0.576	0.612	0.617	0.621	0.626
400	0.563	0.568	0.572	0.578	0.611	0.616	0.620	0.624
500	0.566	0.571	0.575	0.579	0.609	0.613	0.617	0.621
600	0.569	0.573	0.576	0.580	0.608	0.612	0.615	0.618
700	0.571	0.575	0.578	0.582	0.607	0.610	0.613	0.617
800	0.572	0.576	0.579	0.582	0.606	0.609	0.612	0.615
900	0.574	0.577	0.580	0.583	0.605	0.608	0.611	0.614
1000	0.575	0.578	0.580	0.584	0.605	0.608	0.610	0.613

*Table B13. Upper and lower percentage points of the standardized version,
Y, of DAgostino's D (from D'Agostino, 1972; used with permission of the
Biometrika Trustees)*

	percentage points				
n	0.5%	1%	2.5%	5%	10%
10	-4.66	-4.06	-3.25	-2.62	-1.99
12	-4.63	-4.02	-3.20	-2.58	-1.94
14	-4.57	-3.97	-3.16	-2.53	-1.90
16	-4.52	-3.92	-3.12	-2.50	-1.87
18	-4.47	-3.87	-3.08	-2.47	-1.85
20	-4.41	-3.83	-3.04	-2.44	-1.82
22	-4.36	-3.78	-3.01	-2.41	-1.81
24	-4.32	-3.75	-2.98	-2.39	-1.79
26	-4.27	-3.71	-2.96	-2.37	-1.77
28	-4.23	-3.68	-2.93	-2.35	-1.76
30	-4.19	-3.64	-2.91	-2.33	-1.75
32	-4.16	-3.61	-2.88	-2.32	-1.73
34	-4.12	-3.59	-2.86	-2.30	-1.72
36	-4.09	-3.56	-2.85	-2.29	-1.71
38	-4.06	-3.54	-2.83	-2.28	-1.70
40	-4.03	-3.51	-2.81	-2.26	-1.70
42	-4.00	-3.49	-2.80	-2.25	-1.69
44	-3.98	-3.47	-2.78	-2.24	-1.68
46	-3.95	-3.45	-2.77	-2.23	-1.67
48	-3.93	-3.43	-2.75	-2.22	-1.67
50	-3.91	-3.41	-2.74	-2.21	-1.66
60	-3.81	-3.34	-2.68	-2.17	-1.64
70	-3.73	-3.27	-2.64	-2.14	-1.61
80	-3.67	-3.22	-2.60	-2.11	-1.59
90	-3.61	-3.17	-2.57	-2.09	-1.58
100	-3.57	-3.14	-2.54	-2.07	-1.57

Table B13 (cont'd). Upper and lower percentage points of the standardized version, Y, of DAgostino's D

n	90%	95%	97.5%	99%	99.5%
10	0.149	0.235	0.299	0.356	0.385
12	0.237	0.329	0.381	0.440	0.479
14	0.308	0.399	0.460	0.515	0.555
16	0.367	0.459	0.526	0.587	0.613
18	0.417	0.515	0.574	0.636	0.667
20	0.460	0.565	0.628	0.690	0.720
22	0.497	0.609	0.677	0.744	0.775
24	0.530	0.648	0.720	0.783	0.822
26	0.559	0.682	0.760	0.827	0.867
28	0.586	0.714	0.797	0.868	0.910
30	0.610	0.743	0.830	0.906	0.941
32	0.631	0.770	0.862	0.942	0.983
34	0.651	0.794	0.891	0.975	1.02
36	0.669	0.816	0.917	1.00	1.05
38	0.686	0.837	0.941	1.03	1.08
40	0.702	0.857	0.964	1.06	1.11
42	0.716	0.875	0.986	1.09	1.14
44	0.730	0.892	1.01	1.11	1.17
46	0.742	0.908	1.02	1.13	1.19
48	0.754	0.923	1.04	1.15	1.22
50	0.765	0.937	1.06	1.18	1.24
60	0.812	0.997	1.13	1.26	1.34
70	0.849	1.05	1.19	1.33	1.42
80	0.878	1.08	1.24	1.39	1.48
90	0.902	1.12	1.28	1.44	1.54
100	0.923	1.14	1.31	1.48	1.59

Table B14. Lower 5% percentage points of the sample entropy tests, K_{mn} (from Vasicek, 1976; copyright 1976 Blackwell Publishers. Used with permission)

			m		
n	1	2	3	4	5
3	0.99	-	-	-	-
4	1.05	-	-	-	-
5	1.19	1.70	-	-	-
6	1.33	1.77	-	-	-
7	1.46	1.87	1.97	-	-
8	1.57	1.97	2.05	-	-
9	1.67	2.06	2.13	-	-
10	1.76	2.15	2.21	-	-
12	1.90	2.31	2.36	-	-
14	2.01	2.43	2.49	-	-
16	2.11	2.54	2.60	2.57	-
18	2.18	2.62	2.69	2.67	-
20	2.25	2.69	2.77	2.76	-
25	-	2.83	2.93	2.93	2.91
30	-	2.93	3.04	3.06	3.05
35	-	3.00	3.13	3.16	3.16
40	-	-	3.19	3.24	3.24
45	-	-	3.25	3.29	3.30
50	-	-	3.29	3.34	3.35

Table B15. *Upper percentage points of modified empirical distribution function tests (from Stephens, 1974; reproduced with permission from the Journal of the American Statistical Association. Copyright 1974 by the American Statistical Association. All rights reserved.)*

		percentage points		
test	90%	95%	97.5%	99%
D^*	0.819	0.895	0.955	1.035
V^*	1.386	1.489	1.585	1.693
W^{2*}	0.104	0.126	0.148	0.178
U^{2*}	0.096	0.116	0.136	0.163
A^{2*}	0.656	0.787	0.918	1.092

Table B16. *Upper percentage points of the modified Anderson-Darling test of normalized spacings for selected left (p) and right (q) censoring proportions (from Lockhart, O'Reilly and Stephens, 1986; copyright 1986 Blackwell Publishers. Used with permission)*

		percentage points			
p	q	90%	95%	97.5%	99%
0	1	1.422	1.798	2.191	2.728
0	0.75	1.592	2.026	2.479	3.100
0	0.50	1.667	2.129	2.612	3.273
0	0.25	1.730	2.215	2.722	3.416
0.25	0.75	1.800	2.306	2.835	3.559
0.25	0.50	1.889	2.430	2.996	3.770

Table B17. Upper percentage points of Dixon tests for outliers (partially from Dixon, 1951; used with permission of the Institute of Mathematical Statistics. Remainder of table based on simulation of 20,000 samples)

percentage points

r_{10}

n	90%	95%	99%
4	0.679	0.765	0.889
5	0.557	0.642	0.780
6	0.482	0.560	0.698
7	0.434	0.507	0.637
8	0.399	0.468	0.590
9	0.370	0.437	0.555
10	0.349	0.412	0.527
12	0.318	0.376	0.482
14	0.294	0.349	0.450
16	0.277	0.329	0.426
18	0.263	0.313	0.407
20	0.252	0.300	0.391

$max(r_{10}, r_{10}')$

n	90%	95%	99%
5	0.641	0.708	0.817
6	0.563	0.630	0.745
7	0.507	0.569	0.683
8	0.465	0.522	0.633
9	0.439	0.496	0.599
10	0.413	0.469	0.576
12	0.376	0.426	0.521
14	0.349	0.397	0.488
16	0.331	0.376	0.464
18	0.315	0.360	0.444
20	0.300	0.342	0.423
22	0.287	0.330	0.409
24	0.282	0.322	0.399
26	0.272	0.311	0.387
28	0.265	0.305	0.378
30	0.260	0.300	0.370

Table B17 (cont'd). Upper percentage points of Dixon tests for outliers

		r_{20}	
n	90%	95%	99%
5	0.782	0.845	0.929
6	0.670	0.736	0.836
7	0.596	0.661	0.778
8	0.545	0.607	0.710
9	0.505	0.565	0.667
10	0.474	0.531	0.632
12	0.429	0.481	0.579
14	0.395	0.445	0.538
16	0.370	0.418	0.508
18	0.350	0.397	0.484
20	0.333	0.372	0.464
22	0.320	0.358	0.447
24	0.309	0.347	0.434
26	0.300	0.338	0.422
28	0.292	0.330	0.412
30	0.285	0.322	0.402

		r_{21}	
5	0.952	0.976	0.995
6	0.821	0.872	0.951
7	0.725	0.780	0.885
8	0.650	0.710	0.829
9	0.594	0.657	0.776
10	0.551	0.612	0.726
12	0.490	0.546	0.642
14	0.448	0.501	0.593
16	0.416	0.467	0.557
18	0.391	0.440	0.529
20	0.371	0.419	0.506
22	0.356	0.402	0.487
24	0.343	0.388	0.471
26	0.331	0.376	0.457
28	0.320	0.365	0.444
30	0.312	0.355	0.433

Table B17 (cont'd). Upper percentage points of Dixon tests for outliers

		r_{22}	
n	90%	95%	99%
10	0.620	0.682	0.791
12	0.543	0.600	0.704
14	0.492	0.546	0.641
16	0.454	0.507	0.595
18	0.424	0.475	0.561
20	0.401	0.450	0.535
22	0.382	0.430	0.514
24	0.367	0.413	0.497
26	0.354	0.399	0.486
28	0.342	0.387	0.469
30	0.332	0.376	0.457
35	0.311	0.355	0.431
40	0.296	0.337	0.408
45	0.284	0.323	0.393
50	0.273	0.312	0.386

Table B18. *Lower percentage points of the Grubbs' test for k outliers in a specified tail, L_k (from Tietjen and Moore, 1972; reproduced with permission from Technometrics. Copyright 1972 by the American Statistical Association. All rights reserved.)*

1% percentage points

k

n	1	2	3	4	5
3	0.000	-	-	-	-
4	0.011	0.000	-	-	-
5	0.045	0.004	-	-	-
6	0.091	0.021	0.002	-	-
7	0.148	0.047	0.010	-	-
8	0.202	0.076	0.028	0.008	-
9	0.235	0.112	0.048	0.018	-
10	0.280	0.142	0.070	0.032	0.012
11	0.327	0.178	0.098	0.052	0.026
12	0.371	0.208	0.120	0.070	0.038
13	0.400	0.233	0.147	0.094	0.056
14	0.424	0.267	0.172	0.113	0.072
15	0.450	0.294	0.194	0.132	0.090
16	0.473	0.311	0.219	0.151	0.108
17	0.480	0.338	0.237	0.171	0.126
18	0.502	0.358	0.260	0.192	0.140
19	0.508	0.366	0.272	0.201	0.154
20	0.533	0.387	0.300	0.231	0.175
25	0.603	0.468	0.377	0.308	0.246
30	0.650	0.526	0.434	0.369	0.312
35	0.690	0.574	0.484	0.418	0.364
40	0.722	0.608	0.522	0.460	0.408
45	0.745	0.636	0.558	0.498	0.444
50	0.768	0.668	0.592	0.531	0.483

Table B18 (cont'd). Lower percentage points of the Grubbs' test for k outliers in a specified tail, L_k

1% percentage points

k

n	6	7	8	9	10
12	0.019	-	-	-	-
13	0.033	-	-	-	-
14	0.046	0.027	-	-	-
15	0.057	0.037	-	-	-
16	0.072	0.049	0.030	-	-
17	0.091	0.064	0.044	-	-
18	0.104	0.076	0.053	0.036	-
19	0.118	0.088	0.064	0.046	-
20	0.136	0.104	0.078	0.058	0.042
25	0.204	0.168	0.144	0.112	0.092
30	0.268	0.229	0.196	0.166	0.142
35	0.321	0.282	0.250	0.220	0.194
40	0.364	0.324	0.292	0.262	0.234
45	0.399	0.361	0.328	0.296	0.270
50	0.438	0.400	0.368	0.336	0.308

Table B18 (cont'd). Lower percentage points of the Grubbs' test for k outliers in a specified tail, L_k

	5% percentage points				
n	1	2	3	4	5
3	0.003	-	-	-	-
4	0.051	0.001	-	-	
5	0.125	0.018	-	-	
6	0.203	0.055	0.010	-	-
7	0.273	0.106	0.032	-	-
8	0.326	0.146	0.064	0.022	-
9	0.372	0.194	0.099	0.045	-
10	0.418	0.233	0.129	0.070	0.034
11	0.454	0.270	0.162	0.098	0.054
12	0.489	0.305	0.196	0.125	0.076
13	0.517	0.337	0.224	0.150	0.098
14	0.540	0.363	0.250	0.174	0.122
15	0.556	0.387	0.276	0.197	0.140
16	0.575	0.410	0.300	0.219	0.159
17	0.594	0.427	0.322	0.240	0.181
18	0.608	0.447	0.337	0.259	0.200
19	0.624	0.462	0.354	0.277	0.209
20	0.639	0.484	0.377	0.299	0.238
25	0.696	0.550	0.450	0.374	0.312
30	0.730	0.599	0.506	0.434	0.376
35	0.762	0.642	0.554	0.482	0.424
40	0.784	0.672	0.588	0.523	0.468
45	0.802	0.696	0.618	0.556	0.502
50	0.820	0.722	0.646	0.588	0.535

*Table B18 (cont'd). Lower percentage points of the Grubbs' test for k out-
liers in a specified tail, L_k*

		5% percentage points			
n	6	7	8	9	10
12	0.042	-	-	-	-
13	0.060	-	-	-	-
14	0.079	0.050	-	-	-
15	0.097	0.066	-	-	-
16	0.115	0.082	0.055	-	-
17	0.136	0.100	0.072	-	-
18	0.154	0.116	0.086	0.062	-
19	0.168	0.130	0.099	0.074	-
20	0.188	0.150	0.115	0.088	0.066
25	0.262	0.222	0.184	0.154	0.126
30	0.327	0.283	0.245	0.212	0.183
35	0.376	0.334	0.297	0.264	0.235
40	0.421	0.378	0.342	0.310	0.280
45	0.456	0.417	0.382	0.350	0.320
50	0.490	0.450	0.414	0.383	0.356

Table B18 (cont'd). Lower percentage points of the Grubbs' test for k out-liers in a specified tail, L_k

	10% percentage points				
n	1	2	3	4	5
3	0.011	-	-	-	-
4	0.098	0.003	-	-	-
5	0.200	0.038	-	-	-
6	0.280	0.091	0.020	-	-
7	0.348	0.148	0.056	-	-
8	0.404	0.200	0.095	0.038	-
9	0.448	0.248	0.134	0.068	-
10	0.490	0.287	0.170	0.098	0.051
11	0.526	0.326	0.208	0.128	0.074
12	0.555	0.361	0.240	0.159	0.103
13	0.578	0.388	0.270	0.186	0.126
14	0.600	0.416	0.298	0.212	0.150
15	0.611	0.436	0.322	0.236	0.172
16	0.631	0.458	0.342	0.260	0.194
17	0.648	0.478	0.364	0.282	0.216
18	0.661	0.496	0.384	0.302	0.236
19	0.676	0.510	0.398	0.316	0.251
20	0.688	0.530	0.420	0.339	0.273
25	0.732	0.588	0.489	0.412	0.350
30	0.766	0.637	0.523	0.472	0.411
35	0.792	0.673	0.586	0.516	0.458
40	0.812	0.702	0.622	0.554	0.499
45	0.826	0.724	0.648	0.586	0.533
50	0.840	0.744	0.673	0.614	0.562

Table B18 (cont'd). Lower percentage points of the Grubbs' test for k outliers in a specified tail, L_k

			10% percentage point		
n	6	7	8	9	10
12	0.062	-	-	-	-
13	0.082	-	-	-	-
14	0.104	0.068	-	-	-
15	0.124	0.086	-	-	-
16	0.144	0.104	0.073	-	-
17	0.165	0.125	0.092	-	-
18	0.184	0.142	0.108	0.080	-
19	0.199	0.158	0.124	0.094	-
20	0.220	0.176	0.140	0.110	0.085
25	0.296	0.251	0.213	0.180	0.152
30	0.359	0.316	0.276	0.240	0.210
35	0.410	0.365	0.328	0.294	0.262
40	0.451	0.408	0.372	0.338	0.307
45	0.488	0.447	0.410	0.378	0.348
50	0.518	0.477	0.442	0.410	0.380

Table B19. Lower percentage points of the Grubbs' test for one outlier in each tail, $S_{1,n}^2/S^2$ (based on simulation of 20,000 samples)

			percentage points		
n	0.5%	1%	2.5%	5%	10%
10	0.096	0.117	0.158	0.197	0.246
15	0.222	0.254	0.300	0.345	0.394
20	0.328	0.363	0.407	0.444	0.488
25	0.416	0.444	0.485	0.519	0.557
30	0.481	0.503	0.541	0.573	0.607
35	0.532	0.554	0.589	0.615	0.646
40	0.566	0.586	0.621	0.647	0.675
45	0.602	0.623	0.653	0.677	0.703
50	0.631	0.648	0.676	0.698	0.723
60	0.671	0.692	0.715	0.736	0.757
70	0.709	0.724	0.747	0.765	0.784
80	0.738	0.753	0.772	0.788	0.804
90	0.758	0.774	0.791	0.805	0.820
100	0.777	0.791	0.808	0.821	0.835
250	0.895	0.900	0.907	0.913	0.919
500	0.941	0.944	0.948	0.951	0.954

Table B20. Lower percentage points of the Grubbs' test for k outliers in one or both tails, E_k (from Tietjen and Moore, 1972; reproduced with permission from Technometrics. Copyright 1972 by the American Statistical Association. All rights reserved.)

1% percentage points

k

n	1	2	3	4	5
3	0.000	-	-	-	-
4	0.004	0.000	-	-	-
5	0.029	0.002	-	-	-
6	0.068	0.012	0.001	-	-
7	0.110	0.028	0.006	-	-
8	0.156	0.050	0.014	0.004	-
9	0.197	0.078	0.026	0.009	-
10	0.235	0.101	0.048	0.006	-
11	0.274	0.134	0.064	0.030	0.012
12	0.311	0.159	0.083	0.042	0.020
13	0.337	0.181	0.103	0.056	0.031
14	0.374	0.207	0.123	0.072	0.042
15	0.404	0.238	0.146	0.090	0.054
16	0.422	0.263	0.166	0.107	0.068
17	0.440	0.290	0.188	0.122	0.079
18	0.459	0.306	0.206	0.141	0.094
19	0.484	0.323	0.219	0.156	0.108
20	0.499	0.339	0.236	0.170	0.121
25	0.571	0.418	0.320	0.245	0.188
30	0.624	0.482	0.386	0.308	0.250
35	0.669	0.533	0.435	0.364	0.299
40	0.704	0.574	0.480	0.408	0.347
45	0.728	0.607	0.518	0.446	0.386
50	0.748	0.636	0.550	0.482	0.424

Table B20 (cont'd). Lower percentage points of the Grubbs' test for k outliers in one or both tails, E_k

	1% percentage points				
n	6	7	8	9	10
12	0.008	-	-	-	-
13	0.014	-	-	-	-
14	0.022	0.012	-	-	-
15	0.032	0.018	-	-	-
16	0.040	0.024	0.014	-	-
17	0.052	0.032	0.018	-	-
18	0.062	0.041	0.026	0.014	-
19	0.074	0.050	0.032	0.020	-
20	0.086	0.058	0.040	0.026	0.017
25	0.146	0.110	0.087	0.066	0.050
30	0.204	0.166	0.132	0.108	0.087
35	0.252	0.211	0.177	0.149	0.124
40	0.298	0.258	0.220	0.190	0.164
45	0.336	0.294	0.258	0.228	0.200
50	0.376	0.334	0.297	0.264	0.235

Table B20 (cont'd). Lower percentage points of the Grubbs' test for k outliers in one or both tails, E_k

	5% percentage points				
n	1	2	3	4	5
3	0.001	-	-	-	-
4	0.025	0.001	-	-	-
5	0.081	0.010	-	-	-
6	0.146	0.034	0.004	-	-
7	0.208	0.065	0.016	-	-
8	0.265	0.099	0.034	0.010	-
9	0.314	0.137	0.057	0.021	-
10	0.356	0.172	0.083	0.037	0.014
11	0.386	0.204	0.107	0.055	0.026
12	0.424	0.234	0.133	0.073	0.039
13	0.455	0.262	0.156	0.092	0.053
14	0.484	0.293	0.179	0.112	0.068
15	0.509	0.317	0.206	0.134	0.084
16	0.526	0.340	0.227	0.153	0.102
17	0.544	0.362	0.248	0.170	0.116
18	0.562	0.382	0.267	0.187	0.132
19	0.581	0.398	0.287	0.203	0.146
20	0.597	0.416	0.302	0.221	0.163
25	0.652	0.493	0.381	0.298	0.236
30	0.698	0.549	0.443	0.364	0.298
35	0.732	0.596	0.495	0.417	0.351
40	0.758	0.629	0.534	0.458	0.395
45	0.778	0.658	0.567	0.492	0.433
50	0.797	0.684	0.599	0.529	0.468

Table B20 (cont'd). Lower percentage points of the Grubbs' test for k outliers in one or both tails, E_k

			5% percentage points		
n	6	7	8	9	10
12	0.018	-	-	-	-
13	0.028	-	-	-	-
14	0.039	0.021	-	-	-
15	0.052	0.030	-	-	-
16	0.067	0.041	0.024	-	-
17	0.078	0.050	0.032	-	-
18	0.091	0.062	0.041	0.026	-
19	0.105	0.074	0.050	0.033	-
20	0.119	0.085	0.059	0.041	0.028
25	0.186	0.146	0.114	0.089	0.068
30	0.246	0.203	0.166	0.137	0.112
35	0.298	0.254	0.214	0.181	0.154
40	0.343	0.297	0.259	0.223	0.195
45	0.381	0.337	0.299	0.263	0.233
50	0.417	0.373	0.334	0.299	0.268

Table B20 (cont'd). Lower percentage points of the Grubbs' test for k out-liers in one or both tails, E_k

		10% percentage points			
n	1	2	3	4	5
3	0.003	-	-	-	-
4	0.050	0.002	-	-	-
5	0.127	0.022	-	-	
6	0.204	0.056	0.009	-	-
7	0.268	0.094	0.027	-	-
8	0.328	0.137	0.053	0.016	-
9	0.377	0.175	0.080	0.032	-
10	0.420	0.214	0.108	0.052	0.022
11	0.449	0.250	0.138	0.073	0.036
12	0.485	0.278	0.162	0.094	0.052
13	0.510	0.309	0.189	0.116	0.068
14	0.538	0.337	0.216	0.138	0.086
15	0.558	0.360	0.240	0.160	0.105
16	0.578	0.384	0.263	0.182	0.122
17	0.594	0.406	0.284	0.198	0.140
18	0.610	0.424	0.304	0.217	0.156
19	0.629	0.442	0.322	0.234	0.172
20	0.644	0.460	0.338	0.252	0.188
25	0.693	0.528	0.417	0.331	0.264
30	0.730	0.582	0.475	0.391	0.325
35	0.763	0.624	0.523	0.443	0.379
40	0.784	0.657	0.562	0.486	0.422
45	0.803	0.684	0.593	0.522	0.459
50	0.820	0.708	0.622	0.552	0.492

Table B20 (cont'd). Lower percentage points of the Grubbs' test for k outliers in one or both tails, E_k

	10% percentage points				
n	6	7	8	9	10
12	0.026	-	-	-	-
13	0.038	-	-	-	-
14	0.052	0.029	-	-	-
15	0.067	0.040	-	-	-
16	0.082	0.053	0.032	-	-
17	0.095	0.064	0.042	-	-
18	0.110	0.076	0.051	0.034	-
19	0.124	0.089	0.062	0.042	-
20	0.138	0.102	0.072	0.051	0.035
25	0.210	0.168	0.132	0.103	0.080
30	0.270	0.224	0.186	0.154	0.126
35	0.324	0.276	0.236	0.202	0.172
40	0.367	0.320	0.278	0.243	0.212
45	0.406	0.360	0.320	0.284	0.252
50	0.440	0.396	0.355	0.319	0.287

Table B21. The proportion of box plot outside values in a normal population (from Hoaglin, Iglewicz and Tukey, 1986; reproduced with permission from the Journal of the American Statistical Association. Copyright 1986 by the American Statistical Association. All rights reserved.)

n	f = 1.5	f = 3
5	8.60	3.34
6	3.48	0.797
7	4.30	0.717
8	1.94	0.262
9	5.04	1.089
10	2.83	0.362
11	3.47	0.324
12	2.00	0.184
13	3.63	0.417
14	2.23	0.159
15	2.68	0.175
16	1.75	0.0919
17	2.93	0.258
18	1.97	0.124
19	2.39	0.106
20	1.66	0.074
30	1.48	0.0363
40	1.24	0.0180
50	1.15	0.0106
75	1.14	0.00267
100	0.946	0.00200
200	0.786	0.00150
300	0.746	0.00067
∞	0.698	0.0002342

Table B22. *Upper percentage points of the extreme studentized deviate test for multiple outliers (from Jain, 1981; reproduced with permission from Technometrics. Copyright 1981 by the American Statistical Association. All rights reserved.)*

			percentage points				
		k = 2			k = 3		
n	test	90%	95%	99%	90%	95%	99%
20	ESD_1	2.69	2.83	3.09	2.76	2.88	3.13
	ESD_2	2.41	2.52	2.76	2.47	2.60	2.83
	ESD_3	-	-	-	2.34	2.45	2.68
30	ESD_1	2.89	3.05	3.35	2.97	3.12	3.41
	ESD_2	2.55	2.67	2.92	2.61	2.73	3.01
	ESD_3	-	-	-	2.44	2.56	2.75
40	ESD_1	3.01	3.17	3.52	3.07	3.22	3.58
	ESD_2	2.64	2.77	2.98	2.69	2.81	3.03
	ESD_3	-	-	-	2.52	2.62	2.82
50	ESD_1	3.10	3.27	3.61	3.18	3.34	3.68
	ESD_2	2.72	2.85	3.08	2.76	2.89	3.15
	ESD_3	-	-	-	2.58	2.68	2.89
60	ESD_1	3.15	3.34	3.70	3.26	3.42	3.75
	ESD_2	2.77	2.90	3.17	2.83	2.95	3.20
	ESD_3	-	-	-	2.64	2.73	2.95
80	ESD_1	3.28	3.45	3.80	3.32	3.49	3.85
	ESD_2	2.85	2.97	3.23	2.90	3.03	3.27
	ESD_3	-	-	-	2.71	2.81	3.01
100	ESD_1	3.34	3.52	3.87	3.44	3.60	3.97
	ESD_2	2.92	3.03	3.28	2.97	3.10	3.34
	ESD_3	-	-	-	2.77	2.86	3.06

Table B22 (cont'd). Upper percentage points of the extreme studentized deviate test for multiple outliers

			k = 4			k = 5	
n	test	90%	95%	99%	90%	95%	99%
20	ESD_1	2.81	2.95	3.20	2.85	2.97	3.18
	ESD_2	2.51	2.63	2.83	2.55	2.65	2.89
	ESD_3	2.38	2.49	2.68	2.40	2.51	2.69
	ESD_4	2.29	2.39	2.58	2.33	2.42	2.61
	ESD_5	-	-	-	2.27	2.37	2.57
30	ESD_1	3.02	3.16	3.48	3.05	3.19	3.48
	ESD_2	2.65	2.77	3.02	2.67	2.78	3.03
	ESD_3	2.48	2.59	2.79	2.51	2.60	2.80
	ESD_4	2.39	2.49	2.70	2.42	2.51	2.74
	ESD_5	-	-	-	2.35	2.45	2.62
40	ESD_1	3.14	3.32	3.64	3.16	3.31	3.63
	ESD_2	2.74	2.86	3.10	2.76	2.88	3.13
	ESD_3	2.57	2.67	2.87	2.59	2.69	2.89
	ESD_4	2.45	2.55	2.74	2.46	2.55	2.74
	ESD_5	-	-	-	2.39	2.47	2.65
50	ESD_1	3.24	3.40	3.74	3.28	3.45	3.77
	ESD_2	2.81	2.93	3.18	2.84	2.96	3.21
	ESD_3	2.62	2.72	2.92	2.65	2.74	2.94
	ESD_4	2.50	2.59	2.78	2.52	2.61	2.79
	ESD_5	-	-	-	2.44	2.52	2.70
60	ESD_1	3.31	3.48	3.82	3.34	3.51	3.81
	ESD_2	2.85	2.98	3.20	2.88	3.01	3.24
	ESD_3	2.67	2.77	2.97	2.68	2.77	2.96
	ESD_4	2.54	2.63	2.82	2.56	2.65	2.83
	ESD_5	-	-	-	2.48	2.56	2.72
80	ESD_1	3.40	3.57	3.91	3.44	3.61	3.93
	ESD_2	2.94	3.05	3.31	2.98	3.11	3.36
	ESD_3	2.74	2.84	3.04	2.77	2.86	3.08
	ESD_4	2.61	2.69	2.87	2.63	2.72	2.89
	ESD_5	-	-	-	2.54	2.62	2.76
100	ESD_1	3.47	3.64	3.96	3.53	3.70	4.01
	ESD_2	3.00	3.13	3.34	3.04	3.16	3.42
	ESD_3	2.79	2.89	3.06	2.81	2.91	3.10
	ESD_4	2.66	2.74	2.90	2.68	2.77	2.93
	ESD_5	-	-	-	2.59	2.67	2.84

Table B23. *Upper percentage points of the sequential kurtosis procedure for multiple outliers (from Jain, 1981; reproduced with permission from Technometrics. Copyright 1981 by the American Statistical Association. All rights reserved.)*

			percentage points				
		k = 2			k = 3		
n	test	90%	95%	99%	90%	95%	99%
20	KUR_1	4.03	4.56	5.77	4.22	4.76	6.04
	KUR_2	3.30	3.66	4.45	3.45	3.83	4.72
	KUR_3	-	-	-	3.15	3.48	4.23
30	KUR_1	3.96	4.48	5.64	4.16	4.70	5.91
	KUR_2	3.28	3.56	4.25	3.37	3.69	4.43
	KUR_3	-	-	-	3.07	3.31	3.87
40	KUR_1	3.90	4.31	5.43	3.99	4.42	5.61
	KUR_2	3.24	3.50	4.07	3.31	3.58	4.16
	KUR_3	-	-	-	3.04	3.28	3.73
50	KUR_1	3.86	4.26	5.17	3.96	4.37	5.43
	KUR_2	3.25	3.49	4.01	3.29	3.56	4.14
	KUR_3	-	-	-	3.04	3.23	3.69
60	KUR_1	3.79	4.18	5.05	3.88	4.27	5.19
	KUR_2	3.22	3.46	3.93	3.28	3.51	4.04
	KUR_3	-	-	-	3.04	3.23	3.60
80	KUR_1	3.72	4.04	4.78	3.75	4.06	4.92
	KUR_2	3.20	3.39	3.80	3.24	3.46	3.88
	KUR_3	-	-	-	3.03	3.19	3.53
100	KUR_1	3.63	3.91	4.57	3.71	4.00	4.66
	KUR_2	3.19	3.37	3.74	3.24	3.41	3.76
	KUR_3	-	-	-	3.03	3.17	3.48

Table B23 (cont'd). Upper percentage points of the sequential kurtosis procedure for multiple outliers

n	test	k = 4			k = 5		
		90%	95%	99%	90%	95%	99%
20	KUR_1	4.37	4.97	6.37	4.52	5.10	6.40
	KUR_2	3.55	3.92	4.73	3.63	4.02	5.00
	KUR_3	3.25	3.57	4.26	3.32	3.62	4.36
	KUR_4	3.07	3.39	4.13	3.17	3.46	4.09
	KUR_5	-	-	-	3.08	3.37	4.10
30	KUR_1	4.24	4.78	6.11	4.29	4.80	6.14
	KUR_2	3.44	3.75	4.46	3.47	3.78	4.48
	KUR_3	3.14	3.40	3.97	3.18	3.42	3.98
	KUR_4	2.97	3.22	3.77	3.03	3.25	3.75
	KUR_5	-	-	-	2.91	3.14	3.62
40	KUR_1	4.12	4.61	5.81	4.14	4.61	5.80
	KUR_2	3.39	3.64	4.24	3.40	3.68	4.27
	KUR_3	3.11	3.30	3.76	3.11	3.34	3.78
	KUR_4	2.94	3.13	3.55	2.94	3.14	3.56
	KUR_5	-	-	-	2.84	3.02	3.44
50	KUR_1	4.05	4.44	5.57	4.10	4.51	5.67
	KUR_2	3.34	3.59	4.15	3.38	3.61	4.16
	KUR_3	3.08	3.28	3.73	3.11	3.31	3.72
	KUR_4	2.92	3.10	3.49	2.95	3.12	3.48
	KUR_5	-	-	-	2.84	2.99	3.31
60	KUR_1	3.95	4.36	5.32	3.98	4.37	5.28
	KUR_2	3.32	3.54	4.00	3.33	3.57	4.07
	KUR_3	3.07	3.25	3.63	3.08	3.27	3.66
	KUR_4	2.91	3.08	3.42	2.94	3.09	3.45
	KUR_5	-	-	-	2.84	2.98	3.29
80	KUR_1	3.83	4.14	4.87	3.87	4.22	5.05
	KUR_2	3.28	3.47	3.90	3.32	3.53	3.96
	KUR_3	3.05	3.21	3.54	3.08	3.26	3.57
	KUR_4	2.91	3.05	3.35	2.94	3.09	3.36
	KUR_5	-	-	-	2.84	2.97	3.25
100	KUR_1	3.75	4.03	4.69	3.79	4.11	4.80
	KUR_2	3.26	3.44	3.81	3.29	3.48	3.83
	KUR_3	3.05	3.20	3.42	3.07	3.22	3.51
	KUR_4	2.92	3.05	3.33	2.94	3.07	3.33
	KUR_5	-	-	-	2.84	2.97	3.20

Table B24. Lower percentage points of a sequential sum of squares test for multiple outliers (from Prescott, 1979; copyright 1979 Blackwell Publishers. Used with permission)

	sequence percentage points for 2 outliers							
	1%		2.5%		5%		10%	
n	$\lambda_1(\beta)$	$\lambda_2(\beta)$	$\lambda_1(\beta)$	$\lambda_2(\beta)$	$\lambda_1(\beta)$	$\lambda_2(\beta)$	$\lambda_1(\beta)$	$\lambda_2(\beta)$
10	0.205	0.220	0.255	0.280	0.310	0.335	0.360	0.385
11	0.235	0.285	0.285	0.330	0.345	0.370	0.410	0.430
12	0.280	0.325	0.325	0.370	0.380	0.420	0.430	0.475
13	0.320	0.380	0.365	0.430	0.410	0.460	0.465	0.510
14	0.350	0.400	0.400	0.460	0.440	0.495	0.490	0.540
15	0.385	0.435	0.430	0.480	0.465	0.520	0.515	0.565
16	0.415	0.465	0.450	0.505	0.495	0.545	0.540	0.590
17	0.430	0.475	0.470	0.530	0.515	0.565	0.555	0.605
18	0.445	0.505	0.490	0.545	0.535	0.585	0.575	0.625
19	0.465	0.535	0.515	0.570	0.555	0.605	0.590	0.645
20	0.475	0.560	0.525	0.590	0.565	0.620	0.605	0.660
25	0.550	0.630	0.600	0.660	0.630	0.690	0.665	0.715
30	0.615	0.675	0.645	0.710	0.675	0.735	0.705	0.765
35	0.655	0.715	0.690	0.740	0.715	0.770	0.745	0.790
40	0.685	0.745	0.720	0.770	0.745	0.795	0.770	0.810
45	0.715	0.770	0.740	0.795	0.765	0.810	0.790	0.830
50	0.735	0.785	0.760	0.810	0.785	0.830	0.805	0.845

Table B24 (cont'd). Lower percentage points of a sequential sum of squares test for multiple outliers

<table>
<tr><th colspan="7">sequence percentage points for 3 outliers</th></tr>
<tr><th></th><th colspan="3">1%</th><th colspan="3">2.5%</th></tr>
<tr><th>n</th><th>$\lambda_1(\beta)$</th><th>$\lambda_2(\beta)$</th><th>$\lambda_3(\beta)$</th><th>$\lambda_1(\beta)$</th><th>$\lambda_2(\beta)$</th><th>$\lambda_3(\beta)$</th></tr>
<tr><td>15</td><td>0.375</td><td>0.430</td><td>0.455</td><td>0.410</td><td>0.455</td><td>0.470</td></tr>
<tr><td>16</td><td>0.390</td><td>0.450</td><td>0.470</td><td>0.430</td><td>0.485</td><td>0.500</td></tr>
<tr><td>17</td><td>0.405</td><td>0.470</td><td>0.495</td><td>0.445</td><td>0.505</td><td>0.530</td></tr>
<tr><td>18</td><td>0.425</td><td>0.490</td><td>0.515</td><td>0.465</td><td>0.530</td><td>0.555</td></tr>
<tr><td>19</td><td>0.440</td><td>0.515</td><td>0.535</td><td>0.485</td><td>0.550</td><td>0.580</td></tr>
<tr><td>20</td><td>0.455</td><td>0.530</td><td>0.560</td><td>0.500</td><td>0.570</td><td>0.600</td></tr>
<tr><td>25</td><td>0.535</td><td>0.620</td><td>0.655</td><td>0.580</td><td>0.645</td><td>0.680</td></tr>
<tr><td>30</td><td>0.605</td><td>0.670</td><td>0.710</td><td>0.635</td><td>0.695</td><td>0.735</td></tr>
<tr><td>35</td><td>0.650</td><td>0.705</td><td>0.745</td><td>0.675</td><td>0.730</td><td>0.770</td></tr>
<tr><td>40</td><td>0.680</td><td>0.735</td><td>0.775</td><td>0.705</td><td>0.760</td><td>0.790</td></tr>
<tr><td>45</td><td>0.705</td><td>0.765</td><td>0.795</td><td>0.730</td><td>0.780</td><td>0.810</td></tr>
<tr><td>50</td><td>0.725</td><td>0.785</td><td>0.820</td><td>0.755</td><td>0.800</td><td>0.825</td></tr>
</table>

Table B24 (cont'd). Lower percentage points for a sequential sum of squares test for multiple outliers

	sequence percentage points for 3 outliers					
	5%			10%		
n	$\lambda_1(\beta)$	$\lambda_2(\beta)$	$\lambda_3(\beta)$	$\lambda_1(\beta)$	$\lambda_2(\beta)$	$\lambda_3(\beta)$
15	0.445	0.495	0.505	0.490	0.540	0.555
16	0.465	0.520	0.540	0.510	0.565	0.580
17	0.490	0.540	0.565	0.530	0.585	0.605
18	0.510	0.560	0.590	0.550	0.605	0.625
19	0.525	0.585	0.610	0.570	0.625	0.645
20	0.545	0.605	0.630	0.585	0.640	0.660
25	0.610	0.675	0.705	0.650	0.705	0.730
30	0.665	0.725	0.755	0.695	0.750	0.770
35	0.700	0.760	0.785	0.730	0.780	0.805
40	0.730	0.785	0.810	0.755	0.805	0.825
45	0.755	0.805	0.825	0.780	0.825	0.845
50	0.770	0.820	0.840	0.795	0.835	0.855

*Table B25. Upper percentage points of the R-statistic many outlier proce-
dure (from Jain, 1981; reproduced with permission from Technometrics.
Copyright 1981 by the American Statistical Association. All rights reserved.)*

					percentage points		
			k = 2			k = 3	
n	test	90%	95%	99%	90%	95%	99%
20	RST_1	4.66	5.17	6.29	5.94	6.62	8.33
	RST_2	3.50	3.86	4.64	4.56	5.05	6.25
	RST_3	-	-	-	3.77	4.11	5.01
30	RST_1	4.23	4.63	5.44	4.98	5.47	6.50
	RST_2	3.30	3.55	4.08	3.91	4.25	4.93
	RST_3	-	-	-	3.33	3.54	4.10
40	RST_1	4.06	4.37	5.10	4.58	4.95	5.75
	RST_2	3.23	3.44	3.83	3.66	3.92	4.42
	RST_3	-	-	-	3.19	3.39	3.80
50	RST_1	3.98	4.28	4.89	4.43	4.76	5.51
	RST_2	3.21	3.42	3.81	3.58	3.79	4.30
	RST_3	-	-	-	3.14	3.31	3.70
60	RST_1	3.90	4.21	4.81	4.33	4.63	5.28
	RST_2	3.20	3.38	3.77	3.53	3.73	4.14
	RST_3	-	-	-	3.12	3.27	3.60
80	RST_1	3.89	4.13	4.66	4.16	4.43	5.04
	RST_2	3.19	3.36	3.71	3.45	3.63	4.02
	RST_3	-	-	-	3.09	3.22	3.51
100	RST_1	3.83	4.08	4.59	4.14	4.39	4.94
	RST_2	3.21	3.35	3.66	3.43	3.59	3.95
	RST_3	-	-	-	3.09	3.21	3.46

Table B25 (cont'd). Upper percentage points of the R-statistic many outlier procedure

		k = 4			k = 5		
n	test	90%	95%	99%	90%	95%	99%
20	RST_1	7.58	8.54	11.09	9.80	11.31	14.91
	RST_2	5.84	6.54	8.08	7.69	8.75	11.39
	RST_3	4.88	5.42	6.65	6.44	7.32	9.43
	RST_4	4.17	4.60	5.71	5.54	6.28	8.09
	RST_5	-	-	-	4.81	5.46	7.05
30	RST_1	5.76	6.33	7.66	6.63	7.33	8.91
	RST_2	4.57	4.97	5.81	5.30	5.76	6.74
	RST_3	3.93	4.23	4.88	4.57	4.95	5.85
	RST_4	3.47	3.73	4.30	4.04	4.38	5.16
	RST_5	-	-	-	3.62	3.91	4.58
40	RST_1	5.18	5.60	6.56	5.72	6.20	7.47
	RST_2	4.15	4.44	5.04	4.60	4.95	5.69
	RST_3	3.61	3.84	4.30	4.02	4.29	4.84
	RST_4	3.24	3.43	3.83	3.61	3.85	4.37
	RST_5	-	-	-	3.29	3.50	3.92
50	RST_1	4.87	5.24	6.06	5.35	5.71	6.65
	RST_2	3.94	4.18	4.69	4.29	4.56	5.14
	RST_3	3.46	3.66	4.08	3.80	4.02	4.46
	RST_4	3.14	3.30	3.65	3.43	3.62	4.00
	RST_5	-	-	-	3.16	3.33	3.65
60	RST_1	4.69	5.05	5.76	5.04	5.40	6.10
	RST_2	3.81	4.03	4.46	4.11	4.34	4.87
	RST_3	3.38	3.56	3.91	3.63	3.83	4.22
	RST_4	3.07	3.23	3.52	3.33	3.51	3.86
	RST_5	-	-	-	3.08	3.22	3.54
80	RST_1	4.48	4.76	5.33	4.75	5.06	5.71
	RST_2	3.69	3.88	4.29	3.93	4.13	4.58
	RST_3	3.31	3.46	3.76	3.51	3.68	4.01
	RST_4	3.04	3.16	3.41	3.23	3.37	3.62
	RST_5	-	-	-	3.02	3.14	3.37
100	RST_1	4.36	4.63	5.16	4.61	4.90	5.47
	RST_2	3.36	3.82	4.15	3.83	4.02	4.41
	RST_3	3.26	3.41	3.64	3.43	3.58	3.85
	RST_4	3.02	3.13	3.35	3.18	3.30	3.54
	RST_5	-	-	-	2.98	3.09	3.33

Table B26. Lower percentage points of the Shapiro-Francia test for censored samples (from Verrill and Johnson, 1988; reproduced with permission from the Journal of the American Statistical Association. Copyright 1988 by the American Statistical Association. All rights reserved.)

		proportion of uncensored observations				
n	100α	0.2	0.4	0.6	0.8	1.0
20	10%	0.89085	0.92061	0.93992	0.95375	0.96052
30	-	0.90544	0.93900	0.95588	0.96559	0.97039
40	-	0.91876	0.94922	0.96410	0.97291	0.97700
60	-	0.93502	0.96200	0.97359	0.98075	0.98377
80	-	0.94766	0.97028	0.98004	0.98547	0.98709
100	-	0.95404	0.97446	0.98273	0.98790	0.98949
150	-	0.96583	0.98135	0.98782	0.99133	0.99258
250	-	0.97634	0.98796	0.99217	0.99465	0.99538
500	-	0.98614	0.99301	0.99572	0.99709	0.99754
20	5%	0.86464	0.89905	0.92641	0.94407	0.95140
30	-	0.88206	0.92431	0.94512	0.95704	0.96359
40	-	0.89538	0.93701	0.95586	0.96696	0.97217
60	-	0.92081	0.95080	0.96691	0.97618	0.98010
80	-	0.93427	0.96247	0.97494	0.98190	0.98459
100	-	0.94160	0.96709	0.97854	0.98484	0.98728
150	-	0.95612	0.97618	0.98439	0.98927	0.99100
250	-	0.97077	0.98493	0.99016	0.99328	0.99448
500	-	0.98249	0.99142	0.99474	0.99652	0.99709
20	1%	0.77801	0.84473	0.89574	0.91664	0.92476
30	-	0.83040	0.88869	0.91559	0.93782	0.94504
40	-	0.84912	0.90721	0.93531	0.95311	0.95772
60	-	0.88454	0.92799	0.95154	0.96310	0.97171
80	-	0.90535	0.94059	0.96123	0.97234	0.97802
100	-	0.91061	0.94717	0.96603	0.97736	0.98245
150	-	0.93116	0.96318	0.97609	0.98406	0.98742
250	-	0.95228	0.97514	0.98458	0.98991	0.99262
500	-	0.97363	0.98735	0.99214	0.99507	0.99599

Table B27. Upper percentage points of W^2 and A^2 for complete or Type 2 right censored data (from Stephens, 1986; reprinted by courtesy of Marcel Dekker, Inc.)

| | | percentage points | | | |
| | | W^2 | | | |
n	p	90%	95%	97.5%	99%
20	0.2	0.006	0.010	0.016	0.024
40	-	0.008	0.013	0.021	0.041
60	-	0.008	0.014	0.021	0.039
80	-	0.008	0.013	0.021	0.035
100	-	0.008	0.013	0.020	0.032
10	0.4	0.017	0.028	0.041	0.057
20	-	0.024	0.037	0.055	0.090
40	-	0.026	0.038	0.057	0.089
60	-	0.026	0.036	0.052	0.077
80	-	0.026	0.036	0.049	0.073
100	-	0.026	0.035	0.047	0.071
10	0.6	0.040	0.054	0.075	0.109
20	-	0.044	0.060	0.080	0.113
40	-	0.047	0.061	0.077	0.106
60	-	0.047	0.060	0.077	0.105
80	-	0.048	0.061	0.077	0.103
100	-	0.048	0.061	0.076	0.101
10	0.8	0.062	0.078	0.094	0.115
20	-	0.067	0.083	0.100	0.122
40	-	0.070	0.084	0.101	0.124
60	-	0.070	0.086	0.103	0.125
80	-	0.071	0.087	0.105	0.127
100	-	0.072	0.089	0.106	0.129
10	0.9	0.076	0.093	0.110	0.137
20	-	0.079	0.097	0.113	0.137
40	-	0.082	0.099	0.116	0.142
60	-	0.082	0.100	0.118	0.141
80	-	0.084	0.102	0.120	0.144
100	-	0.085	0.103	0.122	0.146

Table B27 (cont'd). Upper percentage points of W^2 and A^2 for complete or Type 2 right censored data

n	p	90%	95%	97.5%	99%
10	0.95	0.084	0.103	0.122	0.145
20	-	0.087	0.106	0.124	0.147
40	-	0.089	0.108	0.126	0.154
60	-	0.089	0.109	0.128	0.152
80	-	0.090	0.110	0.130	0.154
100	-	0.091	0.112	0.132	0.156
10	1.0	0.098	0.119	0.141	0.167
20	-	0.100	0.121	0.142	0.171
40	-	0.101	0.122	0.141	0.169
60	-	0.101	0.123	0.144	0.171
80	-	0.101	0.124	0.146	0.173
100	-	0.102	0.125	0.146	0.174

A^2

n	p	90%	95%	97.5%	99%
20	0.2	0.061	0.092	0.131	0.182
40	-	0.079	0.112	0.158	0.253
60	-	0.084	0.114	0.160	0.246
80	-	0.087	0.116	0.159	0.236
100	-	0.089	0.119	0.158	0.228
10	0.4	0.121	0.172	0.236	0.319
20	-	0.162	0.220	0.297	0.439
40	-	0.177	0.236	0.316	0.433
60	-	0.174	0.228	0.299	0.410
80	-	0.178	0.229	0.292	0.395
100	-	0.182	0.231	0.288	0.385
10	0.6	0.233	0.304	0.405	0.592
20	-	0.259	0.339	0.437	0.607
40	-	0.278	0.348	0.430	0.570
60	-	0.275	0.348	0.430	0.557
80	-	0.278	0.350	0.429	0.548
100	-	0.280	0.351	0.429	0.541

Table B27 (cont'd). Upper percentage points of W^2 and A^2 for complete or Type 2 right censored data

n	p	90%	95%	97.5%	99%
10	0.8	0.352	0.440	0.542	0.698
20	-	0.380	0.473	0.574	0.743
40	-	0.394	0.478	0.575	0.711
60	-	0.396	0.482	0.574	0.705
80	-	0.401	0.489	0.580	0.707
100	-	0.405	0.494	0.585	0.709
10	0.9	0.425	0.530	0.642	0.825
20	-	0.453	0.549	0.654	0.807
40	-	0.473	0.566	0.669	0.814
60	-	0.472	0.571	0.675	0.805
80	-	0.478	0.579	0.683	0.811
100	-	0.483	0.585	0.689	0.818
10	0.95	0.474	0.584	0.696	0.840
20	-	0.500	0.600	0.708	0.853
40	-	0.519	0.624	0.721	0.865
60	-	0.520	0.626	0.733	0.874
80	-	0.528	0.633	0.744	0.885
100	-	0.534	0.640	0.753	0.893
10	1.0	0.578	0.700	0.818	0.964
20	-	0.601	0.714	0.831	0.993
40	-	0.611	0.723	0.833	0.981
60	-	0.614	0.734	0.847	0.993
80	-	0.616	0.740	0.853	1.000
100	-	0.618	0.742	0.857	1.005

Table B28. Upper percentage points of Watson's $_rU^2$ test for Type 2 right censored data (from Pettitt, 1976; used with permission of the Biometrika Trustees)

		percentage points		
r	90%	95%	97.5%	99%
0.5	0.030	0.037	0.043	0.052
0.75	0.064	0.078	0.092	0.111
0.9	0.086	0.104	0.123	0.147

Table B29. Upper and lower percentage points of Tiku's test of spacings for complete and singly and doubly censored data (from Tiku, 1980; copyright 1980 Blackwell Publishers. Used with permission)

			percentage points			
n	a^*	b^*	2.5%	5%	95%	97.5%
10	0	0	0.65	0.71	1.29	1.35
20	0	0	0.77	0.81	1.19	1.23
20	0	4	0.74	0.78	1.22	1.26
20	4	4	0.66	0.72	1.28	1.34
50	0	0	0.86	0.88	1.12	1.14
50	0	10	0.84	0.87	1.13	1.16
50	10	10	0.80	0.83	1.17	1.20

* a and b are the lower and upper observation numbers of the ordered uncensored data, respectively

Table B30. Upper percentage points of a probability plot correlation test for multivariate data (from Tsai and Koziol, 1988; reprinted by courtesy of Marcel Dekker, Inc.)

		percentage points			
n	m	90%	95%	97.5%	99%
10	2	0.9882	0.9911	0.9929	0.9946
20	-	0.9899	0.9921	0.9936	0.9949
30	-	0.9914	0.9931	0.9943	0.9954
40	-	0.9922	0.9937	0.9948	0.9958
50	-	0.9928	0.9942	0.9952	0.9960
60	-	0.9932	0.9945	0.9955	0.9963
10	3	0.9881	0.9909	0.9929	0.9946
20	-	0.9907	0.9926	0.9939	0.9950
30	-	0.9922	0.9937	0.9947	0.9957
40	-	0.9930	0.9944	0.9954	0.9962
50	-	0.9939	0.9950	0.9957	0.9965
60	-	0.9943	0.9953	0.9960	0.9967
10	4	0.9879	0.9908	0.9927	0.9946
20	-	0.9909	0.9927	0.9940	0.9953
30	-	0.9926	0.9940	0.9950	0.9958
40	-	0.9937	0.9948	0.9956	0.9964
50	-	0.9943	0.9953	0.9960	0.9967
60	-	0.9948	0.9958	0.9964	0.9970
10	5	0.9876	0.9905	0.9925	0.9943
20	-	0.9911	0.9928	0.9940	0.9951
30	-	0.9929	0.9942	0.9951	0.9959
40	-	0.9939	0.9950	0.9958	0.9965
50	-	0.9946	0.9955	0.9962	0.9969
60	-	0.9951	0.9960	0.9965	0.9971
10	6	0.9870	0.9900	0.9920	0.9941
20	-	0.9910	0.9927	0.9940	0.9952
30	-	0.9939	0.9950	0.9958	0.9965
40	-	0.9940	0.9951	0.9958	0.9965
50	-	0.9948	0.9957	0.9963	0.9970
60	-	0.9953	0.9961	0.9967	0.9972

Table B30 (cont'd). Upper percentage points of a probability plot correlation test for multivariate data

n	m	90%	95%	97.5%	99%
10	7	0.9866	0.9898	0.9920	0.9941
20	-	0.9909	0.9926	0.9939	0.9950
30	-	0.9930	0.9942	0.9951	0.9959
40	-	0.9941	0.9951	0.9958	0.9965
50	-	0.9948	0.9957	0.9963	0.9970
60	-	0.9954	0.9962	0.9967	0.9973
10	8	0.9865	0.9897	0.9918	0.9938
20	-	0.9907	0.9925	0.9937	0.9949
30	-	0.9929	0.9942	0.9951	0.9960
40	-	0.9941	0.9951	0.9958	0.9965
50	-	0.9950	0.9958	0.9964	0.9970
60	-	0.9955	0.9963	0.9968	0.9973
10	9	0.9863	0.9895	0.9916	0.9937
20	-	0.9906	0.9924	0.9937	0.9949
30	-	0.9929	0.9942	0.9950	0.9959
40	-	0.9942	0.9952	0.9959	0.9965
50	-	0.9950	0.9958	0.9964	0.9970
60	-	0.9956	0.9963	0.9968	0.9974
20	10	0.9905	0.9923	0.9936	0.9928
30	-	0.9929	0.9940	0.9950	0.9959
40	-	0.9942	0.9953	0.9962	0.9965
50	-	0.9952	0.9961	0.9967	0.9970

Table B31. Upper percentage points of the multivariate Anderson-Darling test (from Paulson, Roohan and Sullo, 1987; copyright 1987 Taylor and Francis Ltd. Used with permission)

				percentage points				
n	m	90%	95%	99%	m	90%	95%	99%
10	1	1.08	1.37	2.11	4	1.05	1.24	1.66
20	-	1.14	1.46	2.22	-	0.98	1.19	1.67
24	-	1.15	1.47	2.23	-	0.98	1.19	1.68
30	-	1.16	1.48	2.24	-	0.97	1.19	1.70
40	-	1.17	1.49	2.25	-	0.97	1.20	1.73
60	-	1.18	1.50	2.26	-	0.98	1.21	1.75
120	-	1.09	1.50	2.26	-	0.99	1.22	1.76
∞	-	1.20	1.51	2.26	-	1.00	1.23	1.77
10	2	0.96	1.19	1.73	5	1.21	1.41	1.76
20	-	1.02	1.26	1.84	-	1.00	1.20	1.67
24	-	1.03	1.27	1.85	-	0.98	1.19	1.67
30	-	1.04	1.28	1.87	-	0.97	1.18	1.67
40	-	1.05	1.29	1.90	-	0.97	1.18	1.67
60	-	1.06	1.31	1.93	-	0.97	1.19	1.69
120	-	1.07	1.33	1.95	-	0.97	1.20	1.71
∞	-	1.08	1.35	1.97	-	0.98	1.21	1.73
10	3	0.95	1.14	1.54				
20	-	0.96	1.19	1.69				
24	-	0.97	1.20	1.71				
30	-	0.98	1.21	1.73				
40	-	0.99	1.21	1.75				
60	-	1.01	1.23	1.78				
120	-	1.02	1.25	1.81				
∞	-	1.03	1.27	1.84				

Table B31 (cont'd). Upper percentage points of the multivariate Anderson-Darling test

	asymptotic percentage points		
m	90%	95%	99%
6	0.973	1.194	1.707
7	0.964	1.182	1.686
8	0.957	1.172	1.670
9	0.952	1.165	1.658
10	0.947	1.159	1.648
11	0.943	1.154	1.640
12	0.940	1.150	1.633
13	0.938	1.146	1.627
14	0.935	1.143	1.622
15	0.933	1.140	1.617
16	0.933	1.140	1.616
17	0.932	1.138	1.613
18	0.930	1.136	1.610
19	0.929	1.134	1.607
20	0.928	1.133	1.605
21	0.928	1.133	1.604
22	0.927	1.132	1.602
23	0.926	1.131	1.600
24	0.926	1.129	1.598
25	0.925	1.128	1.597

Table B32. Upper percentage points of the multivariate extreme deviate test (from Jennings and Young, 1988; reprinted by courtesy of Marcel Dekker, Inc.)

10% percentage points

			m			
n	2	3	4	5	6	7
5	3.134					
6	3.884	4.117				
7	4.524	4.928	5.107			
8	5.039	5.612	5.947	6.094		
9	5.525	6.194	6.662	6.944	7.082	
10	5.912	6.692	7.278	7.694	7.954	8.076
12	6.609	7.580	8.338	8.928	9.403	9.725
14	7.132	8.255	9.140	9.900	10.544	11.064
16	7.648	8.826	9.869	10.704	11.480	12.137
18	7.966	9.309	10.444	11.400	12.233	13.004
20	8.365	9.805	10.918	11.928	12.893	13.743
25	9.114	10.592	11.941	13.094	14.257	15.259
30	9.608	11.259	12.808	13.992	15.236	16.337
35	10.174	11.823	13.334	14.736	16.031	17.309
40	10.426	12.272	13.951	15.272	16.637	17.927
45	10.836	12.679	14.415	15.820	17.327	18.634
50	11.058	13.024	14.707	16.332	17.687	19.071
75	12.198	14.333	16.181	17.773	19.276	21.045
100	12.980	15.220	17.042	18.828	20.535	22.240
200	14.540	17.083	19.143	21.159	22.903	24.742
500	16.747	19.204	21.491	23.599	25.597	27.414

Table B32 (cont'd). Upper percentage points of the multivariate extreme deviate test

10% percentage points						
n	8	9	10	12	15	20
12	9.956	10.064				
14	11.469	11.774	11.963			
16	12.665	13.135	13.512	13.970		
18	13.675	14.261	14.772	15.540	16.043	
20	14.488	15.217	15.815	16.820	17.812	
25	16.150	16.999	17.851	19.293	21.047	22.830
30	17.355	18.375	19.328	21.027	23.334	26.212
35	18.339	19.523	20.433	22.384	24.967	28.633
40	19.224	20.335	21.382	23.451	26.239	30.448
45	19.739	21.036	22.237	24.350	27.417	31.866
50	20.423	21.566	22.873	25.061	28.435	33.051
75	22.376	23.895	25.263	27.773	31.466	37.075
100	23.761	25.276	26.645	29.416	33.365	39.494
200	26.513	28.030	29.666	32.831	37.146	44.133
500	29.334	31.273	32.859	36.154	41.017	48.679

Table B32 (cont'd). *Upper percentage points of the multivariate extreme deviate test*

			5% percentage points			
n	2	3	4	5	6	7
5	3.166					
6	3.987	4.142				
7	4.709	5.005	5.125			
8	5.310	5.763	6.011	6.109		
9	5.848	6.413	6.790	7.006	7.096	
10	6.319	6.989	7.471	7.815	8.007	8.088
12	7.093	8.019	8.665	9.185	9.576	9.837
14	7.733	8.753	9.579	10.248	10.832	11.291
16	8.305	9.390	10.395	11.177	11.911	12.465
18	8.703	10.023	11.090	11.986	12.737	13.478
20	9.169	10.473	11.592	12.522	13.486	14.258
25	9.963	11.467	12.787	13.887	14.936	15.936
30	10.483	12.170	13.751	14.938	16.071	17.128
35	11.113	12.794	14.313	15.749	16.951	18.207
40	11.521	13.302	14.959	16.280	17.627	18.939
45	11.879	13.758	15.504	16.921	18.427	19.781
50	12.122	14.075	15.923	17.478	18.796	20.266
75	13.311	15.546	17.538	19.078	20.555	22.294
100	14.229	16.447	18.354	20.197	22.002	23.594
200	15.846	18.447	20.554	22.709	24.367	26.320
500	18.116	20.655	22.989	25.044	27.181	29.084

Table B32 (cont'd). *Upper percentage points of the multivariate extreme deviate test*

			5% percentage points			
n	8	9	10	12	15	20
12	10.003	10.075				
14	11.627	11.866	12.007			
16	12.925	13.337	13.648	14.005		
18	14.063	14.568	15.023	15.674	16.049	
20	14.959	15.646	16.147	17.073	17.883	
25	16.779	17.647	18.396	19.772	21.371	22.893
30	18.172	19.172	20.067	21.654	23.859	26.501
35	19.226	20.441	21.295	23.210	25.678	29.143
40	20.254	21.264	22.359	24.386	27.129	31.146
45	20.848	22.108	23.296	25.394	28.391	32.637
50	21.554	22.675	24.012	26.136	29.458	33.880
75	23.625	25.257	26.617	29.041	32.787	38.353
100	25.110	26.757	28.100	30.907	34.857	41.034
200	28.112	29.545	31.393	34.513	39.004	45.950
500	31.051	32.965	34.525	38.003	41.788	50.631

Table B32 (cont'd). Upper percentage points of the multivariate extreme deviate test

	1% percentage points					
n	2	3	4	5	6	7
5	3.193					
6	4.110	4.163				
7	4.951	5.098	5.139			
8	5.670	5.950	6.084	6.122		
9	6.319	6.783	6.975	7.077	7.108	
10	6.915	7.439	7.771	7.978	8.068	8.098
12	7.873	8.664	9.183	9.576	9.817	9.980
14	8.919	9.682	10.371	10.863	11.285	11.626
16	9.639	10.514	11.361	12.008	12.548	12.996
18	10.095	11.255	12.229	12.978	13.579	14.160
20	10.739	11.860	12.931	13.780	14.541	15.327
25	11.689	13.319	14.448	15.488	16.400	17.349
30	12.425	14.282	15.645	16.844	17.879	18.721
35	13.140	14.806	16.197	17.924	19.038	20.099
40	13.733	15.481	17.115	18.318	19.739	20.993
45	14.283	15.946	17.821	19.121	20.671	21.860
50	14.494	16.385	18.360	19.992	21.182	22.591
75	15.792	18.068	20.112	21.896	23.145	24.848
100	16.915	19.198	21.320	22.766	24.879	26.347
200	19.275	21.202	23.621	26.056	27.611	29.866
500	21.338	23.953	26.376	28.233	30.650	32.586

Table B32 (cont'd). Upper percentage points of the multivariate extreme deviate test

			1% percentage points			
n	8	9	10	12	15	20
12	10.057	10.082				
14	11.852	11.979	12.049			
16	13.374	13.664	13.832	14.045		
18	14.707	15.070	15.389	15.854	16.054	
20	15.822	16.351	16.768	17.450	17.981	
25	18.045	18.760	19.466	20.647	21.910	22.978
30	19.790	20.588	21.426	22.909	24.803	26.953
35	20.973	22.187	23.084	24.914	27.154	30.029
40	22.403	23.247	24.328	26.175	28.738	32.421
45	22.913	24.157	25.450	27.402	30.335	34.164
50	24.136	25.096	26.311	28.473	31.592	35.686
75	26.617	28.147	29.361	31.830	35.452	41.186
100	28.153	29.610	31.269	33.862	37.966	44.199
200	31.688	33.002	34.922	38.018	42.747	49.793
500	34.611	38.744	38.876	42.184	46.993	55.241

Table B33. Lower percentage points of the multivariate Wilks' tests for two or three outliers (from Fung, 1988; copyright 1988 Taylor and Francis Ltd. Used with permission)

		percentage points Λ^2		
n	m	1%	5%	10%
10	2	0.038	0.066	0.090
11	-	0.058	0.095	0.118
12	-	0.078	0.120	0.145
13	-	0.097	0.148	0.172
14	-	0.115	0.171	0.199
15	-	0.133	0.194	0.226
16	-	0.157	0.217	0.250
17	-	0.178	0.239	0.273
18	-	0.200	0.261	0.296
19	-	0.219	0.281	0.315
20	-	0.237	0.300	0.334
22	-	0.268	0.335	0.373
24	-	0.296	0.369	0.406
26	-	0.324	0.399	0.433
28	-	0.354	0.424	0.456
30	-	0.378	0.445	0.476
35	-	0.431	0.494	0.525
40	-	0.475	0.536	0.565
45	-	0.514	0.572	0.602
50	-	0.546	0.604	0.630
100	-	0.724	0.762	0.778
200	-	0.840	0.860	0.870
300	-	0.886	0.900	0.906

Table B33 (cont'd). Lower percentage points of the multivariate Wilks' tests for two or three outliers

		Λ^2		
n	m	1%	5%	10%
10	3	0.012	0.025	0.035
11	-	0.023	0.040	0.054
12	-	0.036	0.059	0.076
13	-	0.049	0.077	0.099
14	-	0.062	0.101	0.123
15	-	0.079	0.125	0.146
16	-	0.097	0.144	0.168
17	-	0.115	0.164	0.189
18	-	0.132	0.182	0.212
19	-	0.149	0.203	0.232
20	-	0.167	0.221	0.251
22	-	0.198	0.253	0.284
24	-	0.228	0.284	0.316
26	-	0.259	0.314	0.344
28	-	0.285	0.343	0.373
30	-	0.311	0.372	0.400
35	-	0.366	0.425	0.456
40	-	0.415	0.475	0.501
45	-	0.456	0.512	0.538
50	-	0.492	0.545	0.570
100	-	0.688	0.724	0.738
200	-	0.819	0.838	0.848
300	-	0.872	0.885	0.891

Table B33 (cont'd). *Lower percentage points of the multivariate Wilks'*
tests for two or three outliers

		Λ^2		
n	m	1%	5%	10%
10	4	0.003	0.007	0.011
11	-	0.007	0.016	0.022
12	-	0.014	0.027	0.036
13	-	0.024	0.043	0.054
14	-	0.034	0.056	0.070
15	-	0.047	0.073	0.091
16	-	0.061	0.093	0.111
17	-	0.075	0.109	0.128
18	-	0.089	0.124	0.145
19	-	0.104	0.143	0.167
20	-	0.117	0.158	0.182
22	-	0.145	0.189	0.217
24	-	0.176	0.220	0.253
26	-	0.204	0.251	0.279
28	-	0.231	0.279	0.308
30	-	0.254	0.305	0.339
35	-	0.315	0.369	0.393
40	-	0.365	0.414	0.441
45	-	0.407	0.457	0.482
50	-	0.444	0.494	0.518
100	-	0.653	0.686	0.705
200	-	0.799	0.819	0.829

Table B33 (cont'd). Lower percentage points of the multivariate Wilks' tests for two or three outliers

n	m	Λ^2 1%	5%	10%
10	5	0.000	0.001	0.002
11	-	0.002	0.005	0.007
12	-	0.006	0.010	0.014
13	-	0.010	0.020	0.028
14	-	0.017	0.032	0.042
15	-	0.026	0.042	0.054
16	-	0.035	0.057	0.069
17	-	0.046	0.072	0.085
18	-	0.058	0.087	0.102
19	-	0.070	0.101	0.120
20	-	0.082	0.114	0.134
22	-	0.108	0.146	0.167
24	-	0.136	0.173	0.199
26	-	0.163	0.204	0.228
28	-	0.187	0.232	0.256
30	-	0.211	0.257	0.284
35	-	0.265	0.316	0.345
40	-	0.315	0.367	0.395
45	-	0.359	0.413	0.437
50	-	0.400	0.451	0.473
100	-	0.624	0.658	0.675
200	-	0.782	0.801	0.811
40	10	0.168	0.200	0.216
50	-	0.247	0.282	0.302
60	-	0.310	0.358	0.375
80	-	0.428	0.462	0.477
100	-	0.503	0.538	0.555

Table B33 (cont'd). Lower percentage points of the multivariate Wilks'
tests for two or three outliers

		Λ^3		
n	m	1%	5%	10%
15	2	0.066	0.098	0.118
16	-	0.081	0.117	0.138
17	-	0.096	0.134	0.157
18	-	0.111	0.149	0.176
19	-	0.127	0.168	0.194
20	-	0.141	0.186	0.213
22	-	0.169	0.220	0.246
24	-	0.198	0.251	0.278
26	-	0.226	0.281	0.306
28	-	0.252	0.305	0.335
30	-	0.276	0.329	0.359
35	-	0.326	0.383	0.412
40	-	0.373	0.432	0.457
45	-	0.416	0.472	0.496
50	-	0.455	0.505	0.529
100	-	0.654	0.690	0.705
200	-	0.791	0.813	0.823
300	-	0.850	0.865	0.871

*Table B33 (cont'd). Lower percentage points of the multivariate Wilks'
tests for two or three outliers*

		Λ^3		
n	m	1%	5%	10%
15	3	0.029	0.048	0.061
16	-	0.039	0.062	0.075
17	-	0.050	0.076	0.091
18	-	0.062	0.090	0.107
19	-	0.075	0.105	0.124
20	-	0.088	0.118	0.137
22	-	0.112	0.146	0.166
24	-	0.134	0.174	0.194
26	-	0.157	0.201	0.222
28	-	0.182	0.227	0.250
30	-	0.206	0.252	0.276
35	-	0.261	0.306	0.332
40	-	0.309	0.353	0.379
45	-	0.349	0.394	0.419
50	-	0.385	0.431	0.455
100	-	0.607	0.640	0.655
200	-	0.762	0.784	0.793
300	-	0.828	0.844	0.851

*Table B33 (cont'd). Lower percentage points of the multivariate Wilks'
tests for two or three outliers*

		Λ^3		
n	m	1%	5%	10%
15	4	0.014	0.023	0.030
16	-	0.020	0.032	0.040
17	-	0.028	0.041	0.051
18	-	0.036	0.051	0.062
19	-	0.044	0.061	0.074
20	-	0.052	0.072	0.087
22	-	0.071	0.096	0.114
24	-	0.091	0.120	0.141
26	-	0.113	0.144	0.166
28	-	0.136	0.168	0.190
30	-	0.157	0.190	0.214
35	-	0.206	0.248	0.267
40	-	0.252	0.295	0.316
45	-	0.293	0.338	0.360
50	-	0.332	0.376	0.398
100	-	0.560	0.595	0.614
200	-	0.735	0.757	0.767

Table B33 (cont'd). Lower percentage points of the multivariate Wilks'
tests for two or three outliers

		Λ^3		
n	m	1%	5%	10%
15	5	0.005	0.010	0.013
16	-	0.008	0.015	0.020
17	-	0.012	0.023	0.028
18	-	0.017	0.029	0.037
19	-	0.024	0.038	0.045
20	-	0.031	0.046	0.055
22	-	0.046	0.064	0.076
24	-	0.064	0.084	0.098
26	-	0.082	0.106	0.121
28	-	0.100	0.126	0.140
30	-	0.117	0.148	0.166
35	-	0.161	0.199	0.218
40	-	0.204	0.246	0.267
45	-	0.247	0.287	0.306
50	-	0.289	0.327	0.347

Table B34. Upper percentage points of the generalized gap test for multivariate outliers (from Rohlf, 1975; copyright 1975. Used with permission of the International Biometric Society)

	5% percentage points					
n	$\eta = 0.1$	0.2	0.3	0.4	0.5	0.6
10	9.680	8.439	7.382	6.604	6.020	5.567
20	16.387	12.387	10.163	8.760	7.789	7.071
40	23.904	16.383	12.893	10.843	9.478	8.497
60	28.394	18.683	14.443	12.015	10.424	9.292
80	31.573	20.287	15.518	12.825	11.077	9.840
100	34.024	21.514	16.337	13.442	11.574	10.257
200	41.505	25.231	18.795	15.315	13.068	11.498
n	$\eta = 0.7$	0.8	0.9	1.0	1.5	2.0
10	5.205	4.908	4.660	4.450	3.733	3.311
20	6.517	6.075	5.712	5.409	4.410	3.843
40	7.752	7.168	6.693	6.301	5.031	4.327
60	8.441	7.773	7.237	6.794	5.373	4.592
80	8.915	8.190	7.614	7.133	5.608	4.774
100	9.275	8.507	7.898	7.390	5.786	4.913
200	10.334	9.481	8.757	8.165	6.321	5.328
n	$\eta = 2.5$	3.0	3.5	4.0	4.5	
10	3.028	2.823	2.666	2.541	2.439	
20	3.471	3.205	3.003	2.845	2.716	
40	3.871	3.549	3.307	3.117	2.965	
60	4.091	3.737	3.473	3.266	3.100	
80	4.241	3.866	3.586	3.368	3.192	
100	4.355	3.963	3.672	3.445	3.262	
200	4.698	4.257	3.930	3.676	3.473	
n	$\eta = 5.0$	6.0	8.0	10.0	12.0	
10	2.354	2.218	2.033	1.909	1.821	
20	2.609	2.441	2.213	2.063	1.955	
40	2.838	2.639	2.373	2.198	2.074	
60	2.962	2.747	2.459	2.272	2.139	
80	3.047	2.821	2.518	2.322	2.182	
100	3.112	2.877	2.563	2.360	2.216	
200	3.305	3.044	2.697	2.473	2.315	

Table B34 (cont'd). Upper percentage points of the generalized gap test for multivariate outliers

			1% percentage points			
n	$\eta = 0.1$	0.2	0.3	0.4	0.5	0.6
10	9.946	9.343	8.520	7.784	7.175	6.675
20	18.396	14.871	12.437	10.784	9.598	8.705
40	28.986	20.430	16.114	13.512	11.763	10.500
60	35.302	23.484	18.088	14.960	12.910	11.448
80	39.662	25.538	19.411	15.928	13.676	12.083
100	42.941	27.066	20.393	16.648	14.247	12.556
200	52.528	31.556	23.189	18.824	15.912	13.874
n	$\eta = 0.7$	0.8	0.9	1.0	1.5	2.0
10	6.260	5.911	5.614	5.358	4.469	3.934
20	8.007	7.446	6.982	6.594	5.308	4.576
40	9.537	8.784	8.169	7.663	6.027	5.124
60	10.351	9.483	8.793	8.226	6.406	5.412
80	10.893	9.953	9.228	8.603	6.660	5.606
100	11.298	10.303	9.544	8.884	6.851	5.752
200	12.367	11.438	10.486	9.711	7.411	6.181
n	$\eta = 2.5$	3.0	3.5	4.0	4.5	
10	3.572	3.308	3.106	2.945	2.814	
20	4.096	3.754	3.495	3.292	3.128	
40	4.542	4.132	3.826	3.586	3.394	
60	4.777	4.332	4.000	3.742	3.534	
80	4.935	4.466	4.117	3.846	3.629	
100	5.054	4.566	4.206	3.925	3.700	
200	5.407	4.865	4.468	4.159	3.912	
n	$\eta = 5.0$	6.0	8.0	10.0	12.0	
10	2.704	2.530	2.292	2.134	2.021	
20	2.991	2.777	2.488	2.300	2.165	
40	3.235	2.986	2.654	2.439	2.286	
60	3.363	3.097	2.742	2.512	2.350	
80	3.450	3.171	2.801	2.562	2.394	
100	3.515	3.228	2.845	2.600	2.426	
200	3.709	3.393	2.977	2.709	2.523	

Table B35. *Upper percentage points of the standardized difference between component means for testing for a two component normal mixture in multivariate samples (from Everitt, 1988; copyright 1988 Taylor and Francis Ltd. Used with permission)*

n	%age point	2	3	4	5	6
20	95	4.5	5.1	5.6	6.3	6.8
40	-	3.4	3.8	4.2	4.4	4.8
60	-	3.0	3.3	3.6	3.9	4.1
80	-	2.8	3.1	3.3	3.4	3.7
100	-	2.6	2.8	3.1	3.3	3.5
200	-	2.2	2.3	2.6	2.7	2.8
500	-	1.8	1.9	1.9	2.1	2.3
20	99	5.5	5.9	6.7	7.3	7.9
40	-	3.8	4.3	4.6	4.9	5.1
60	-	3.3	3.7	3.9	4.1	4.4
80	-	3.1	3.4	3.6	3.8	4.0
100	-	2.9	3.0	3.3	3.4	3.6
200	-	2.6	2.5	2.8	2.8	2.9
500	-	1.9	2.1	2.2	2.4	2.4

n	%age point	7	8	9	10
20	95	7.3	7.9	8.6	9.2
40	-	5.1	5.3	5.5	5.7
60	-	4.3	4.5	4.7	4.9
80	-	3.9	4.1	4.2	4.3
100	-	3.6	3.7	3.9	4.0
200	-	2.9	3.1	3.2	3.3
500	-	2.3	2.4	2.5	2.5
20	99	8.6	9.3	10.1	11.2
40	-	5.5	5.8	6.0	6.2
60	-	4.5	4.8	5.0	5.1
80	-	4.2	4.4	4.4	4.5
100	-	3.9	3.9	4.1	4.2
200	-	3.1	3.2	3.4	3.4
500	-	2.5	2.5	2.6	2.7

Table B36. Values of z as a function of zη for estimating gamma parameters (from Greenwood and Durand, 1960; reproduced with permission from Technometrics. Copyright 1960 by the American Statistical Association. All rights reserved.)

z	zη	z	zη	z	zη
0.00	0.50000000	-	-	-	-
0.01	0.50166108	0.31	0.54597209	0.61	0.58061573
0.02	0.50331088	0.32	0.54726833	0.62	0.58163629
0.03	0.50494925	0.33	0.54855377	0.63	0.58264927
0.04	0.50657603	0.34	0.54982850	0.64	0.58365479
0.05	0.50819111	0.35	0.55109265	0.65	0.58465291
0.06	0.50979438	0.36	0.55234633	0.66	0.58564375
0.07	0.51138573	0.37	0.55358965	0.67	0.58662738
0.08	0.51296508	0.38	0.55482273	0.68	0.58760389
0.09	0.51453237	0.39	0.55604568	0.69	0.58857337
0.10	0.51608755	0.40	0.55725862	0.70	0.58953591
0.11	0.51763057	0.41	0.55846165	0.71	0.59049158
0.12	0.51916141	0.42	0.55965491	0.72	0.59144048
0.13	0.52068005	0.43	0.56083849	0.73	0.59238267
0.14	0.52218619	0.44	0.56201253	0.74	0.59331825
0.15	0.52368074	0.45	0.56317712	0.75	0.59424729
0.16	0.52516282	0.46	0.56433238	0.76	0.59516987
0.17	0.52663277	0.47	0.56547842	0.77	0.59608607
0.18	0.52809061	0.48	0.56661536	0.78	0.59699595
0.19	0.52953640	0.49	0.56774331	0.79	0.59789961
0.20	0.53097019	0.50	0.56886236	0.80	0.59879710
0.21	0.53239205	0.51	0.56997264	0.81	0.59968850
0.22	0.53380204	0.52	0.57107426	0.82	0.60057389
0.23	0.53520025	0.53	0.57216730	0.83	0.60145333
0.24	0.53658674	0.54	0.57325189	0.84	0.60232689
0.25	0.53796161	0.55	0.57432812	0.85	0.60319464
0.26	0.53932495	0.56	0.57539610	0.86	0.60405665
0.27	0.54067685	0.57	0.57645593	0.87	0.60491298
0.28	0.54201741	0.58	0.57750771	0.88	0.60576369
0.29	0.54334674	0.59	0.57855154	0.89	0.60660886
0.30	0.54466493	0.60	0.57958751	0.90	0.60744854

Table B36 (cont'd). Values of z as a function of zη for estimating gamma parameters

z	zη	z	zη	z	zη
0.91	0.60828279	1.21	0.63106046	3.6	0.7337487
0.92	0.60911168	1.22	0.63175276	3.8	0.7391165
0.93	0.60993527	1.23	0.63244119	4.0	0.7442040
0.94	0.61075362	1.24	0.63312580	4.2	0.7490353
0.95	0.61156678	1.25	0.63380662	4.4	0.7536319
0.96	0.61237481	1.26	0.63448370	4.6	0.7580126
0.97	0.61317777	1.27	0.63515706	4.8	0.7621942
0.98	0.61397571	1.28	0.63582675	5.0	0.7661916
0.99	0.61476869	1.29	0.63649279	5.2	0.7700182
1.00	0.61555677	1.30	0.63715523	5.4	0.7736861
1.01	0.61633999	1.31	0.63781410	5.6	0.7772060
1.02	0.61711841	1.32	0.63846943	5.8	0.7805879
1.03	0.61789208	1.33	0.63912126	6.0	0.7838406
1.04	0.61866105	1.34	0.63976962	6.2	0.7869722
1.05	0.61942537	1.35	0.64041454	6.4	0.7899901
1.06	0.62018509	1.36	0.64105606	6.6	0.7929012
1.07	0.62094026	1.37	0.64169420	6.8	0.7957116
1.08	0.62169093	1.38	0.64232900	7.0	0.7984270
1.09	0.62243714	1.39	0.64296049	7.2	0.8010527
1.10	0.62317895	1.4	0.6435887	7.4	0.8035935
1.11	0.62391639	1.6	0.6555077	7.6	0.8060539
1.12	0.62464951	1.8	0.6663340	7.8	0.8084382
1.13	0.62537836	2.0	0.6762335	8.0	0.8107500
1.14	0.62610298	2.2	0.6853382	8.2	0.8129931
1.15	0.62682342	2.4	0.6937540	8.4	0.8151707
1.16	0.62753971	2.6	0.7015677	8.6	0.8172861
1.17	0.62825190	2.8	0.7088511	8.8	0.8193420
1.18	0.62896003	3.0	0.7156639	9.0	0.8213412
1.19	0.62966414	3.2	0.7220568	9.2	0.8232862
1.20	0.63036427	3.4	0.7280727	9.4	0.8251795

Table B36 (cont'd). Values of z as a function of zη for estimating gamma parameters

z	zη	z	zη	z	zη
9.6	0.8270232	13.6	0.8560277	17.6	0.8755080
9.8	0.8288196	13.8	0.8571778	17.8	0.8763207
10.0	0.8305705	14.0	0.8583059	18.0	0.8771208
10.2	0.8322778	14.2	0.8594127	-	-
10.4	0.8339433	14.4	0.8604989	-	-
10.6	0.8355687	14.6	0.8615651	-	-
10.8	0.8371555	14.8	0.8626118	-	-
11.0	0.8387052	15.0	0.8636397	-	-
11.2	0.8402193	15.2	0.8646493	-	-
11.4	0.8416989	15.4	0.8656411	-	-
11.6	0.8431455	15.6	0.8666156	-	-
11.8	0.8445602	15.8	0.8675733	-	-
12.0	0.8459441	16.0	0.8685146	-	-
12.2	0.8472983	16.2	0.8694401	-	-
12.4	0.8486239	16.4	0.8703502	-	-
12.6	0.8499219	16.6	0.8712452	-	-
12.8	0.8511931	16.8	0.8721255	-	-
13.0	0.8524384	17.0	0.8729916	-	-
13.2	0.8536588	17.2	0.8738438	-	-
13.4	0.8548549	17.4	0.8746825	-	-

References

D'Agostino, R.B. (1972). Small sample probability points for the D test of normality. Biometrika 59, 219-221.

David, H.A., Hartley, H.O., and Pearson, E.S. (1954). The distribution of the ratio, in a single normal sample, of the range to the standard deviation. Biometrika 41, 482-493.

Dixon, W. (1951). Ratios involving extreme values. Annals of Mathematical Statistics 22, 68-78.

Everitt, B.S. (1988). A test of multivariate normality against the alternative that the distribution is a mixture. Journal of Statistical Computation and Simulation 30, 103-115.

Fung, W.-K. (1988). Critical values for testing in multivariate statistical outliers. Journal of Statistical Computation and Simulation 30, 195-212.

Greenwood, J.A., and Durand, D. (1960). Aids for fitting the gamma distribution by maximum likelihood. Technometrics 2, 55-65.

Grubbs, F., and Beck, G. (1972). Extension of sample sizes and percentage points for significance tests of outlying observations. Technometrics 14, 847-859.

Hoaglin, D.C., Iglewicz, B., and Tukey, J.W. (1986). Performance of some resistant rules for outlier labeling. Journal of the American Statistical Association 81, 991-999.

Jain, R.B. (1981). Percentage points of many-outlier detection procedures. Technometrics 23, 71-76.

Jennings, L.W., and Young, D.M. (1988). Extended values of the multivariate extreme deviate test for detecting a single spurious observation. Communications in Statistics - Simulation and Computation 17, 1359-1373.

LaBreque, J. (1977). Goodness-of-fit tests based on nonlinearity in probability plots. Technometrics 19, 293-306.

Locke, C., and Spurrier, J.D. (1981). On the distribution of a new measure of heaviness of tails. Communications in Statistics - Theory and Methods A10, 1967-1980.

Lockhart, R.A., O'Reilly, F.J., and Stephens, M.A. (1986). Tests of fit based on normalized spacings. Journal of the Royal Statistical Society B 48, 344-352.

Pearson, E.S., and Hartley, H.O., eds. (1976). **Biometrika Tables for Statisticians, Volume I**. Biometrika Trust, London, England.

Paulson, A.S., Roohan, P., and Sullo, P. (1987). Some empirical distribution function tests for multivariate normality. Journal of Statistical Computation and Simulation 28, 15-30.

Pettitt, A.N. (1976). Cramer-von Mises statistics for testing normality with censored samples. Biometrika 63, 475-481.

Prescott, P. (1979). Critical values for a sequential test for many outliers. Applied Statistics 28, 36-39.

Rohlf, F.J. (1975). Generalization of the gap test for the detection of multivariate outliers. Biometrics 31, 93-101.

Shapiro, S.S., and Wilk, M.B. (1965). An analysis of variance test for normality (complete samples). Biometrika 52, 591-611.

Spiegelhalter, D.J. (1977). A test for normality against symmetric alternatives. Biometrika 64, 415-418.

Spiegelhalter, D.J. (1980). An omnibus test for normality for small samples. Biometrika 67, 493-496.

Stephens, M.A. (1974). EDF statistics for goodness of fit and some comparisons. Journal of the American Statistical Association 69, 730-737.

Stephens, M.A. (1986). Tests based on EDF statistics. In D'Agostino, R.B., and Stephens, M.A., eds., **Goodness of Fit Techniques**, Marcel Dekker, New York.

Tietjen, G.L., and Moore, R.H. (1972). Some Grubbs-type statistics for the detection of outliers. Technometrics 14, 583-597.

Tiku, M.L. (1980). Goodness of fit statistics based on the spacings of complete or censored samples. Australian Journal of Statistics 22, 260-275.

Tsai, K.-T., and Koziol, J.A. (1988). A correlation procedure for assessing multivariate normality. Communications in Statistics - Simulation and Computation 17, 637-651.

Vasicek, O. (1976). A test for normality based on sample entropy. Journal of the Royal Statistical Society B 38, 54-59.

Verrill, S., and Johnson, R.A. (1988). Tables and large sample distribution theory for censored data correlation statistics for testing normality. Journal of the American Statistical Association 83, 1192-1197.

APPENDIX C

FUNCTION OPTIMIZATION COMPUTER SUBROUTINE

VARIABLE METRIC METHOD

```
      SUBROUTINE VARMET(B,P0,X,N,GRAD,MAXLIK,ITR)
      IMPLICIT DOUBLE PRECISION (A-H,O-Z)
C
C VARMET IS A VARIABLE METRIC (QUASI-NEWTON) FUNCTION MINIMIZATION
C SUBROUTINE BASED ON AN ALGORITHM BY NASH (1979), ADAPTED FOR THE NORMAL
C MIXTURE PROBLEM
C
C SUBROUTINE CHARACTERISTICS WHICH MAY BE CHANGED BASED ON
C HARDWARE CAPABILITIES ARE THE STEP SIZE (W) AND TOLERANCE (TOL)
C
C B     = VECTOR OF STARTING POINTS (INPUT) AND PARAMETER
C          ESTIMATES (OUTPUT)
C P0    = VALUE OF THE FUNCTION AFTER CONVERGENCE (OUTPUT)
C X     = VECTOR OF DATA VALUES OF SIZE N (INPUT)
C N     = NUMBER OF OBSERVATIONS (INPUT)
C GRAD = SUBROUTINE NAME WHICH CALCULATES THE GRADIENTS
C          (PARTIAL DERIVATIVES) OF THE FUNCTION (INPUT)
C MAXLIK = SUBROUTINE NAME WHICH CALCULATES THE VALUE OF THE
C            FUNCTION FOR A GIVEN SET OF PARAMETER VALUES (INPUT)
C ITR   = MAXIMUM NUMBER OF ITERATIONS SET FOR CONVERGENCE (INPUT)
C
      DOUBLE PRECISION MAXLIK,K
      REAL X(N)
      DIMENSION B(4),BB(4,4),T(4),C(4),XX(4),G(4)
      IG=0
      IFN=0
      ITER=0
C
C NASH SUGGESTS USING STEP SIZE PARAMETER W = 0.2 AND TOLERANCE .0001
C
      W=.2D0
      TOL=.0001D0
C
C COMPUTE LIKELIHOOD AT STARTING PARAMETER VALUES
C
    1 P0=MAXLIK(B,X,N,IER)
      IF(IER.EQ.1)STOP 'NO ML ON INITIALIZATION'
      IFN=IFN+1
C
C COMPUTE GRADIENT AT STARTING VALUE
C
    2 GR= GRAD(G,B,X,N)
      IG=IG+1
```

```
C
C  INITIALIZE GRADIENT TRANSFORMATION MATRIX TO THE UNIT MATRIX
C
    3 DO 21 I=1,4
      DO 20 J=1,4
   20 BB(I,J)=0.D0
   21 BB(I,I)=1.D0
      ILAST=IG
C
C  BEGIN ITERATION LOOP
C
    4 CONTINUE
      ITER=ITER+1
      IF(ITER.GT.ITR)GOTO 18
      DO 30 I=1,4
      XX(I)=B(I)
   30 C(I)=G(I)
    5 D1=0.D0
      DO 40 I=1,4
      S=0.D0
      DO 50 J=1,4
   50 S=S-BB(I,J)*G(J)
      T(I)=S
   40 D1=D1-S*G(I)
C
C  PERFORM LINEAR SEARCH (D1>0) OR END BECAUSE OF CONVERGENCE (IG=ILAST);
C    OTHERWISE, RESTART ITERATION
C
    6 IF(D1.GT.0.D0)GOTO 7
      IF(ILAST.EQ.IG)GOTO 18
      GOTO 3
    7 K=1
    8 NCOUNT=0
      DO 60 I=1,4
      B(I)=XX(I)+K*T(I)
      IF(B(I).EQ.XX(I))NCOUNT=NCOUNT+1
   60 CONTINUE
C
C  TEST FOR CONVERGENCE
C
    9 IF(NCOUNT.LT.4)GOTO 10
      IF(ILAST.EQ.IG)GOTO 18
      GOTO 3
   10 P=MAXLIK(B,X,N,IER)
      IFN=IFN+1
      IF(IER.NE.1)GOTO 11
      K=W*K
      GOTO 8
```

```
C
C   ACCEPTANCE TEST ON NEW POINT
C
   11   IF(P.LT.P0-D1*K*TOL)GOTO 12
        K=W*K
        GOTO 8
C
C   SAVE NEW VALUE OF LIKELIHOOD; COMPUTE NEW GRADIENT
C
   12   P0=P
        GR= GRAD(G,B,X,N)
        IG=IG+1
C
C   BEGIN MATRIX UPDATE
C
   13   D1=0.D0
        DO 70 I=1,4
        T(I)=K*T(I)
        C(I)=G(I)-C(I)
   70   D1=D1+T(I)*C(I)
   14   IF(D1.LE.0.D0)GOTO 3
   15   D2=0.D0
        DO 80 I=1,4
        S=0.D0
        DO 90 J=1,4
   90   S=S+BB(I,J)*C(J)
        XX(I)=S
   80   D2=D2+S*C(I)
   16   D2=1.D0+D2/D1
        DO 100 I=1,4
        DO 100 J=1,4
  100   BB(I,J)=BB(I,J)-(T(I)*XX(J)+XX(I)*T(J)-D2*T(I)*T(J))/D1
C
C   START A NEW ITERATION
C
   17   GOTO 4
C
C   ALGORITHM HAS CONVERGED OR EXCEEDED MAXIMUM NO. OF ITERATIONS
C
   18   IERR=0
        IF(ITER.GT.ITR)IERR=-1
        RETURN
        END
```

Index

Milton Keynes UK
Ingram Content Group UK Ltd.
UKHW020008071024
449327UK00031B/2703